If Then

IF THEN

HOW THE SIMULMATICS CORPORATION INVENTED THE FUTURE

JILL LEPORE

THORNDIKE PRESS
A part of Gale, a Cengage Company

**LIBRARY OF CONGRESS CIP DATA ON FILE.
CATALOGUING IN PUBLICATION FOR THIS BOOK
IS AVAILABLE FROM THE LIBRARY OF CONGRESS.**

ISBN-13: 978-1-4328-8614-1 (hardcover alk. paper).

Published in 2021 by arrangement with Liveright Publishing
Corporation, a wholly owned subsidiary of W.W. Norton & Company,
Inc.

Printed in Mexico
Print Number: 01   Print Year: 2021

to T.R.L., always

The Company proposes to engage principally in estimating probable human behavior by the use of computer technology.

— Simulmatics Corporation,
stock offering, 1961

CONTENTS

PROLOGUE:
WHAT IF?

The mystery surrounding Simulmatics started with its name.
— Statement to Simulmatics Corporation stockholders, 1966

The geodesic dome in Wading River, Long Island, where Simulmatics met in 1961, with the Greenfields' house to the right.

The scientists of the Simulmatics Corporation spent the summer of 1961 on a beach on Long Island beneath a geodesic dome that looked as if it had landed there, amid the dunes, a spaceship gone to ground.[1] Inside, they wrote mathematical formulas on blackboards. Chalk dusted their fingertips. Reams of perforated computer printouts unfurled across the floor.

The Simulmatics Corporation, Cold War America's Cambridge Analytica, claimed credit for having gotten John F. Kennedy elected president of the United States in November 1960. Months later, its scientists spent a summer at the beach planning new projects for their invention: a computer program designed to predict and manipulate human behavior, all sorts of human behavior, from buying a dishwasher to countering an insurgency to casting a vote. They called it the People Machine.[2]

Hardly anyone, almost no one, remembers Simulmatics anymore. But beneath that honeycombed dome, the scientists of this long-vanished American corporation helped build the machine in which humanity would, by the twenty-first century, find itself trapped and tormented: stripped bare, driven to distraction, deprived of its senses, interrupted, exploited, directed, connected and disconnected, bought and sold, alienated and coerced, confused, misinformed, and even

12

governed. They never meant to hurt anyone.

They were young men, the best and the brightest, fatally brilliant, Icaruses with wings of feathers and wax, flying to the sun. "The scientists are from the Massachusetts Institute of Technology, Yale, Harvard, Columbia and Johns Hopkins," the *New York Times* reported. "They are preparing to work with electronic computers, the giant question-answering devices in use for some years, but are using social and economic data and their own knowledge to work out new programs for computer simulation, the name given to the technique of acting out, so to speak, all the probabilities that might flow from a given set of circumstances."[3] They wrote in a new language, FORTRAN, using an expression known as an IF/THEN statement to instruct a computer to simulate possible actions and calculate their consequences, under different conditions, again and again and again. IF this, THEN that. IF this, THEN that. IF this, THEN that, an infinity of outcomes.

To the beach that summer, they brought their wives and their children. The men wore bathing trunks and polo shirts and pondered punch cards; the women wore summer dresses and sandals and made potato salad and tuna salad and barbecue and macaroni salad and ham salad and pots of stew and piles of corn on the cob; their children — seventeen of them in all — waded in the

13

ocean and built sandcastles, Camelots-by-the-sea, and sailed one-masted Sunfishes and chased a black poodle named Sputnik up and down the beach and over the creek. The children got so badly sunburned that at night their mothers doused them with vinegar to cool their skin: they smelled like pickles. On rainy days, they played Monopoly, hopscotching from Park Place to the B. & O. Railroad, collecting two hundred dollars every time they passed Go, and trying, as all monopolists must, to keep out of jail. The wives traded paperback copies of *Peyton Place,* a steamy novel about sex and female rebellion, its pages wilted from the humidity.[4] And everything, and everyone, was covered with sand, as if, if they'd stayed there long enough, they'd have been buried, like ancient Egyptians.

The sun rises, the sun sets, and still no one ever really knows what will happen next. In a world of endless uncertainty, the forecasting of the future began with the very oldest human societies. The Greeks built a shrine to the Oracle of Delphi; the Incas built a temple to the Oracle at Pachacamac. Buddhists, Muslims, Christians, Jews, every religion, every culture: all have had their prophets and their temples, their diviners, their readers of omens, their seers. Time passed, centuries, millennia. And then, beginning in the middle decades of the twentieth century, Americans

14

began building machines meant to serve as their oracles, new seers, electronic prophets, diviners of data.

Founded in 1959, the Simulmatics Corporation established offices in New York, Washington, Cambridge, and, eventually, Saigon before it declared bankruptcy, in 1970. The company wore a cloak of intrigue. This was, in part, unintentional. "The mystery surrounding Simulmatics started with its name," its president once explained to the company's stockholders. "We were a contraction of two words — 'simulation' and 'automatic.' "[5] Its founders hoped the name would become a watchword, a byword, like "cybernetics." It did not. The obscurity of the word "simulmatics" is a measure of their failure. But its meaning is a measure of their ambition: to automate the simulation of human behavior.

The scientists of the Simulmatics Corporation acted on the proposition that if they could collect enough data about enough people and feed it into a machine, everything, one day, might be predictable, and everyone, every human mind, simulated, each act anticipated, automatically, and even driven and directed, by targeted messages as unerring as missiles. Facebook, Palantir, Cambridge Analytica, Amazon, the Internet Research Agency, Google — they were all incubated there, beneath that honeycombed dome by the edge of the gray-green sea, like

15

so many eggs.

Simulmatics' scientists were known as the What-If Men. They believed that by simulating human behavior, their People Machine could help the human race avert each and every disaster. It could defeat communism. It could counter insurgencies. It could win elections. It could sell mouthwash. It could accelerate news, like so much amphetamine. It could calm agitated wives. It could win the war in Vietnam by targeting hearts and minds. It could predict race riots, and even plagues. It could end chaos. The scientists of Simulmatics believed they had invented "the A-bomb of the social sciences."[6] They did not predict that it would take decades to detonate, like a long-buried grenade.

Still, even at the time, the People Machine seemed to many people to be a species of madness, a harbinger of a coming dystopia. In 1964, the Simulmatics Corporation served as the subject of two ominous novels. In Eugene Burdick's political thriller *The 480,* a barely disguised "Simulations Enterprises," equipped with hulking, sinster IBM computers, meddles with the 1964 U.S. presidential election. In Daniel F. Galouye's *Simulacron-3,* science fiction set in the year 2033, specialists in the field of "simulectronics" build a people machine — "a total environment simulator" — only to discover that they themselves don't exist and are instead merely

16

the ethereal, Escherian inventions of yet another people machine.[7] After that, Simulmatics lived on, in fiction and film, an anonymous avatar. In 1973, the avant-garde German filmmaker Rainer Werner Fassbinder adapted *Simulacron-3* into *World on a Wire,* a terrifying futurist tour de force, a forerunner to the 1999 film *The Matrix* in which all of humanity lives in a simulation, locked, trapped, deluded, and dehumanized; *Matrix*'s main character, trying to set humanity free, hides stolen software inside a hollowed-out copy of Jean Baudrillard's 1981 book, *Simulacra and Simulation,* a metatext about the meaningless "hell of simulation."[8]

In fiction and film, Dr. Frankenstein yielded to Dr. Jekyll and, finally, to Dr. Strangelove, as mad science moved from biology to chemistry to physics. But Simulmatics' fiction-and-film avatar — the mad scientist of computer science — is wildly outsized, the lengthening shadow of a very small man. The Simulations Enterprises of *The 480* is a megacorporation, and the simulectronics specialists in *Simulacron-3* are technical geniuses. The real Simulmatics Corporation was a tiny, struggling company, its technicians bumbling, its accounts disastrous. It soared and then it sank, like a helium balloon. The geodesic dome became a Space Burger, a drive-through hamburger joint.

And yet Simulmatics' legacy endures in

17

predictive analytics, what-if simulation, and behavioral data science: it lurks behind the screen of every device. Simulmatics, notwithstanding its own failure, helped invent the data-mad and near-totalitarian twenty-first century, in which the only knowledge that counts is prediction and, before and after the coming of the coronavirus, corporations extract wealth by way of the collection of data and the manipulation of attention and the profit of prophecy. In a final irony, Simulmatics, whose very past has been all but erased, helped invent a future obsessed with the future, and yet unable to improve it.

Simulmatics' own origins lie still further back in time, in the early-twentieth-century science of psychological warfare: the control of people's minds by assault, interruption, and distraction. Simulmatics' scientists carried that work into the 1950s, the age of the modern computer, and into electoral politics, with a commission from the Democratic National Committee during the 1960 presidential election, and then into targeted advertising. Later, they flew that work across an ocean, to Vietnam, until student protesters called them war criminals.

It would be easier, more comforting, less unsettling, if the scientists of Simulmatics were villains. But they weren't. They were midcentury white liberals in an era when white liberals were not expected to understand people who weren't white or liberal. They were husbands and fathers in an age when men were not expected to understand women and children. By "human behavior," they meant the behavior of men; by "artificial intelligence," they meant their own intelligence — a fantasy of their own intelligence — which they intended to graft onto a machine. They did not consider the intelligence of women to be intelligence; they did not consider a female understanding of human behavior to be knowledge.

They built a machine to control and predict what they could not. They are the long-dead, white-whiskered grandfathers of Mark Zuckerberg and Sergey Brin and Jeff Bezos and Peter Thiel and Marc Andreessen and Elon Musk. The Simulmatics Corporation is a missing link in the history of technology, a clasp that fastens the first half of the twentieth century to the beginning of the twenty-first, a future in which humanity's every move is predicted by algorithms that attempt to direct and influence our each and every decision through the simulation of our very selves, this particular hell.

If, then, in the 1950s and 1960s, things had gone differently, this future might have been averted. If, then, history had taken a different course, humanity might not have been demoted, humanistic knowledge might still be cherished, and democracy might have grown stronger, not weaker. Or very little might have been different. It is not possible to know. No machine can run an IF/THEN program backward, calculating possible pasts. History cannot answer "What if?" But it can explain what happened, and why.

The future invented by Simulmatics has a past, a history washed away, like a sandcastle, by the tide of time. It can only be pieced back together grain by grain, each parapet and battlement, each rampart and turret, every last feature of its towering audacity.

The new underworld is made up of innocent and well-intentioned people who work with slide rules and calculating machines and computers which can retain an almost infinite number of bits of information as well as sort, categorize, and reproduce this information at the press of a button. Most of these people are highly educated, many of them are Ph.D.s, and none that I have met have malignant political designs on the American public. They may, however, radically reconstruct the American political system, build a new politics, and even modify revered and venerable American institutions — facts of which they are blissfully innocent.

— Eugene Burdick, *The 480,* 1964

■ ■ ■ ■

PART ONE: THE SOCIAL NETWORK

■ ■ ■ ■

Given that A knows B, what is the probability that B knows *n* persons in the circle of acquaintances of A?
— Ithiel de Sola Pool, 1956

CHAPTER 1
MADLY FOR ADLAI

It did Adlai Stevenson great harm, not having a wife, and trying to be funny all the time, too. Great harm.

— Gore Vidal, *The Best Man,* 1960

Ed and a pregnant Patty Greenfield, with Michael, 1954.

Ed Greenfield collected people the way other men collect comic books or old stamps or vintage cars. "Ed Greenfield," he'd say, flashing a made-for-TV, Dean Martin grin, slapping a back, clasping a hand, offering a vodka and tonic, palming a business card, *Edward L. Greenfield, President, Edward L. Greenfield & Co., 501 Madison Ave.* He was like a ten-million-volt Looney Tunes electric magnet, a giant red-handled iron U that pulled everyone toward him, *plink, plink, plink.*

Greenfield founded the Simulmatics Corporation in 1959 and became its president, but the company was years in the making, and the use of computer technology to estimate probable human behavior, like any starry-eyed idea, involved scores of people. To pull off a big bank heist, you need a munitions expert, a surveillance guy, a computer whiz, a security team, a money man, and an all-around huckster. To pull off the computer prediction of human behavior, you need a political theorist, a mathematician, a behavioral scientist, a market researcher, a computer scientist, and an all-around huckster. Greenfield was the huckster. "If you see a frog sitting on top of a flag pole, you know it didn't get up there by itself," a very wise man once said.[1] Greenfield collected the men who figured out how to get the frog up there, watched them build the tools to do it, and then, when they were done, roped in a crowd, pointed to the top of the flag pole, and shouted, "Look, a frog!"

Ed Greenfield had thick, wavy black hair, a big nose, and jug-handle ears. He had wide shoulders and narrow hips and toothpick legs, misfit parts disguised by his custom-made suits. He was warm, loving, and affectionate and he was charming and he was sweetly funny and impossibly fun and he was sexy, in the way of a certain very dapper, animal of a man. He smoked Pall Malls, except when he was smoking a pipe that smelled like a campfire, of pine and night sky. He drank Scotch, in glasses made of crystal as clear as ice.

The future president of the Simulmatics Corporation was a Madison Avenue ad man — a "mad man" — and like all ad men, he sold nothing so well as himself. Born in Chicago in 1927, he was the only child of Jacob Greenfield, an insurance salesman who used to be a Communist, and Theodora Rubenstein, the daughter of a rabbi. He had a neat little résumé: "Edward L. Greenfield, Public Relations, N.Y.C. Formerly Univ. of Chicago, Yale Law School." Most of these credentials were fake. He never graduated from the University of Chicago, or from Yale Law School, either. He went to Wabash College in Indiana for a year, in 1945, and then dropped out, and although he liked to tell people that he still owed the University of

27

Chicago a library book, the University of Chicago has no record of his ever enrolling there. Neither does Yale Law School.[2] Still, he'd been to Yale and must've sat in on a seminar there, because once, when he applied for membership at the Yale Club of New York, Harold Lasswell, a dome-headed, grim-faced, world-famous Yale professor and expert on propaganda, vouched that Greenfield had studied with him in 1950. "Greenfield is a very convivial and attractive human being who has a very wide net of personal acquaintances in this country and abroad," Lasswell testified.[3] Very convivial. Very wide net. A fisherman, fisher of men.

When Ed Greenfield was in his early twenties, he met the very smart and very pretty and sometimes terribly sad Patricia Safford, a talented pianist and dancer who'd studied with Martha Graham. Patty Safford was born in 1928; her mother was a Vienna-trained Freudian analyst and her much older and much-married father, Frank Safford, was an eminent neurosurgeon and patron of the arts. She spent her summers at the family's forty acres of seaside and hills, a little village of cottages on Wading River, Long Island, with her father's friends: artists, writers, and intellectuals who included the painters Willem and Elaine de Kooning, the poet Edwin Denby, and the novelist Richard Wright. Denby and the de Koonings made silent films

28

starring the Safford children, Patty in a swimsuit, her little brother in a rowboat, abducted by pirates, kidnapped by a witch, rescued at the last minute: fairy tales in black and white, tales of sorcery.[4]

In 1951, when Ed Greenfield and Patty Safford got married, Patty's father gave them, as a wedding gift, a rambling old wooden Victorian house on the beach. It had a fireplace made of stones from Long Island Sound. Next door, another of Frank Safford's friends, the visionary and eccentric architect Buckminster Fuller, would build for the Saffords one of his early geodesic domes, a shell of struts of aluminum and triangles of glass and porcelain, intricately balanced, a feat of engineering, a marvel, out of this world: the future summer headquarters of the Simulmatics Corporation.[5]

Ed Greenfield had big ideas and big ideals, big liberal ideas. For all his hucksterism, he was much more than an ad man: he was a philanthropist earnestly dedicated to mid-century American liberalism. He raised money for liberal causes, especially civil rights and civil liberties; he grabbed checks out of thin air, like a magician who pulls a nickel out from behind your ear. He served on the boards of the Fund for the Republic (which fought for the freedom of speech), the American Freedom of Residence Fund (which fought for desegregated housing), and Opera-

29

tion Crossroads Africa (a precursor to the Peace Corps). The civil rights attorney Harris Wofford, who would serve as John F. Kennedy's special assistant for civil rights and help found the Peace Corps in 1961, once advised Martin Luther King Jr., "Let me suggest that some time soon you try to talk with a good friend of mine, a very astute public relations man, Ed Greenfield."[6]

Very astute, my good friend, knows everyone. Very convivial. *So* fun. Warmhearted, witty. And that *laugh.* You'll love him. Very wide net. He was here, he was there, he was everywhere, the beautiful and clever Patty on his arm.

But for all his interests and acquaintances, Ed Greenfield's real passion was politics, Democratic politics, presidential politics. A huckster's game.

The People Machine began as a glint in Ed Greenfield's eye during the election of 1952, the first presidential election waged in the television age, the first presidential contest whose outcome was predicted by a computer, and the first presidential campaign orchestrated by a mass advertising firm. It was also, notably, a devastating loss for the Democratic Party.

Democrats had held the White House for two decades, since Franklin Delano Roosevelt's historic election in 1932. By 1952,

30

liberalism, which crossed party lines, appeared triumphant, unassailable. This turned out to have been an illusion, but at the time, it was hardly questioned. In the 1930s, Democrats and Republicans had fought over Roosevelt's New Deal, with Democrats arguing for the regulation of business and banking, and Republicans opposed. And they'd fought over the United States' entry into the Second World War, with Roosevelt in favor and Republican isolationists opposed. But starting in 1941, in the let's-all-come-together war years, and after 1945, in the isn't-life-grand postwar years, the distance between the parties had narrowed. What did they have left to fight over? After the Cold War began, in 1949, opposition to a common enemy tended to smooth over whatever differences remained. Republicans were still the party of business and Democrats the party of labor but both parties were liberal, and by 1952, Americans could hardly tell Democrats and Republicans apart: Tweedledee and Tweedledum.

Given that the differences between the parties were so few, the contest in 1952 seemed less likely to turn on policy issues than on the two candidates' personalities, which made it a perfect campaign for Madison Avenue. This posed a problem for Democrats, since, with rare exceptions, Republicans made much better use of ad men than did Democrats, a problem that got a whole lot worse when the Democratic nominee, Adlai E. Stevenson, decided to run on a platform that included opposition to the influence of advertising agencies on American politics.

Advertising was booming. In 1935, the Manhattan phone book listed ten public relations firms. By the middle of the 1950s, that same phone book list covered seven columns and contained the names of more than seven hundred firms, including Edward L. Greenfield & Co.[7] During the Second World War, American manufacturers had churned out for the Allies not only arms and ammunitions but clothing and food and more. After the war, hoping not to close shop but instead to find new markets for consumer goods, manufacturers churned out everything from dishwashers to hair curlers to Barbie dolls. To sell these products — many of which no one had ever thought to make or own before — manufacturers turned to advertising agencies, whose industry, between 1950 and 1955, grew from $6 billion to $9 billion. "We don't sell lipstick," one manufacturer explained. "We buy customers."[8]

Political campaigns had begun turning to advertising agencies, too, saying, in effect, "We don't sell candidates, we buy voters." Shrewd observers greeted this development with alarm. In 1951, the fearless muckraker Carey McWilliams published an explosive three-part series in the *Nation,* a profile of a married couple, Clem Whitaker and Leone Baxter, who ran a California company called Campaigns, Inc., the first political consulting firm in the history of the world.[9] They'd opened shop in 1933, chiefly running political campaigns for Republican candidates. For a long time, they'd taken only California clients. But beginning in 1949, they'd engaged in a national campaign, and they'd won: retained by the American Medical Association, they'd defeated a national health insurance plan proposed by the Democratic president, Harry S. Truman — the last, unfinished work of the New Deal. The AMA paid Campaigns, Inc., $3.5 million. "This must be a campaign to arouse and alert the American people in every walk of life, until it generates a great public crusade and a fundamental fight for freedom," Whitaker and Baxter's Plan of Campaign began. "Any other plan of action, in view of the drift towards socialization and despotism all over the world, would invite disaster."[10] When Whitaker and Baxter claimed that national health insurance amounted to socialized medicine, Truman fumed. Nothing in his bill, he insisted, "came any closer to socialism than the payments the American Medical Association makes to the advertising firm of Whitaker and Baxter to misrepresent my health program."[11] Whitaker and Baxter, McWilliams concluded, represented the new, cynical future of American politics. "This is expert political management," he wrote. "This is government by Whitaker and Baxter."[12]

For the Republican presidential nominee in 1952, Clem Whitaker and Leone Baxter liked the amiable and avuncular Dwight D. Eisenhower, former supreme commander of the Allied Expeditionary Force in Europe, a greatly admired war hero who had never been known to belong to either party. Eisenhower agreed to run out of a sense of duty to the country, even though he didn't think military men should occupy the Oval Office. He didn't enter the New Hampshire primary, but he won anyway, as a write-in, upsetting the conservative candidate, Robert Taft, an Ohio senator and the son of former president William Taft.

In 1952, most states didn't hold primaries and instead chose their nominees at state nominating conventions. Primaries weren't binding, and party leaders tended to ignore them. Instead, party leaders used polls to gauge the prospects of their candidates, and polls, of course, can be driven by advertising. Eisenhower won five primaries; Taft won six. But Eisenhower led in the polls.[13]

That summer, delegates to the Republican National Convention met in the International Amphitheatre, on the South Side of Chicago. Like a lot of big halls, it had originally been built for livestock shows: promenades of cattle. The convention was televised live, coast to coast, a first. Eisenhower won on the first ballot. To balance the ticket, party leaders connived to anoint as his running mate the shovel-jawed young California senator Richard M. Nixon, a Whitaker and Baxter protégé. Eisenhower was sixty-two; Nixon, thirty-nine. Eisenhower was a liberal, Nixon a ferocious anti-Communist. Whitaker and Baxter ran the Eisenhower-Nixon campaign in California.

The Republicans had put together a formidable ticket. The Democrats were vulnerable. Truman had assumed the presidency in 1945 with FDR's death and had been elected in 1948. He campaigned for a Fair Deal. But in 1952 he was unpopular, not least because voters blamed him for the United States' involvement in the Korean War. He also faced a challenge from within his own party from slender Tennessee senator Estes Kefauver, who'd made a national name for himself by heading a sensational investigation into organized crime. "I'm running on my own," Kefauver said, distancing himself from the party of Truman. Asked if he was a New Dealer and a Fair Dealer, he said, "Well, I don't classify myself on the dealers. I believe in progress."[14] Kefauver entered the Democratic primary in New Hampshire, where he campaigned in a coonskin hat by dogsled.

When Kefauver won, Truman announced that he would not seek another term and instead urged his former secretary of commerce, Averell Harriman, a wealthy New York businessman, to seek the nomination. Harriman hired Edward L. Greenfield & Co. to help manage his campaign. Harriman won only one primary. Kefauver entered fifteen primaries and won twelve. But at the Democratic National Convention, held in the same Chicago amphitheater as the Republican convention, the unruly delegates of the Democratic Party drafted into the contest Adlai E. Stevenson, the governor of Illinois, who hadn't run in a single primary.

Stevenson would become the Hamlet of American presidential politics. A moderate and a party loyalist who had served in both the Roosevelt and the Truman administrations, he was best known in 1952 for the role he'd played in establishing the United Nations. He was famously eloquent and learned. He also enjoyed nearly universal support among intellectuals, including the historian Arthur Schlesinger Jr. and the economist John Kenneth Galbraith, and among some of the nation's finest political writers, including *The New Yorker*'s Richard Rovere and John Hersey. (Rovere, wry and fair-minded, wrote the *The New Yorker*'s regular "Letter from Washington," and Hersey, one of the most acclaimed political reporters of the twentieth century, had written a breathtaking account of the bombing of Hiroshima and its aftermath.) Schlesinger, Galbraith, Rovere, and Hersey all wrote speeches for Stevenson, speeches celebrated for their erudition and elegance. Eisenhower placed his faith in ad men; Stevenson placed his faith in writers.

At the Democratic National Convention in Chicago, Stevenson agreed to enter his name into nomination. He won on the third ballot. His acceptance speech is one of the best in the history of American political rhetoric. He offered himself as a bridge between the democracy of the New Deal and the democracy of a new America. "The workingman, the farmer, the thoughtful businessman, all know that they are better off than ever before, and they all know that the greatest danger to free enterprise in this country died with the Great Depression under the hammer blows of the Democratic Party," he told a crowd that stamped in the stands. The Democratic Party, Stevenson argued, had rescued the country from the Depression and ushered in an age of abundance. And yet a danger lurked. Something evil stalked the land. The political savage. Senator Joseph McCarthy, a brawny Republican from Wisconsin, had in 1950 begun a campaign against supposed Communist subversives in the United States, a campaign of nearly unrivaled demagoguery. He stoked fear. He fought phantoms. He incited panic. He persecuted the weak. He lied. And people believed him. That night in Chicago, Adlai E. Stevenson of Illinois presented himself, not to his party but to the nation, as a political savior who could rescue Americans from the malice and vulgarity of modern American politics.

"I hope and pray that we Democrats, win or lose, can campaign not as a crusade to exterminate the opposing Party, as our opponents seem to prefer," Stevenson said, "but as a great opportunity to educate and elevate a people whose destiny is leadership, not alone of a rich and prosperous, contented country, as in the past, but of a world in ferment." Promising to "talk sense to the American people," Stevenson spoke with a forcefulness and a dedication to principle not often heard in American politics, before or after. Truman talked with something of a whine. Eisenhower stammered. Nixon raged. McCarthy seethed and sweated and spat. Stevenson spoke with the precision of a scholar and the power of a poet. "Better we lose the election than mislead the people," he said, "and better we lose than misgovern the people."[15] He ran against dishonesty. He ran against demagoguery itself.

In the late, candlelit decades of the unruly eighteenth century, American political philosophers had thought a great deal about the dangers of demagoguery. "Men of factious tempers, of local prejudices, or of sinister designs may, by intrigue, by corruption, or by other means, first obtain the suffrages, and then betray the interests, of the people," James Madison warned in 1787.[16] Madison drafted a Constitution designed to thwart such men by the nature and very structure of government, its separation of powers, its checks and balances. But the framers hadn't anticipated the electrified, neon-glowing, vacuum-tubed twentieth century's methods and machines of mass advertising and political manipulation, methods and machines so powerful that they sparked a panic about something well beyond demagoguery and into mind control.

Adlai Stevenson worried about all of that. McCarthy troubled him, led him to fear for the republic. But he had another concern, too, one that had to do with the cost Democrats and Democratic policies had paid for Republicans' willingness to engage the services of advertising agencies. In 1952, after the Republican National Convention, Eisenhower's national campaign retained one of the largest advertising agencies in the United States, Batten, Barton, Durstine & Osborn, better known as BBDO. (Straight-talking Harry S. Truman said BBDO really stood for "Bunko, Bull, Deceit, and Obfuscation.") BBDO's ad men sold Eisenhower like a laundry detergent. The most-watched television advertisement of 1952 was a Disney-produced animated short of little people marching in a parade, led by an elephant, singing a jingle written by Irving Berlin: "You like Ike, I like Ike, everybody likes Ike." Eisenhower became the first presidential candidate to appear in televised ads, including one called "The Man from Abilene": it borrowed its graphics from the television version of *Superman.*[17]

The Eisenhower campaign hired ad men; the Stevenson campaign denounced them. To run against demagoguery is to commit to a campaign of restraint, a campaign of decorum, a campaign of understatement. For Stevenson, running against dishonesty in American politics meant running a campaign almost entirely without the aid, or at least without the seeming aid, of Madison Avenue. Forty-two-year-old George Ball, a principled New Dealer and former law partner of Stevenson's, headed Volunteers for Stevenson. Some people "like Elvis Presley, and I like Marilyn Monroe," Ball said, in a much-reported speech, "but I doubt that is sufficient reason for electing either president." Ball dubbed the Eisenhower operation the Cornflakes Campaign.[18]

Behind the Stevenson campaign's criticism of the role of mass advertising in American politics lay another fear. "Brainwashing" entered the American lexicon in 1951 with the publication of *Brain-Washing in Red China: The Calculated Destruction of Men's Minds,* by the journalist Edward Hunter. Hunter promised to reveal "the terrifying methods that have put an entire nation under hypnotic control."[19] "Brain-washing" was Hunter's translation of the Chinese *hsi nao,* and he used it to describe Communist China's methods of Maoist indoctrination. (At the end of the Korean War, American psychologists would be charged with interviewing former prisoners of war to determine if they had been brainwashed, too, a story that became the plot of the 1959 novel *The Manchurian Candidate.*)[20] If McCarthy tapped into a fear that Communists were secretly controlling Americans' minds, animus against mass advertising tapped into a growing fear that someone else was tampering with the American mind. Not Chairman Mao or the Communist Party or the Soviets but American ad agencies.

For these reasons and more, Stevenson balked at campaigning on television. He found the idea of appearing in television advertisements — advertising himself — undignified and insulting to the office of president. He refused. Eisenhower had no such qualms or, at least, he'd had those qualms assuaged by Rosser Reeves, of the

41

Ted Bates advertising agency.

Reeves was the unquestioned top dog of Madison Avenue, later famous for the campaign he'd devise for M&M's ("Melts in your mouth, not in your hand"). "I think of a man in the voting booth who hesitates between two levers as if he were pausing between competing tubes of toothpaste in a drugstore," Reeves explained. "The brand that has made the highest penetration on his brain will win his choice."[21] Speeches were long and boring; candidates had to be willing to get their message across in under a minute. For the Eisenhower campaign, Reeves proposed to make a series of short television ads, called spots. "Is there a new way of campaigning that can guarantee victory for Eisenhower in November?" Reeves asked. "The answer is: 'Yes!' . . . Most people do not know the power of spots. However, here are the cold facts. THE HUMBLE RADIO OR TV 'SPOT' CAN DELIVER MORE LISTENERS FOR LESS MONEY THAN ANY OTHER FORM OF ADVERTISING. Let us repeat that!! THE HUMBLE RADIO OR TV SPOT CAN DELIVER MORE LISTENERS FOR LESS MONEY THAN ANY OTHER FORM OF ADVERTISING."[22]

Reeves also pioneered targeted political advertising. To win in 1952, Republicans needed to flip the vote in forty-nine counties and twelve states they'd lost in the last election. Reeves made their spots for those counties. He titled his series of spots "Eisenhower Answers America." After George Gallup conducted polls to establish what Americans cared about the most, Reeves wrote scripts for Eisenhower, answers to questions raised by voters, about those issues. Eisenhower read the answers from cue cards, and then Reeves got ordinary Americans, off the streets of New York — tourists waiting in line outside Radio City Music Hall — to come inside and read the questions off the cue cards.

"General, the Democrats are telling me I never had it so good," says a young black man in a suit.

Answers Eisenhower: "Can that be true when America is billions in debt, when prices have doubled, when taxes break our backs and we are still fighting in Korea? It's tragic. And it's time for a change."[23]

Stevenson supporters found Eisenhower's spots embarrassing. Eisenhower was "a plodding five-star general uttering pedestrian language written by some journalistic hack with all the grace of a gun carriage being hauled across cobblestones," George Ball complained. Stevenson, Ball said, "was a man of culture and intellect seeking not only to educate the country but also to elevate its taste."[24] And that, alas, was the heart of the problem. The country didn't much want to be educated and elevated. It wanted slogans.

Melts in your mouth, not in your hand! I like Ike! It's time for a change!

Meanwhile, Nixon went after Stevenson, viciously, which is the way Nixon always campaigned, not educating and elevating the country but misinforming and degrading it. He called Stevenson a "weakling, a hustler, and a small-caliber Truman." He dubbed him "Adlai the Appeaser." McCarthy attacked Stevenson as a Communist; Nixon merely slyly *hinted* that Stevenson was a Communist. (Stevenson called Nixonism "McCarthyism in a white collar.")[25] And Nixon used television even more effectively than Eisenhower, answering seemingly career-ruining charges of corruption in a televised speech that ended when an aw-shucks Nixon admitted that he had indeed taken a campaign gift: he'd accepted the gift of Checkers, a black-and-white spaniel, and "we're going to keep it." Ball later said that watching the Checkers speech was like watching a Geritol commercial.[26] But Geritol commercials sell a lot of Geritol.

Stevenson all but boycotted television advertising, agreeing only to have his speeches televised — very long speeches, which few stations were willing to broadcast and few people bothered to watch no matter how well-written they were. In 1952, Republicans spent $1.5 million on television advertising to the Democrats' puny $77,000.[27]

44

This seemed, somehow, unfair. Ball asked the Federal Communications Commission to look into the legality of the Republican television spots, suspecting them of violating provisions of the 1934 Communications Act (requiring equal time for candidates) and of the Corrupt Practices Act (governing campaign spending). But the FCC said there was nothing to be done.[28] American politics never looked back.

Plenty of other problems riddled Stevenson's campaign. To Eisenhower's "I like Ike," Stevenson's campaign offered a pathetic rejoinder: "Madly for Adlai." The fact that he was divorced hurt him, and badly. He was also saddled with a terrible lower ticket as a consequence of the splintering of the Democratic Party in the last election. In 1948, southerners had bolted from the Democratic convention over civil rights and held their own convention, as the Dixiecrat Party, under this platform: "We stand for the segregation of the races and the racial integrity of each race." In 1952, in a Faustian bargain, the Democratic convention chose as Stevenson's running mate the conservative John Sparkman of Alabama, a leading segregationist. The party was trying to stitch itself back together by betraying civil rights. It didn't work. Meanwhile, Republicans, tapping into the deep well of anti-intellectualism in American life, made Stevenson's learnedness and eloquence into liabilities, mocking the bald Stevenson as an "egghead." Stevenson's weakness for obscure allusions and bad puns didn't help, either. "Eggheads of the world unite!" he once said. "You have nothing to lose but your yolks!"[29]

Many Republicans admired Stevenson; newspaper columnist Stewart Alsop asked a Connecticut Republican whether they'd also vote for him. "Sure," the Republican answered. "All the eggheads love Stevenson. But how many eggheads do you think there are?"[30]

Still, the race looked close, too close to call. On Election Day, Stevenson followed the returns from the governor's mansion in Illinois while Eisenhower waited for the results at the Commodore Hotel in New York. On television that night, for the first time ever, the networks broadcast coast to coast.

In 1952, television news executives hoped to undo the damage done on Election Night in 1948, when not only had the coverage been painfully boring (there was very little to watch except commentators who had very little information), but the networks, like the newspapers, had also made a colossal error, calling the election for the Republican Thomas E. Dewey. (A jubilant president had later held up the front page of the *Chicago Daily Tribune*, displaying its headline, "DEWEY

DEFEATS TRUMAN.")

For Election Night 1952, CBS had a plan, a secret plan known as Project X. Like the rest of America, Ed Greenfield watched, rapt. "Balmy weather over most of the United States today, and a record turnout apparently throughout the United States," said CBS newsman Walter Cronkite, early in the evening, speaking to the nation from CBS's Studio 41, just above Grand Central Station. They'd outfitted the studio with a giant wall map of the United States, to be filled in as the returns came in. States that went for Stevenson would be colored black and states that went for Eisenhower would get stripes, like so many zebras (it was the age of black-and-white television, too soon for red and blue).

But the real star of the evening wasn't Cronkite or the map. It was a computer, the first computer most Americans had ever seen. The Universal Automatic Computer, or UNIVAC, had been built in 1951 by a company named Remington Rand, for the U.S. Census Bureau. In 1952, CBS had arranged to hire a UNIVAC for Election Night, to tally returns and predict the winner. "A ROBOT COMPUTER WILL GIVE CBS THE FASTEST REPORTING IN HISTORY," read the headline.[31] The UNIVAC would stay in Philadelphia — it was too big to move — but in New York, CBS would install a fake, a console lit, from

47

the inside, by a string of Christmas lights. The first computer most Americans ever saw was an empty shell: a stunt.

On Election Night, at Studio 41 in New York, legendary CBS newsman Charles Collingwood sat at the fake UNIVAC terminal to offer the predictions he'd be getting, by telephone, from an actual UNIVAC, in Philadelphia. Cronkite smiled at the camera and addressed the audience:

"And now for perhaps a prediction on how this voting is going, what the vote that is in so far means, let's turn so far to that miracle of the modern age, the electronic brain, UNIVAC, and Charles Collingwood."

The camera panned to the corner of the studio and zoomed in on Collingwood, at a console kitted up with lights on timers, to make it look as though it was doing something.

"This is the face of a UNIVAC," purred Collingwood. "A UNIVAC is a fabulous electronic machine which we have borrowed to help us predict this election from the basis of the early returns as they come in. UNIVAC is going to try to predict the winner for us just as early as we can possibly get the returns in. . . . This is not a joke or a trick. It's an experiment, and we think it's going to work. We don't know, we hope it will work . . ."

It didn't work. Or at least it didn't work

48

well. Cronkite kept throwing to Collingwood — "And now to find out perhaps what this all means, at least in the electronic age, let's turn to that electronic miracle, the electronic brain, UNIVAC, with a report from Charles Collingwood" — but Collingwood kept having to explain that no prediction had yet come in.

"UNIVAC, our fabulous mathematical brain, is down in Philadelphia mulling over the returns that we've sent him so far," Collingwood said, filling time. "He's sitting there in his corner, humming away. A few minutes ago I asked him what his prediction was, and he sent me back a very caustic answer, for a machine. He said that if we continue to be so late in sending him the results, it's going to take him a few minutes to find out just what the prediction is going to be. So he's not ready yet with his prediction but we're going to go to him in just a little while."

Not long after midnight, CBS turned to the man from Remington Rand, in Philadelphia with the actual UNIVAC, for an explanation. He claimed that UNIVAC had predicted an Eisenhower landslide early in the evening, but he'd been too nervous about it to pass the projection along to Collingwood. "When UNIVAC made its first prediction with only three million votes in, it gave five states for Stevenson, 43 for Eisenhower, 93 electoral votes for Stevenson, 438 for Eisenhower," he said. "We just plain didn't believe it." In the end, Eisenhower won 442 Electoral College votes to Stevenson's 89 and 55.2% of the popular vote to Stevenson's 44.3%. A rout. UNIVAC had been right.

Ed Greenfield was mesmerized. Two new machines, the television and the computer, were transforming American politics. The influence of the first was much easier to see than the influence of the second. But the way Greenfield figured it, Republicans had made better use of television in 1952, by way of advertising, which meant that Democrats ought to figure out how to make better use of the computer, and fast, because what could be more valuable to a campaign than a computer that could predict the vote?

Eyeing the election of 1956, Greenfield began collecting men, the very best.[32] Not Madison Avenue men but scientists. He knew everyone. Very convivial. Very astute. The man was a magnet. Ed Greenfield, University of Chicago, Yale. He cast his net, fishing for men, the best and the brightest scientists of the mind and minders of machines, gigantic brains. He went to California, where he found Eugene Burdick in the ocean, surfing.

CHAPTER 2
IMPOSSIBLE MAN

$f + h = p$
(fear plus hate equals power)
— Advertising copy for Eugene Burdick,
The Ninth Wave, 1956

Eugene Burdick as the "Ale man," c. 1961.

Everyone assumed that Eugene Burdick was a spy, which meant, of course, that he wasn't. A man can't drink whiskey with Marlon Brando and eat dinner with Ingrid Bergman and teach political theory to PhD students and advise American presidents and write best-selling novels that get made into big-budget Hollywood films and also steal secrets, undercover, incognito, unknown, as if he were nobody.

Or maybe he could. If anyone could pull it off, it would've been Burdick. He was an impossible man, exactly the sort of man he was forever inventing in his fiction, Cold War thrillers starring dashing men of mystery, rugged, daring, and brilliant, who defended Americans against the age of automation. Once, when Burdick sent a short story to *The New Yorker* — "Happy Man in Berlin" — his editor wrote back, "I understand that Pico is a skillful man, but sometimes he seems almost too miraculous."[1] Burdick, too.

Eugene Burdick walked with the rubbery gait of a surfer and wore owl's-eye glasses and smoked a pipe and liked to be photographed sitting at his typewriter, an old Royal, well-used and well-oiled. He had two selves and two costumes, both disguises: he wore scuba gear; he wore tweed suits. He was James Bond; he was Ernest Hemingway. He didn't write like Hemingway, but he looked like him, with the same hair, straight and thin and flat, and the same head, too: as boxy as a biscuit tin. Wallace Stegner once said Burdick was "as energetic as a bulldozer and as persistent as an Egyptian fly."[2]

He'd been born in Sheldon, Iowa, a railroad town, in 1918, at the end of the century of the railroad, the great machine of the nineteenth century, whose coal-black, smoke-puffing locomotives chugged from town to town on tracks that reached across the continent like the legs of a giant arachnid. He was named after Indiana-born Eugene V. Debs, the onetime railroad worker, founder of the American Railway Union, and self-proclaimed socialist who, in 1920, when Burdick was not yet two, ran for president for the fifth and final time, from prison in Atlanta, as Convict No. 9653, with the campaign slogan "From the Jail House to the White House." It earned him nearly a million votes.[3]

Democracy is a mystery. In the 1950s, Eugene Burdick wanted to solve that mystery. He wanted to know whether or not it could be solved by the great machine of the twentieth century, the computer, whose whirring, blinking parts were housed in gray boxes and stored in gleaming, temperature-controlled rooms, as if they were streamlined steel coffins, standing on end, in a hospital morgue. He decided to immerse himself in a new field that became known as behavioral science,

53

which was why, in the late summer of 1954, he drove his Jaguar up the winding, redwood-lined roads of Palo Alto, California, to a paradise sometimes called Lotus Land, a Buddhist-style, monastic retreat of cedar and glass perched on top of a hill overlooking San Francisco Bay: the Center for Advanced Study in the Behavioral Sciences.[4] It was exactly the sort of place where Ed Greenfield would turn up, looking for men to staff a new department of Edward L. Greenfield & Co., a Social Science Division, with which he hoped to help get a Democrat elected president in 1956, by a method other than plastering the candidate's face on boxes of corn-flakes or hawking him like canned soup in one-minute television spots, as if the Oval Office were a kitchen pantry.

Eugene Burdick was a beach boy. After Iowa, he'd grown up in Los Angeles. His father died when he was four. His mother went to work in a waffle shop; she sent the children out to live with neighbors, little vagabonds, wandering the beaches and dunes and board-walks. Burdick started smoking cigars and sleeping with one of his teachers when he was fourteen, a story he told in his first, partly autobiographical novel, *The Ninth Wave*.[5] (The novel's lead character counts the women he's slept with the way other people count sheep, ticking off "each pinkening breast,

54

each exhalation of breath, the twist of a thigh, the feel of hair and flesh and moisture.")[6] He had no money for college and worked at a life insurance company to pay for a year at Santa Barbara State College before going to Stanford, where he supported himself by waiting tables while studying psychology, from Freud to Fromm.

He was interested in everything. "He read anthropology, sociology, psychology, mathematics, philosophic, ethics, history, and logic," he wrote in *The Ninth Wave.* "He stole books from the library, bought used books in Palo Alto, borrowed books and bought still more books. Some books he glanced at, threw under his bed and never opened again. Some books he read twice."[7] He mulled over the political theory of the great American poet Carl Sandburg: "The people choose and the people's choice more often than not is one more washout."[8]

The DC comic book superhero Aquaman debuted in 1941, an emblem of American naval forces in the Pacific, but the Protector of the Deep had nothing on Eugene Burdick, hero of the high seas. He graduated from Stanford in 1942, married Carol Warren, the daughter of a San Francisco school superintendent, and shipped out to Treasure Island with the navy. The next year, patrolling off Guadalcanal, he dove off his own boat to swim through burning oil, exploding am-

55

munition, and falling wreckage, to rescue the survivors of a torpedoed battleship.[9]

After the war, Burdick went back to Palo Alto, where trees on Stanford's nine thousand wooded acres were being felled to make way for laboratories. Five million people moved to California in the 1950s, including African Americans, Asian Americans, Mexican immigrants, and a great many young, white veterans like Burdick, men who during the war had learned how to fly planes or fix tanks or use radar, training from which servicemen of color had typically been barred, as had nearly all women. Wartime training in hand, men like Burdick went to California to study engineering and to find jobs in the Defense Department–funded, $8 billion-a-year electronics industry. Palo Alto, a land of farms and orchards, a paradise of fruit trees, was becoming a boomtown, a land of electronics factories and microwave labs, a valley of silicon.[10]

Burdick had a different idea. He planned to become a political scientist. But he'd read a lot of novels in the navy, and he also wanted to be a writer, so he enrolled in a Stanford fiction-writing class taught by the novelist Wallace Stegner, whose courses would one day become legendary. Burdick gave Stegner a story he'd written about his service in the Pacific, a story he'd dug out of his navy foot locker. "It wasn't very finished," Stegner later remembered, "but it ran over the class like a tank."[11] That was Burdick, all over.

Burdick revised the story and sent it to *Harper's.* "Rest Camp on Maui" won second prize for the O. Henry Award, and the decorated navy gunner became a celebrated fiction writer. He won a fellowship at the Bread Loaf Writers' Conference. The plaudits kept coming. He telephoned Stegner. "For the love of Mike, you know what? I'm a Rhodes Scholar!"[12]

He didn't much love Oxford, a cloister of stone.[13] England had suffered during the war and, while not defeated, had emerged depleted. Burdick had no patience for that, no sympathy for it, no sense of the lingering of loss. Instead, he wrote about England with a callow disappointment. He wrote about how he'd expected to find Oxford's famous debating society a place where "wildly precocious youths, their eye firmly fixed on the main chance at parliament, debate with cruelly deflating epigrams." But he found it full of nothing so much as "dozens of stuttering youths" who say very little of any interest to anyone, least of all of any interest to an undersea diver who read Machiavelli before breakfast.[14]

After Oxford, he traveled the world and then landed back by the bay, as a professor in the political science department at the University of California, Berkeley. "There is a waiting list for his classes," the *San Francisco Examiner* reported, and "students — most of them female — flock to his office hours."[15] In Political Science 212B, Contemporary Political Theory, Professor Burdick raised a topic at the start of the term: "the crisis in method." Political theory was on the wane, quantification on the rise. But to what end? A lecture hall full of girls in skirts and baby blue eye shadow and midnight black mascara eyes looked up at him adoringly. What was the way out of political science's crisis in method? Burdick directed them to "the new epistemology of Harold Lasswell."[16] Ed Greenfield had studied with Lasswell in 1950. He'd imbibed Lasswell's epistemology like so much wine, had fallen under his intoxicating sway.

Burdick was fascinated by Lasswell, too. But Burdick wasn't really under anyone's sway. Instead, he followed his own curiosity. In 1954, thirty-five, he would spend his sabbatical with Lasswell, studying the mathematics of mass persuasion — the punch-carded computer programs that purported to model the behavior of the American voter, the fickle Coke-or-Pepsi voter, the television-watching, shopping cart–pushing, Sears-Maytag, Eisenhower-Nixon voter. And he meant to do that in the company of the nation's most distinguished social scientists, men in horn-rims and bow ties and V-neck sweaters, at the latest front in the Cold War battle of minds: California.

The year Eugene Burdick began his sabbatical at the Center for Advanced Study in the Behavioral Sciences, the long struggle for civil rights moved to the very center of American politics. African Americans had been fighting for those rights for decades, but in May 1954, in the landmark decision *Brown v. Board of Education,* the Supreme Court denounced the regime of Jim Crow and ruled racially segregated schools unconstitutional. The decision, heralded and celebrated in much of the country, also fueled white supremacy organizations: terrorists attacked black churches and schools and pro—civil rights temples and synagogues, setting fires, igniting bombs, commiting murders.

Burdick had some but not much of this on his mind in the James Bond and bikini summer of 1954, as he drove his Jaguar convertible through stands of palms and cedars and piñons and eucalyptus groves, up to the Center for Advanced Study in the Behavioral Sciences. He meant to spend his year — the year of Elvis Presley's debut, the year of the polio vaccine, the year of Joseph McCarthy's downfall — away from his office, away from his classes on Hobbes and Locke, away from his colleagues and their sniping at his successes, away from his students, the boys he could still beat at squash, the girls in tight sweaters. He was particularly keen to meet Lasswell, fifty-two, bald and jowly, best known for two works from the 1930s, *Psychopathology and Politics* (1930) and *Politics: Who Gets What, When, How* (1936). Lasswell had psychologized political science. He also likely introduced Eugene Burdick to Ed Greenfield. Later, Lasswell would help found Simulmatics, a company in which he held a great deal of stock.

"The American public believes that it 'chooses' the Party candidates for the Presidency and then makes a free and sovereign choice between the two candidates," Burdick would later write. "This is hardly an accurate description of what happens. The American public believes it is sovereign. It is not."[17] Burdick arrived at this understanding of American elections during his year at the center, and mainly from Lasswell.

Harold Lasswell had gone to the University of Chicago when he was only sixteen, completed a PhD in political science, and published his dissertation, *Propaganda Technique in the World War,* in 1927, when he was twenty-five. Then he'd gone to Berlin to be psychoanalyzed by a disciple of Freud's, before accepting a position teaching at the University of Chicago and publishing his two seminal works. If Burdick was a bulldozer,

60

Lasswell was a virus. He didn't lecture people; he infected them. He flared his nostrils and sniffed. He held forth like an oracle, as if he were Aristotle. "His monologues are symposiums," said one student. "Lasswell was not a teacher but a tornado."[18] People treated him like a god. He flirted with boys. He humiliated girls. "If you asked Harold a question, he would say, 'We don't know enough about that,' and you knew he meant the entire body of knowledge of the universe, because if anyone knew it, he knew it," said Naomi Spatz, the woman for whom Ed Greenfield would one day leave his wife. Once, when Spatz had Lasswell over to brunch at her apartment, her cat started rubbing against him. "The cat likes you," Spatz said. Answered Lasswell, nostrils flaring, "He knows where the power is."[19]

Lasswell enjoyed such influence because his work purported to explain *"who* says *what* to *whom* in *which channel* and with *what effect."*[20] He claimed to know how ideas get into people's heads — and how to get them out. During the Second World War, he founded a war communications research project at the Library of Congress and recommended that the United States preserve democracy from authoritarianism by way of systematic, government-run mass manipulation.[21] For a long time, this stuff was known either as propaganda or as psychological

61

warfare (the Nazi version was known as *Weltanschauungskrieg,* or worldview warfare), but after a while, people who worried about how that sounded started calling it the study of "mass communication."[22] With the war over, this sort of work also required a new justification. Who says what to whom with what effect? That's the question that had drawn Ed Greenfield into Lasswell's seminar room at Yale in 1950. In peacetime, there seemed no better place to study mass communication, or psychological warfare, than democratic campaigns and elections, because campaigns produce their own propaganda, a body of political opinion, and elections produce their own data, an inventory of political behavior.

The pioneer in this field was an Austrian émigré named Paul Lazarsfeld who, like Burdick and Lasswell, went to the center in Palo Alto in 1954. "Every four years, the country stages a large-scale experiment in political propaganda and public opinion," Lazarsfeld explained in 1944, the year he founded the Bureau of Applied Social Research at Columbia.[23] The data piled up. In 1948, the University of Michigan launched a massive voter study known as the American National Election Study, one of the biggest social science research projects ever undertaken, and Lazarsfeld and his colleague Bernard Berelson embarked on "a two-phase study of how the voter makes up his mind in congressional and state elections," eventually producing a landmark work, *Voting: A Study of Opinion Formation in a Presidential Campaign.*[24] Their coauthor would be a fellow named William McPhee. One day soon, Ed Greenfield would meet McPhee, too: he would put him in his collection of men, a collection he called "the Group."

In November 1954, Democrats won back the House in an election that amounted to a repudiation of McCarthyism. In December, the Senate formally censured McCarthy, and McCarthyism, as Eisenhower liked to say, became "McCarthywasm." Nixonism, though, was still Nixonism.

Ed Greenfield, meanwhile, was putting together the Social Science Division of his advertising agency. Here was another way to fight demagoguery: with a very big, super-fast calculating machine. To find a way to bring the speed of giant electronic brains into political campaigning, Greenfield needed to gather together scientists from this brave new world, quantitative social scientists, who were engaged in founding a new body of scholarship, a new field of knowledge. Its assumptions would take over universities, spread across the intellectual landscape, and leak into the soil of American culture itself so entirely that people would forget that it had emerged out of the urgency of the Cold War.

Cold Warriors understood themselves as engaged in a battle over the future. To win that battle, they tried to turn a lot of things, including the study of human behavior, into predictive sciences. The more dangerous the Cold War got, the more madly its scientists scrambled to foretell the future. And the more heedlessly and violently they cast aside the past, the knowledge of ages, the humanities, the study of the human condition: history, philosophy, literature.

In 1949, the Soviet Union had tested an atomic bomb and China had become Communist. Recruited into a Manichaean battle between Americanism and communism, social scientists began waging wars on two fronts: protecting American democracy from demagoguery and fighting the spread of communism. They asked two questions: How do voters in a democracy form their opinions? And how can the influence of communism be stopped? Answering these questions led social scientists to gather large bodies of data — not only in the United States, where they collected public opinion surveys, campaign polls, and election returns, but also in Europe, where they interviewed Soviet dissidents and Warsaw Pact refugees and made transcriptions of Soviet radio and television broadcasts. No longer did American political scientists require merely an explanation of who says what to whom with what effect; they required a prediction, based on a mathematical model. If we tell voters X, will they then do Y? For that, they looked to physics. The mathematics of the targeting of messages would be derived from the targeting of missiles and it, too, would require computers.

Anyone who wanted to use computers to study and produce propaganda needed vast sums of money; most of that money came from the Ford Foundation, soon to become the richest philanthropy in the world. Established by Edsel Ford in 1936, it had been taken over in 1947 by his son Henry Ford II, who tapped California lawyer H. Rowan Gaither to help prepare a report about the foundation's postwar priorities. Gaither's report complained that existing ways of studying human behavior — both ancient humanistic fields like philosophy and political theory and modern fields like psychology and sociology — were "polemical, speculative, and pre-scientific." He recommended turning the study of human behavior into a science, like physics, to be based on "experiment, the accumulation of data, the framing of general theories, attempts to verify the theories, and prediction."[25] After all, if a body of knowledge couldn't be used to make predictions, what use was it?

Before working for Ford, Gaither had helped found the RAND Corporation, in Santa Monica. RAND — short for "Research and Development" — had begun as an arm of the U.S. Air Force, part of the Douglas Aircraft Company, but in 1948, newly independent, and with funding from both the Department of Defense and the Ford Foundation, RAND had hired a pioneer in psychological warfare to head its new Social Science Division.[26] Its projects included charting, mathematically, "a general theory of the future."[27]

Prophecy is ancient, a species of mysticism. Prediction as a quantitative social science was new: outside of economics, social scientists had not, till then, generally made predictions. But Ford asked social scientists to make prediction the entire object of their research, to claim knowledge — empirical, probabilistic, mathematical knowledge — of what would happen next. This, however, was no less mystical than the ancient art of prophecy. The attempt to derive a general theory of the future rested on the presumption that human behavior can be predicted, a presumption that, in turn, rested on the conviction that reckoning with people and societies and the human mind itself isn't the stuff of poetry and paintings and philosophy but is, instead, the stuff of numbers and graphs and simulations.[28] The Ford Foundation called this field of study "the behavioral sciences."

In 1951, when the Ford Foundation established a special division for the study of the human mind, it named it the "Behavioral Sciences Division." The next year, Ford agreed to a grant of $3.5 million to establish the Center for Advanced Study in the Behavioral Sciences.[29] But what even were the "behavioral sciences"? The best explanation came from the economist Kenneth Boulding, who said a behavioral science was a science "that gets money from the Ford Foundation."[30] The second best explanation came from Eugene Burdick, who, in a novel, made the question into a sexual tease.

"What does a behavioral scientist do?" he asked.
"I study the behavior of humans," she said, and smiled.[31]

During the center's inaugural years, its first fellows, Eugene Burdick, Harold Lasswell, and Paul Lazarsfeld, took turns offering "confessions," each scholar telling the story of his life.[32] Then began the work. There were no telephones. The fellows were required to follow a strict schedule.[33] Boulding, following that schedule, pondered the imperative of prediction. "In about an hour I shall rise, leave my office, go to a car, drive down to my home, play with the children, have supper, perhaps read a book, go to bed," he wrote from his desk at the center. "I can predict this behavior with a fair degree of accuracy because of the knowledge which I have: the knowledge that I have a home not far away, to which I am accustomed to go. The prediction, of course, may not be fulfilled. There may be an earthquake, I may have an accident with the car on the way home, I may get home to find that my family has been suddenly called away."[34]

Most of the fellows held a deep faith, a worshipful faith, in the predictability of human behavior, believing that it required only knowledge of the world, a theory of behavior, and a sufficiently sophisticated mathematical model. This was why Burdick had come.[35] Trained as a political theorist, he was eager to measure the influence of the quantitative turn in the study of man. He attended an intensive seminar on mathematics. He began assembling a collection of essays on the most promising new research, editing an anthology called *American Voting Behavior.*[36] He was impressed, but he was also worried.

As Burdick saw it, quantitative political science proceeded with very little concern either for political theory or for the actual workings of democracy. Liberal democratic theory posits the rationality of citizens and their interest in and active participation in politics, but the citizen described in the voting behavior research does not meet even a minimum test for rationality and is not very interested in politics, or involved, either. Democratic self-government relies on feelings of belonging and political community, and divisions that include political parties and interest groups, but the voting studies sorted voters into artificial, identity-based categories like "upper-middle-Catholic-urban." Political parties are self-conscious groups, Burdick pointed out, and so are interest groups, like farmers, but "upper-middle-Catholic-urban" people are not; they're just an invention of a punch-card sorter. It was possible they could be *turned into* a group, by virtue of being classed as one, consistently, but that would seem more likely to harm than to benefit the polity. Burdick wondered whether pluralism could survive when the nation's political scientists had dedicated themselves to the project of segmenting the electorate.[37]

He was not alone in his skepticism about the real-world effects of the behavioral sciences. At a holiday party, the fellows sang, to the tune of "I've Been Working on the Railroad," lyrics written by Boulding:

Something is the matter with Science,
Something is the matter with the plan (I've
    got it!)
Something is the matter with Science
For there ain't no science of Man.[38]

They were confident, and they were cocky. But they sometimes wondered if all of it

69

wasn't mainly bullshit.

Eugene Burdick spent most of his year at the Center for Advanced Study in the Behavioral Sciences failing to produce scholarship and instead writing a novel, *The Ninth Wave,* a dystopian political thriller about the problems with the quantitative study of voting behavior. "The Center is not very healthy or sympathetic an environment in which to grow the delicate flower of fiction," he confided to a friend. "But I am giving it a good try."[39]

In *The Ninth Wave,* Burdick created a version of himself in a character called Mike Freesmith, a surfer who studies psychology at Stanford, where, by observation and experiment, he derives a set of cold-blooded and finally monstrous principles of human behavior. According to Freesmith's Fear Principle, "There is one thing that the masses know: real authority. And a real authority is someone who can satisfy their desire to hate and their fear. A good authority works the two of them together."[40] This insight leads Freesmith into politics: he finds a demagogue who preaches hate and fear, and offers to help him get elected to office.

Burdick had observed this sort of thing firsthand in the rise of Richard Nixon, a California boy and, also like Burdick, a naval officer. Nixon had run for a seat in Congress in 1946. Handsome, well-educated war heroes were very electable. John F. Kennedy was elected to Congress in Massachusetts in 1946, too. ("Second place is failure," Kennedy's wealthy, influential, and hard-charging father had always told him.)[41] But Nixon undertook a new style of campaign, a campaign of hate and fear, something he got better at each time he ran. In 1950, when he ran for a seat in the Senate, he allegedly said his opponent, Democrat Helen Gahagan Douglas, was "Pink right down to her underwear."[42] Nixon, like McCarthy, served on the House Un-American Activities Committee, where, like McCarthy, if with more polish and more subtlety, he gleefully accused a lot of other people of wearing pink underwear, too. McCarthy had been exposed, defeated, humiliated, censured. Nixon had been elected vice president. Burdick wondered what a man like Nixon could do, if he also had the latest behavioral science behind him. He tried to picture it.

In *The Ninth Wave,* Freesmith tries to get the demagogue in his pocket elected governor of California. "You start with this," Freesmith says, handing a campaign worker named Georgia a printed and annotated census as he shows her around campaign headquarters, a room that houses a giant mainframe computer. "Tells you how the population breaks down: how many street cleaners, Negroes, veterans, trade union members, truck drivers, fry cooks, Protestants, Jews, Catholics, foreign born, Okies, doctors and teachers there are. Also how much money they make, the size of their houses, the kind of car they drive, the degree of education, lodges they join and a lot more." This, Freesmith explains, is the "Great Beast," the public. From the Great Beast, he goes on, you make a Little Beast, a three-thousand-person sample, and you engage a polling company to ask those three thousand people a series of questions. "Then they punch the answers into the IBM cards and bring them back here and we run them through the machine. We pay them three dollars for each card," Freesmith says, pointing to stacks of cards.

Georgia asks to see a card. Burdick lingered on this description. His readers had likely never seen a punch card.

"It was rectangular and its face was covered with closely printed, black rows of numbers. Some of the numbers had been punched out, leaving tiny slots in the cards. There were no words on the cards." A pollster holds the card up to the light and reports, "Subject is: White. Male. Thirty-four years old. Catholic. Married. Three children. Clerk. Less than four thousand and more than thirty-five hundred a year. No television. In debt."[43]

Georgia learns to feed the punch cards into a reading machine. Burdick writes, "She put the cards from the box into the hopper and pressed the button. The machine began to purr. She looked down at it with pleasure, moved her fingertips lightly over the quivering surface. Then she touched the lever and the cards began to flick through the machine."[44]

Burdick is not a subtle writer of fiction. He's more of an armored tank than a sniper. Georgia, feeding the cards into the machine, looks out the office window, across the street, and into another office, where a dentist is sedating a female patient. When the dentist inserts a syringe into the woman's mouth, her shoes jerk suddenly. Georgia watches as the dentist begins to pull a tooth and — just as Freesmith removes the last of the cards from the punch-card reader — "the dentist stepped away from the woman and a burr in his hand glistened with bright red blood."[45] You can almost hear Burdick hollering, "Get it? The punch card is like a tooth pulled out of your head! It's horrible! Mike has blood on his hands!"

The Ninth Wave, a Book-of-the-Month Club selection, became a bestseller. Even before the book came out, Burdick sold the film rights.[46] Hollywood gossip had it that Frank Sinatra and Marlon Brando had been tapped for the two leads.[47] Burdick contracted to write the script with Stanley Roberts, who in 1954 had been nominated for an Oscar for his screenplay for *The Caine Mutiny,* starring Humphrey Bogart. No film of *The Ninth Wave* was ever made. "I hear California Democrats are hoping to keep it off the screen," wrote one Hollywood reporter.[48] Maybe that was because by then Eugene Burdick was working for the Adlai E. Stevenson campaign, as part of the new Social Science Division of Edward L. Greenfield & Co.

Ed Greenfield had been busy since the election of 1952. In the Baby Boom 1950s, he and Patty had three children, one after another: Michael (1952), Ann (1954), and Susan (1955). Diapers, bottles, strollers. They mainly lived in New York, in a brick townhouse in Chelsea. They spent their summers at Wading River. Greenfield liked to bring clients out there. In 1954, he'd begun working for Averell Harriman again, this time for Harriman's campaign for governor of New York. (Harriman went on to win.) But he'd come to admire Stevenson. Maybe he even wanted to prove to Stevenson that ad men weren't evil, that ad men could make politics better, not worse.

Stevenson formally entered the race on November 15, 1955, under the even more pathetic banner "I'm Still Madly for Adlai." Three days later, Ed Greenfield of Edward L. Greenfield & Co. wrote to Stevenson's campaign manager, offering his firm's services.[49]

"You can only be a virgin once," Stevenson

said about his decision to run again.[50] The second time around, his prospects didn't look terribly good. He wasn't a new face anymore, and there was very little reason to believe that he could beat Eisenhower in 1956: it's always harder to run against a sitting president, especially in a time of prosperity. Also, in the cruel logic of politics, Stevenson was a loser, and voters don't like losers. But one of the many strange things about Stevenson was that he didn't mind losing, if only he could lose well, if only he could deliver the beautiful speeches and say the brilliant things, educate and elevate, and break through the din, the mindless jingles of liking M&M's, Colgate toothpaste, and Ike.

Eisenhower had a heart attack in September 1955, which was, undoubtedly, a vulnerability. But Stevenson, unlike Eisenhower, faced a divided party and a hard fight for the nomination. Estes Kefauver entered the race for the Democratic nomination in December. Stevenson hadn't run in a single primary in 1952, and he didn't want to enter any in 1956, either. "I'm not going to run around like I'm running for sheriff, shaking hands at shopping centers," he told his close aide and law partner, Newton Minow. "You're wrong, Guv," Minow said. Minow hadn't wanted Stevenson to run, thinking Eisenhower was unbeatable; he'd urged him to hold off until 1960. But if he was going to run, Minow thought, he had to actually run.[51] Stevenson conceded the inevitable, entered the primaries, and shook hands at shopping centers. This contest would be decided by the California primary, scheduled to be held on June 5, 1956.[52] Even before the primary race began, Stevenson found himself in a jam. In February 1956, in Fresno, California, Kefauver and Stevenson spoke back to back. Kefauver talked about civil rights forthrightly, while Stevenson spoke "vaguely and metaphysically." Metaphysically, Stevenson had a lot to say about civil rights, and he had a stronger record on civil rights than Kefauver. But Stevenson refused to say anything simply; he considered simplicity a species of vulgarity.[53]

The Constitution had guaranteed equal rights, regardless of race, under the terms of the Fourteenth Amendment, ratified in 1868. *Brown v. Board* had declared racial segregation in public schools unconstitutional. And still change had not come. The promise *of Brown v. Board* had not been met: most southern schools simply refused to abide by it, and unless the federal government was willing to act, there wasn't much that could be done about that. The Democratic Party needed black voters, but much of the party was composed of white, segregationist southerners. Adlai Stevenson could not figure out how to thread this needle.

In 1956, three days after he spoke in

Fresno, Stevenson told a mostly black audience in Los Angeles that desegregation ought to "proceed gradually" and that care must be taken "not to upset, overnight, traditions and habits older than the republic." These traditions, of course, included not only segregation but terrorism, the murder of children, the raping of women, and the lynching of men. Stevenson's stumbles, and his timidity, cost him. The director of the California NAACP switched his support from Stevenson to Kefauver.[54] The time for patience, the time for waiting, was over, if there'd ever been a time for patience, or for waiting. Adam Clayton Powell Jr., an African American congressman from Harlem, one of the most visible black men in the country, and a Democrat, damned his party for its cowardice and endorsed Eisenhower.[55]

The proposal Edward L. Greenfield & Co. submitted to the Stevenson campaign must have seemed appealing. Greenfield, who had strong civil rights credentials, proposed to undertake research for Stevenson in Powell's Harlem, in San Francisco, and in Los Angeles. Greenfield understood that Stevenson had a civil rights problem, and he wanted to help him solve it.

The Stevenson campaign might have looked favorably on that proposal, except that Stevenson had once again made his opposition to political advertising part of his platform. "The men who run the Eisenhower administration evidently believe that the minds of Americans can be manipulated by shows, slogans, and the arts of advertising," he said. "And that conviction will, I dare say, be backed up by the greatest torrent of money ever poured out to influence an American election — poured out by men who fear nothing so much as change and who want everything to stay as it is — only more so. This idea that you can merchandise candidates for high office like breakfast cereal — that you can gather votes like box tops — is, I think, the ultimate indignity to the democratic process."[56]

No doubt Stevenson's distress about the influence of mass advertising, especially television ads, on American politics was genuine. And, to be sure, when Stevenson declared, "I say, my friends, that what this country needs is not propaganda and a personality cult," a lot of voters agreed with him.[57] Whether they were then, or later, in a position to do much about it is harder to determine. And it's also true that, their supposed scruples notwithstanding, plenty of Democrats wanted to hire ad agencies but had trouble getting any of the best ones to work for them. The big advertising agencies had big businesses as clients. Big businesses supported the Republican Party. If the big ad agencies ran campaigns for Democratic candidates, and were known for helping get Democrats elected, they risked losing their biggest clients. So they said no. "Big persuaders cost big money," as the prominent social critic Vance Packard observed. And "the big contributors were mainly on the Republican side."[58] It's this sort of thing that led candidates like Stevenson to the Ed Greenfields of the world. Little outfits, run by liberals. And especially little outfits, run by liberals, who were selling something different.

In any event, the Democratic National Committee had long since conceded the inevitable, retaining for the 1956 political season an up-and-coming ad agency, Norman, Craig & Kummel (NCK), which was best known for its Maidenform bra ads. NCK took on the DNC's $8 million account for the national campaign.[59] But, for the primaries, the Stevenson campaign needed smaller operators. That's where Ed Greenfield came in. By 1955, Greenfield & Co., with offices in New York and San Francisco, boasted dozens of affiliates expert in "all phases of publicity, public relations, and social science research."[60]

For the Stevenson campaign, Greenfield & Co. proposed to conduct what were known as "depth studies," the qualitative and especially quantitative analysis of interviews with panels of consumers or voters — what later would be called focus groups — to test messages. One brand of soap isn't that different from another. To get consumers to choose one over another required bombarding them with messages. But which messages worked best?[61] To critics of advertising agencies, this, too, looked sinister. Aldous Huxley imagined the Dr. Jekyll and Mr. Hyde of an imaginary public relations firm, Dr. Jekyll believing the people to be rational and Mr. Hyde — "or rather a Dr. Hyde, for Hyde is now a Ph.D. in psychology and has a master's degree as well in the social sciences" — knowing them to be eminently persuadable.[62]

Greenfield pitched depth studies as a higher ground, a better, more decent alternative to a Cornflakes Campaign. No glitz, no brainwashing, no fakery. He also brought the best minds, from the Center for Advanced Study in the Behavioral Sciences: Burdick, Lasswell, and Lazarsfeld.

To begin, Burdick, with help from Lasswell and Lazarsfeld, prepared a series of confidential white papers for the Stevenson campaign, summarizing the voting behavior studies. This work did not inspire a great deal of faith in American politics.[63] There was little ideological difference between the two parties. People sorted themselves into parties based on their families and their neighborhoods, not on any particular set of ideas. Voters who were not sorted in this way tended to know very little about politics and were unable to identify the meaning of "liberal" or "conservative" (asked "Would you say that either one of the parties is more *conservative* or more *liberal* than the other?," most Americans interviewed could not answer) or to display any knowledge about "what goes with what" (laissez-faire with free enterprise) or "what the parties stand for" (Democrats for labor, Republicans for business). Voters, it turned out, were not especially rational. Complicated messages were lost on them. Elections turned on the undecided voters, and undecided voters make choices that, election to election, prove extremely difficult to predict.[64]

None of this was any secret. It was the sort of thing Burdick also wrote about in popular magazines. In a magazine story called "How You'll Vote in '56," he explained that "scientific voting studies show that party preferences are determined by an individual's background, the political attitudes of his parents and friends, his guess as to the 'best' party for him, his vision of the future and other factors." The article included a quiz and a chart ("Is your father a Democrat or Republican? Put 5 points under the column of your father's party"; "If you are below thirty years of age put 4 points in the Republican column"), allowing readers to calculate their own partisan preferences and likely vote: "In 9 cases out of 10, it will forecast your ballot now!"[65]

Greenfield hired Burdick to conduct interviews with influential men in California. For instance, Burdick reported to Greenfield, after interviewing a newspaper publisher: "Feels that Stevenson has almost no chance against Ike in November. Only another coronary could help. Dislikes Nixon very intensely, but doesn't see how the campaign could be focused on Nixon."[66] This was more or less Burdick's view, too. He didn't expect Stevenson to win. Mainly, he wanted him to run well, "to wage a campaign designed to rephrase the issues, educate the electorate and set the stage for the next Presidential year." Stevenson's biggest problem, Burdick said, had to do with his personality: his divorce, his sense of humor ("too ironic and satirical"), and "the ambiguity and sophistication of his speeches." By no means should Stevenson talk about actual issues, Burdick advised, because "it is clear from the voting studies that specific issues play almost no role, except with the undecideds."[67]

Rereading Stevenson's speeches from the 1952 campaign, Burdick concluded that the problem with Stevenson's speeches was that they were too good. "There has seldom been anything like them in candor, pure intellectual content, high responsibility, and charm," Burdick wrote. "Stevenson said that he would talk sense to the American people. He stated that he thought his first obligation should be to aid the development of the rational faculties of the voters and to elevate political discourse. He did these things." Still, there was a problem. "Stevenson stated that politics were intricate and difficult, and he emphasized the hopelessness — as well as the vulgarity — of the slick black-or-white approach," Burdick observed. "Yet this difficulty must be faced: <u>how far can one push the rational faculties of the voter without tripping some mechanism of doubt and anxiety</u>."

A candidate could be only so rational when dealing with an irrational electorate. Burdick's recommendation? "The Emphasis upon Complexity Should Be Minimized."[68] Stevenson might be brilliant. The electorate, Burdick concluded, was not.

This conclusion led Eugene Burdick to a crisis of political faith. For all his worry about behavioral science, he glumly recommended to Greenfield that Greenfield & Co. ought to compile California voter data and do its behavioral science best with it.[69] For that, Burdick was not the right man, not at all. Instead, Greenfield turned to another fellow in his collection of men: Ithiel de Sola Pool, a numbers guy who taught at MIT and walked the halls of the Pentagon.

CHAPTER 3
THE QUIET AMERICAN

Perhaps I should have seen that fanatic gleam, the quick response to a phrase, the magic sound of figures: Fifth Column, Third Force, Seventh Day. I might have saved all of us a lot of trouble.
— Graham Greene, *The Quiet American,*
1955

Early in the afternoon of December 7, 1950, the anniversary of Pearl Harbor, Ithiel de Sola Pool, thirty-three, a gentle and soft-spoken man, took a seat before the Industrial Employment Review Board at a meeting held in Room 2E832 in the National Defense Building in Washington, a building constructed during the war on the site of an old airfield and a slum called "Hell's Bottom." It had five sides and five floors and five wings: a pentagon, *the* Pentagon.

He'd typed his speech out on twenty-three pages of white paper, and then he'd scribbled all over them, scratching out words, inserting lines, blotting letters, and slashing whole paragraphs with gashes of black ink. He set his prose as if it were poetry:

I am not,
    and never have been,
        either a member,
            or a sympathizer
of the Communist Party or of Russia.
I do not have,
    and never have had,
        any loyalty to any foreign government,
  or to any instrumentality of one.
My only ~~national~~ loyalty
    is, and always has been
        to the United States of America.
Communism I abhor.
The government of Russia
    is a base,
        brutal,
           vicious,
                reactionary dictatorship.

Ithiel de Sola Pool had a narrow face and short, curly dark hair. His frame was slight, his gaze unwavering. He had a tight voice, clipped and brisk.[1] It had taken him nearly two years to get this hearing, and he was determined not to squander it. He was, understandably, nervous. He'd been denied a security clearance, a suspected Communist, and this meeting served as his only chance to appeal the decision by proving his loyalty. He opened a case holding a file containing the statement that he'd typed so that he could perform each line with the proper emphasis:

Peace,
  freedom,
    and our national survival
      will never be secure
  until this monstrosity called the Soviet
    regime
  is wiped from the face of the earth
    preferably from within,
      but if need be from without.[2]

Pool would win his appeal. In the 1950s, he'd become a familiar figure in the halls of the Pentagon. After working for Edward L. Greenfield & Co. on the Adlai Stevenson campaign in 1956, he'd go on to found Simulmatics with Greenfield in 1959. In the middle decades of the twentieth century, American behavioral scientists worked on political campaigns and they worked on military campaigns. Pool would bring to Simulmatics his expertise as a quantitative behavioral scientist, an expertise he'd gained as a defense analyst.

Later, decades later, Pool would face a different hearing, when students at MIT would plaster the campus with mimeographed posters of his face, posters that would charge the Simulmatics Corporation with having "master-minded . . . a computer-oriented counter-insurgency program, containing data files not only on revolutionary movements abroad, but also on the U.S. student movement, black rebellions, strikes, etc."[3] Student protesters at MIT wouldn't accuse him of being a Communist. They'd accuse him of being a war criminal.

This future had not been predicted in 1950 when an emerging national security state asked scholars and scientists to profess their loyalty and commit themselves to the development of technology for national ends. Still, this future was set in motion then, like a train leaving a station.

The Cold War altered the history of knowledge by distorting the aims and ends of American universities. This began in 1947, with the passage of the National Security Act, which established the Joint Chiefs of Staff, created the Central Intelligence Agency and the National Security Council, and turned the War Department into what would soon be called the Department of Defense, on the back of the belief that defending the nation's security required massive, unprecedented military spending in peacetime. The Defense Department's research and development budget skyrocketed. Most of that money went to research universities — the modern research university was built by the federal government — and the rest went to think tanks, including RAND, the institute of the future. There would be new planes, new bombs, and new missiles. And there would be new tools of psychological warfare: the behavioral science of mass communications.

To staff its Social Science Division, in 1947 RAND held a conference in New York, with Harold Lasswell, after which the head of the division recruited Pool to work at RAND.[4] RAND likely wanted Pool to be part of a top secret study group, soon to be authorized by the National Security Council.[5] But the job came with a catch: the offer stood so long as "the clearance nuisance can be satisfactorily handled."[6] The work Pool was supposed to do at RAND required handling classified military material; he could not be hired without a security clearance.

RAND had staked its reputation on the conviction that national security depended on the expertise of intellectuals. In a national security state, that conviction required intellectuals to agree to conduct their work in secrecy. RAND's security measures at its fortified headquarters included guards at all entrances and exits and the requirement that all visitors wear a badge. Top secret documents could only be checked out of a single top secret office and signed out by hand. Visitors and consultants who did not have top secret clearance — men like Paul Lazarsfeld

— were consigned to a separate building, known as the Isolation Ward.[7] Pool needed to work in the main building. To do that, he had to convince the clearance review board that he wasn't a Communist.

"You have close and sympathetic association with known members of the Communist Party, namely; your parents, Mr. and Mrs. David de Sola Pool," Pool had been told by the board, in a letter of notification.[8] That day in 1950, facing the board in the Pentagon, he'd intended to say something about how he had felt when he read that letter — did they mean for him to disavow his own parents?

> Naturally, when I first read the charges
>   against me
>   I was angry
>   and when I heard my parents maligned
>   I was furious.

Calming himself down, he revised much of this part of his statement to read:

> I realize you have a right to call me here
>   and to interrogate me thoroughly.
> I intend to cooperate to the fullest extent
>   that I can.

And then, thinking still better of it, he seems to have scratched this whole section out.[9]

Pool was the child of rabbis, of learned men and women, for generations, all the way back to medieval Spain. His father, David de Sola Pool, born in London, headed the Orthodox congregation Shearith Israel, in New York. A distinguished scholar and translator with a PhD in ancient languages from Heidelberg University, David de Sola Pool was the most important Sephardic rabbi in the United States. He was also a political activist. "Our religion, Judaism, seeks salvation for all mankind," he said in the 1950s. "The emphasis of our Bible is on social justice for all rather than on the mystic quest of salvation for the individual."[10] Pool's mother, Tamar, the Jerusalem-born daughter of a rabbi, was herself a scholar of languages.[11] They'd named their only son, born in 1917, Ithiel, a name that appears only twice in the Bible — once in the book of Nehemiah and once Proverbs — a name associated with signs and prophecy.

He'd been dyslexic as a boy. A teacher trying to explain to Rabbi and Mrs. Pool what was wrong with their son told them to imagine what it would be like if instead of the Pools they thought they were "the Loops." It was a difficult childhood for the son of lovers of language, to be a boy who stumbled over words.[12] He loved, instead, numbers.

"I do not share all their political views," Pool had intended to tell the review board.

"They are Zionists and I am an anti-Zionist."
Also, "they may have been duped by commie
fronts and hidden commies on a few occa-
sions," but he was sure they were not themselves Communists. "It is easier for me to
imagine Mahatma Gandhi as a member of
the Union League Club than it is to imagine
my Father or my Mother in the Communist
Party," he wrote. He struck all that out, and
decided, instead, to wait for questions.[13]

The second charge against him was harder
to answer: "You have held membership in
Communist Front organizations, namely;
American Civil Liberties Union, Young
Peoples Socialist League, American Student
Union and Consumers Union."[14] None of
this was untrue.

Pool had been a precocious child, a brilliant child; he'd even overcome his dyslexia.
"*Faust* is the life story of the struggle of a
split personality to find meaning and an aim
in life," he'd written in a penetrating and
eerily autobiographical tenth-grade essay, at
the Fieldston School in New York.[15] He
became a socialist when he was sixteen.

I should like,
        gentlemen,
                to outline:
First, how I came to be a socialist.
Second, how I ceased being a socialist.
Third, what my attitude towards communism

91

was <u>then</u>.

And fourth, what relevance I think this has to security.

The Depression had driven him to it, he said. "With every month the breadlines were growing longer," he explained. "I think it is not surprising that a theory which said that misery was bound to increase until the system was changed appealed to me then." He joined the Young People's Socialist League, and he joined the ACLU (which was hardly a Communist front). He went to the University of Chicago in 1933, the year Hitler seized power, and studied political science with Lasswell. He was, at that point, still a socialist.

In 1936, the summer after his junior year, he went to Mexico, which would soon offer asylum to Leon Trotsky, the exiled Russian revolutionary. He talked a lot about the proletariat.[16] In Germany, the atrocities began. Pool, in Chicago, watching the rise of Joseph Goebbels, dedicated himself to the study of propaganda. In September 1939, when Germany invaded Poland, and used propaganda to suggest that the Poles had been the aggressors, Pool enrolled in a course called Political Parties and Propaganda. "Discuss the possibilities of the experimental method in studying political behavior," he was asked on a test. "What difference is there between the authoritarian theory of the state and the totalitarian theory of the state?"[17] Watching the rise of totalitarianism, Pool changed his mind about socialism.

"With the outbreak of the war in Europe in 1939 I concluded that a socialist must be for the Allies against Hitler," he told the review board. A lot of his socialist friends were pacifists: this he found indefensible.

> I came to see that society is far too complex to be tinkered with according to the blue-prints of any idealistic theorist, no matter how idealistic.
> . . .
> In other words, men should not try to play God.
> I saw that when they do that the result is fanaticism,
>   barbarism,
>   and tyranny as in Russia.

Yes, he had been a socialist. Might he not be absolved? In his defense, Pool quoted the 1940 Republican presidential candidate, Wendell Willkie. "Willkie once said, with reference to his own past, that anyone who is not a socialist at 20 has something wrong with his heart and anyone who is still a socialist at 30 has something wrong with his head."[18]

Pool had been questioned about his views

long before he sat in front of a review board in Washington in 1950. After Pearl Harbor, he tried to enlist, only to learn that he had tuberculosis. He wanted to find another way to serve: Lasswell offered him that way. In 1941, when Lasswell started up the Experimental Division for the Study of Wartime Communications at the Library of Congress, Pool, still a graduate student, joined him there, to conduct propaganda by way of the quantitative analysis of words.[19] This field — the counting of words — became known as "content analysis." For Lasswell, Pool worked on a content analysis of Communist propaganda in reaction to frustration. He counted the words, particular words, in Soviet newspapers, searching for patterns.[20] Someone must have questioned whether Pool really ought to be working for the government, in any capacity, because in April 1942, all of twenty-four, he wrote a "To Whom It May Concern" letter attempting to clear up "a number of instances of confusion of my present with past political views," presumably the sort of confusion that was impeding Lasswell's ability to hire him for a government job, permanently. Pool explained in this letter that he'd decided to devote himself to academic research, instead of revolution, and that he'd also become quieter about his opinions.[21] The fiery college student had become a cautious graduate student.

In September 1942, Pool accepted a position teaching political science at Hobart College, in Geneva, in upstate New York, where he moved with his wife, a scientist, and their first baby, a son named Jonathan. Another son, Jeremy, would be born in 1945, the year the war ended, with the destruction of Hiroshima and Nagasaki. Pool was named chairman of Hobart's Division of Social Science.[22] "It is hardly an atmosphere where a violent radical would thrive unnoticed," Pool told the review board.[23] That seemed a reasonable enough statement, but teaching at a small school in a small town, unnoticed by everyone, was just what a former Nazi, a war criminal, disguised as "Charles Rankin," did in Orson Welles's chilling 1946 film, *The Stranger.* "Not only in the College, but also in the town, I was accepted as a reliable, responsible, and loyal citizen," Pool said. So was Charles Rankin. That was the problem with subversives: they found it so easy to hide.

In 1948, when the government denied Pool a security clearance, RAND rescinded its offer of employment. Pool had no intention of spending his career at Hobart College. The next year, he took an unpaid leave of absence from Hobart to accept a job at a Palo Alto think tank at Stanford called the Hoover Institute and Library on War, Revolution and Peace. There, with Lasswell and another young scholar named Daniel Lerner, he continued the work they'd all done during the war in a new form, a Carnegie Corporation–funded project called Revolution and the Development of International Relations, or RADIR. It was meant to be a kind of social science radar. "Radar," short for "radio detection and ranging," had been pioneered at MIT. Radar detects the motion of objects; RADIR was meant to detect the movement of ideas.

At Stanford, Lasswell, Lerner, and Pool were trying to invent a kind of ideological radar that could detect the *bleep, bleep, bleep* of political unrest. (One way to think about this kind of content-analysis work is as a very early version of Google Trends, which also started at Stanford, in the graduate work of Larry Page and Sergey Brin.) In one study, Lasswell, Lerner, and Pool searched editorials in the leading newspapers of Great Britain, Russia, the United States, France, and Germany, from 1890 to 1950, for words like "internationalism" and "security," and then counted these words and sorted, or coded, them into 416 political symbols. Lasswell made the list of symbols. Lerner supervised the coding. And Pool analyzed the results, in hopes of creating a mathematical model — ultimately, a predictive model — of the historical relationship between "ideology" (democracy and capitalism) and "counter-ideology" (authoritarianism and socialism).[24]

They were playing Spot the Revolution. They wanted to know: What will happen next?

All that while, the House Un-American Activities Committee, or HUAC, whose members included a freshman congressman from California named Richard M. Nixon, was playing Spot the Subversive. David de Sola Pool's name turned up in a HUAC document from March 1950, on a list of sponsors of a 1948 Chicago conference of the "American Committee for Protection of Foreign Born," an organization suspected of being a Communist front.[25] Pool's chances at a security clearance were getting worse, not better. By then, he'd realized he'd need outside support. He secured and submitted to the review board an affidavit from Ralph de Toledano, an editor of *Newsweek* and later a founder of *National Review* who'd likely known Pool since they were students together at the Fieldston School.[26] De Toledano also wrote a letter to Nixon, pleading Pool's case. "His record is clean right down the line — anti-Stalinist all the way," de Toledano informed Nixon. "On his parents, the story is a little different. They are not Party members but they've got some bad party front affiliations and they've been pretty involved. If this means that Ithiel is marked lousy, I don't know what can be done. I do know that he sees them practically never — they're in New York and he in California."[27]

In the end, Pool's argument before the review board came down to his personal conviction: socialists are truthful people who state their beliefs in the open; Communists are liars and subversives who hide what they believe. He was an honest man, a former socialist, but he had never been a Communist. And he was no longer a member of the Left.[28]

Two weeks after Pool delivered his statement, the review board reversed its decision and granted him his long-sought security clearance: "The Secretaries of the Army, Navy, and Air Force therefore withdraw any objection to your employment on classified military information."[29] Pool wrote to Nixon to thank him.[30]

Pool had come by his security clearance and his anti-Communist credentials the hard way, and he would cling to them, and burnish them, all his life, like medals earned on a battlefield. Saul Bellow, who'd known Pool since they were undergraduates and fellow Trotskyites at the University of Chicago, later memorialized him as "Ithiel Regler" in a novel called *A Theft.* Bellow must have known about Pool's troubles getting a national security clearance. In the novel, he includes a scene in which Ithiel Regler, asked for an ID, provides a "Pentagon pass." Bellow took pains to remark that this Ithiel could be relied upon: "he wouldn't give out information that might be classified." Ithiel Regler had every president's ear: "If he wanted, he could do with Nixon, Johnson, Kennedy, or Kissinger, with the shah or de Gaulle, what Keynes has done with the Allies at Versailles." Ithiel Regler says oracular things like "Neither the Russians nor the Americans can manage the world. Not capable of organizing the future."[31] Ithiel de Sola Pool was not Ithiel Regler, not exactly, but, like him, he would try and try to organize that future, by the use of numbers and mathematical models, to make human behavior predictable. Fanatics and fascists made men gods, Pool thought. He would place his faith in numbers, patterns, models, and calculated predictions.

Between 1950 and 1952, the budget of the United States Department of Defense grew from $500 million to $1.6 billion. In 1953, Eisenhower committed to a defense strategy known as the "New Look," which placed priority on the precise targeting of weapons, rather than on ground troops. This pumped money into both the electronics and the aeronautics industries and into advanced research into technologies that could speed calculation. In the middle of the 1950s, military spending made up close to three-quarters of the federal budget.[32] In 1952, the National Science Foundation announced that the nation needed one hundred thousand more scientists; universities furiously set about producing them.[33]

In 1951, with his long-sought security clearance in hand, Pool resigned from Hobart College and went back to the Hoover Institute. In nearby Berkeley, or maybe in a meeting at Stanford, he met the mesmerizing, out-of-a-novel-by-Ian-Fleming Eugene Burdick.[34] But the prospects for behavioral science at Stanford were limited. Liberal intellectuals might have concluded that liberalism had triumphed the world over, but liberalism had not triumphed, and one place where it had really not triumphed was Stanford University. Instead, Stanford became an incubator, one among many, of the modern conservative movement. Herbert Hoover, an ex-president, an engineer, a Stanford alum, and the founder of the Hoover Institute, rejected what he called the "squirrel-cage scholasticism" of behavioral science.[35] He was not alone. In 1952, in *God and Man at Yale,* a very young William F. Buckley offered a similar rebuke. To conservatives, behavioral science smacked of the New Deal. Throughout the McCarthyite 1950s, conservatives damned the godlessness and "moral idiocy" of behavioral science, citing its technocratic posture as a species of socialism, the control of the people, even of their very minds, by the state.[36]

There was something more to their critique. Conservatives looked to the past, venerated the past. "A conservative is someone who stands athwart history yelling Stop," Buckley would famously declare in 1955, in *National Review*.[37] Behavioral science looked to the future. That, to conservatives, was a hallmark of its liberalism.

Pool looked for a job somewhere else: MIT. In 1949, when the Russians jammed radio transmissions to the Soviet Union, the State Department turned to university researchers to help solve the problem of "how to get information into Russia." This led in 1950 to Project TROY, a joint operation of scholars from MIT, Harvard, and RAND, which recommended a series of technical means of "perforating the iron curtain." To develop psychological means to the same end, MIT in 1952 established the Center for International Studies, funded by the Ford Foundation and the CIA and headed by Project TROY's Max Millikan, a former assistant director of the CIA. The next year, Millikan hired Pool to direct the center's new International Communications Program. Pool's colleague Daniel Lerner moved to MIT, too.[38]

The departure of Pool and Lerner from the Hoover Institute left Palo Alto without a behavioral sciences research group, a vacancy filled by the Center for Advanced Study, founded the next year.[39] In 1959, Hoover would remake his eponymous institution into a conservative think tank. "The purpose of this institution," its new charter said, "must be, by its research and publications, to demonstrate the evils of the doctrines of Karl Marx."[40]

Disavowing Marx, most midcentury American behavioral scientists flirted with Freud. They submitted themselves to psychoanalysis; it infected their marriages like a virus. They loved women; they found women disgusting; they hated women. They divorced women. Analyzed, Pool left his wife and their two young sons. He found his second wife when, after accepting the position at MIT, he visited the University of Chicago and met a PhD student in psychology, Jean MacKenzie. She had just finished her dissertation, having made a study of the role of research scientists in corporate laboratories.[41] She was glad Pool had already been analyzed.[42] She converted to Judaism and married him in March 1956.[43] Of all the Simulmatics marriages, theirs would be the strongest.

Saul Bellow, who adored the idea of female enthrallment, would describe Ithiel Regler as irresistible to women.[44] "On a scale of one to ten . . . he *was* ten," a female character in Bellow's novel reports. "When Ithiel comes to town and I see him at lunch, I start to flow for him. He used to make me come by stroking my cheek. It can happen when he talks to me. Or even when I see him on TV or just hear his voice. *He* doesn't know it — I think not — and anyway Ithiel wouldn't want to do harm, interfere, dominate or exploit — that's not the way he is."[45] Bellow's heroine (like all Bellow women) "let politics alone, asking no questions." The classified nature of Ithiel Regler's work only added to his allure. "The more hidden his activities, the better she felt about him. Power, danger, secrecy, made him even sexier." In her Manhattan apartment, she cooks for him naked, wearing nothing but clogs, while, "stretched on the bed, Ithiel studied his dangerous documents (all those forbidden facts)."[46] This was less a portrait of Ithiel de Sola Pool than a fantasy of Saul Bellow's, a fantasy of stretching in bed with his very important work, his dangerous documents, his forbidden facts, watching a naked woman cooking for him in the kitchen. Still, it wasn't only Bellow's fantasy; it was the fantasy of an entire generation.

During the years when Pool rose to influence in the Pentagon, the United States exerted its authority around the world. In 1950, when Pool wrote to thank Nixon for his help in getting a security clearance, he urged the new senator to keep hitting Communists hard, especially abroad. "The communists will keep punching wherever they can make a dent, and we must do the same," Pool told Nixon. "If we withdraw it will only be a matter of time before they can mobilize overwhelming force against us from all continents."[47]

The places where communism could make a dent were, at this point, not in Europe but, instead, where people who had freed themselves of European imperial rule were building new nations. Some would be Communist, allied with the Soviet Union and China; others would be capitalist, allied with the United States. American money went where the Pentagon believed it was needed most, to stop the spread of communism. Between 1949 and 1952, three-quarters of U.S. foreign aid went to nations in Europe; between 1953 and 1957, three-quarters of that same budget went to countries outside Europe. Much of that money went to Southeast Asia.

In 1951, Massachusetts congressman John F. Kennedy and his younger brother Robert went on a seven-week tour of the Middle East and Asia. Jack Kennedy intended to run for a seat in the Senate; Bobby would manage his campaign. Fatefully, they stopped in Vietnam, which had declared its independence from French rule at the end of the Second World War. France, though, had refused to recognize Vietnam's independence. "The case of Indochina is perfectly clear," FDR had written in 1944. "France has milked it for one hundred years. The people of Indochina are entitled to something better than that."[48] But in 1951, with France still fighting to regain power in a war against forces led by Ho Chi Minh, the United States, violating a staunch opposition to colonialism, a commitment made by FDR, was providing aid to France.[49]

Could behavioral scientists have predicted anything accurately in the 1950s, much agony would have been averted. The future of war? The United States engaged only in the futility of ill-considered military spending. Despite $2.5 billion in military aid from the United States, France lost the war in 1954, at the Battle of Dien Bien Phu, after which a peace treaty forged at Geneva divided Vietnam at the seventeenth parallel. North Vietnam would be Communist, headed by Ho Chi Minh. South Vietnam would be headed by Ngo Dinh Diem, a U.S.-backed Catholic nationalist. Elections about reunification were to be held in two years. They were never held. Instead, Diem declared South Vietnam a republic, and war broke out between the North and the South. The United States provided, at first, only advisers, then air support, then marines, and finally soldiers, in a war that would not end until 1975, by which time it would have taken the lives of fifty-eight thousand Americans and three million Vietnamese, two million of them civilians.

In 1955, Graham Greene, who had reported from French Indochina for the London *Times,* would lament French colonialism and American advice alike, in a novel called

The Quiet American, in which a hard-bitten English newspaper correspondent on assignment in Saigon watches an agent named Alden Pyle arrive in Vietnam and die there, as deluded about America's role in Indochina at the end the book as he'd been at the beginning. Greene based his Pool-like Pyle on an American aid worker he'd met in Saigon in 1951. "Pyle was very earnest and I had suffered from his lectures on the Far East, which he had known for as many months as I had years," Greene's narrator observes. "Democracy was another subject of his — he had pronounced and aggravating views on what the United States was doing for the world."

Pyle is also often, if wrongly, read as modeled on Colonel Edward Lansdale, a former advertising executive turned CIA psychological warfare expert who spoke neither French nor Vietnamese but nevertheless went to Vietnam in 1954 to support a pro-American government and undermine Vietnamese Communists. Greene denied any connection and insisted that this reading was largely Lansdale's own: arguing that Pyle was modeled on him was a way for Lansdale to damn Greene's anti-Americanism while making himself into a literary celebrity. Greene and Lansdale had in fact met in Vietnam in 1954.[50] In any case, Pyle was younger and more innocent ("I never knew a man who had better motives for all the trouble he caused") and more of a reader than Lansdale. At one point, the narrator, stopping in Pyle's apartment — the two men are fighting for the love of a Vietnamese woman named Phuong — thumbs through the books on Pyle's bookshelf: *The Advance of Red China, The Challenge to Democracy, The Role of the West.*[51] It might have been Pool's bookshelf.

In 1955, the year Greene published *The Quiet American,* Ithiel de Sola Pool, defense intellectual, took on a consulting job with New York ad man Edward L. Greenfield. They might have met earlier, maybe in 1950, when Greenfield was studying, briefly, with Lasswell at Yale. Pool would surely have been in close communication with Lasswell that year, asking for his help in gaining a security clearance. Greenfield and Pool had Lasswell in common, and more, too. They were part of a generational and religious cohort: young, ambitious Jewish men who'd missed the fighting in the Second World War and traded one faith for another, Judaism for liberalism.

"Last night's gathering went a long way towards cementing the group," Greenfield reported to Pool in December 1955, from the offices of Edward L. Greenfield & Co. in New York.[52] Greenfield knew Lasswell. Greenfield knew Burdick. Greenfield, it seemed, knew everyone. Pool, fascinated, started making notes for a new line of research. How many acquaintances do most people have? How many people do any two people know in common? How many degrees of separation separate any one person from any other? How small is the world?

"Given that A knows B, what is the probability that B knows *n* persons in the circle of acquaintances of A," Pool asked. "My first hunch was that perhaps a logarithmic relationship was plausible." The math dealing with "the probability of B knowing C, given that A knows B and A knows C" was fuzzy. But what about "the case where A does not know B, but where they have one acquaintance in common"?[53]

Greenfield collected people. Pool collected data. Pool began to make a list of everyone he knew, including the man who seemed to know everyone: Ed Greenfield. And then he sent that list around, all over the country, to ask people he knew if they knew people he knew: "Check here if you know this person." He included with the request a "Definition of Knowing": If you ran into "Edward L. Greenfield, Public Relations, N.Y.C. Formerly Univ. of Chicago, Yale Law School," on the street, would you recognize him? Would you say hello? Would Edward L. Greenfield know you, too?[54] And then Pool ran the numbers, calculating the chances. He plotted his data and came up with a function to describe that plot. Then he used that function to extrapolate beyond his data. He devised a theory, a theory that would come to undergird all social media companies, including Facebook and Twitter. He called it the theory of "social networks."[55]

In 1956, Greenfield & Co.'s operation in California, installed in an office in San Francisco, reported that Adlai E. Stevenson's prospects in that state's upcoming and must-win June 5 primary looked excellent. As early as April, Greenfield began drafting a press release announcing Stevenson's victory in the state. It began, "In the last twenty-four hours the people of America have overwhelmingly named Adlai Stevenson their choice as the Democratic nominee for President."[56] Meanwhile, Ithiel de Sola Pool and Eugene Burdick were still conducting their research.

"I'm from the Stevenson for President headquarters," Pool had instructed his team of California volunteers to say, when knocking on doors. "Mr. Stevenson would like to know how the voters feel," they were supposed to go on. "He wants to represent the whole American people, but it's pretty hard for a man in his position to know exactly how they feel." Then came the questionnaire, whose answers Pool put on punch cards.[57]

In California, Burdick and his colleagues conducted in-depth interviews with one hundred political "elites." Pool and his volunteers polled 750 voters. In a report on Los Angeles, they concluded that "Stevenson has lost a good deal of badly needed support among Negroes and other minority groups."[58] On May 13, they submitted their final reports to Greenfield. Their leading recommendation had to do with civil rights: "<u>Do not make the mistake of thinking that civil rights and desegregation are important issues only to the Negroes.</u>"[59]

In the June 5 California primary, Stevenson beat Kefauver, 1,139,964 to 680,722, with 62.6 percent of the vote.[60] The senator from Tennessee withdrew from the race. Edward L. Greenfield & Co.'s Social Science Division claimed credit.

The 1956 Democratic National Convention was held in the same stockyard amphitheater on the South Side of Chicago where Stevenson had accepted his party nomination four years before. In Gore Vidal's play *The Best Man,* set at the convention, an unmistakably Stevenson-like presidential candidate named William Russell, played by Henry Fonda in the movie, is forever quoting Shakespeare and the Old Testament and Oliver Cromwell, to the bafflement of reporters and voters alike. Russell doesn't like Gallup. "I don't believe in polls," he says. "Accurate or not." He doesn't like ad men. "This is exactly the kind of thing I went into politics to stop!"

he cries. He's fussy and semantic and meta-physical.

RUSSELL

What is a liberal, Senator?
(*RUSSELL crosses to bedroom,
picks up dictionary, returns to living
room, thumbing pages*)

CARLIN
(*Groans*)

And I thought Adlai Stevenson was a pain in the neck.

Needless to say, Vidal's best man loses.[61]

At the 1956 Democratic National Convention, the actual Adlai Stevenson arranged for John F. Kennedy, now thirty-nine, his star rising, to deliver the nominating speech. Stevenson's speechwriters drafted the speech, but Kennedy rejected it and, with help from his aide Ted Sorensen, wrote his own.[62] He said, from the festooned rostrum, "The American people saw and heard and admired this man for the first time four years ago, when, out of the usual sea of campaign promises and dreary oratory and catchy slogans, there came something new and different — something great and good — a campaign and a candidate dedicated to telling the truth."[63]

Stevenson had been new and different in

'52; in '56, he was neither. And, even though Kefauver had withdrawn, Stevenson still hadn't cinched the nomination. Truman, who'd supported Stevenson in 1952, now supported Averell Harriman, who hadn't entered any of the primaries. Luckily for Stevenson, Truman's support didn't count as much as it used to. Stevenson won the nomination on the first ballot. Stevenson had no intention of hitching himself to his former running mate, the segregationist John Sparkman. Kennedy hoped that Stevenson would name him as the vice-presidential nominee. But Stevenson believed the honor of delivering the nominating address had been enough of a gift for Kennedy. Instead, doing something that had never been done before, Stevenson threw the choice of vice president to the delegates, as a way of demonstrating his commitment to a more democratic process. The delegates chose Kefauver. Stevenson had been nearly certain Kennedy would win. Kennedy never quite forgave him. Nor did his younger brother, closest adviser, and fiercest protector, Bobby, who was famous for bearing grudges, and who believed that Stevenson had double-crossed him.[64]

Stevenson, in his acceptance speech, called for a New America. "I mean a New America where poverty is abolished and our abundance is used to enrich the lives of every family," he said, to a hushed hall. "I mean, my friends, a New America where freedom is made real for all without regard to race or belief or economic condition. I mean a New America which everlastingly attacks the ancient idea that men can solve their differences by killing each other."[65]

He wanted a new America, but he was an old candidate. A lot of Americans were a little sick of Adlai E. Stevenson by 1956. His speeches were beautifully written; his delivery was depressing. After Stevenson delivered his acceptance speech, Arthur Schlesinger ran into Phil Graham, publisher of the *Washington Post*. "You know, I thought that the Democrats really had a chance to win," Graham said, "until Adlai began to speak."[66]

For the general election, Stevenson named George Ball his director of public relations, fairly perversely, since it was Ball who in 1952 dubbed the Eisenhower's approach the Cornflakes Campaign. Ball found the assignment humiliating.[67] Stevenson did agree, this time around, to appear in a series of television ads, except that they were really anti-television ads: their purpose was to warn Americans against believing everything they see on TV. One ad, shot in Stevenson's book-lined study at his house in Libertyville, Illinois, begins with a gaffer snapping a slate to reveal a frame that exposes all the trappings of a television shoot, its clutter of cameras and cables, an entire contrivance. Stevenson's television ad was an indictment of television. "I wish you could see what else is in this room," Stevenson says, directing the viewers' attention to the artificiality of the scene. "Beside the camera, there are lights over here, there are cables all over the floor."[68] It was positively postmodern.

Eisenhower, meanwhile, did nearly all of his campaigning on television. After the heart attack he'd suffered in the fall of 1955, he'd undergone surgery for a bowel obstruction the following spring, which together left him little able to head out on the campaign trail.[69] The GOP's strategy was to build a "political mousetrap": keeping quiet and letting Stevenson do all the talking, on the theory that he'd eventually screw up.[70]

It is difficult to say whether the Stevenson campaign paid much if any attention to the reports that came out of Edward L. Greenfield & Co.'s Social Science Division. In one way, Stevenson appears to have ignored Greenfield's recommendations: he did not strengthen either his position or his rhetoric on civil rights. In another way, he seemed to accept them. Except for civil rights, "the issues, per se, are relatively unimportant," one Greenfield & Co. report advised the Stevenson campaign.[71] And maybe Stevenson took that to heart. Critics called him "the Issueless Candidate."[72]

The eggheads still loved him, but there are

only so many eggheads. Ike won in a rout. In November 1956, thirty-six million out of sixty-one million votes cast went to the general. He won forty-one out of forty-eight states, including California. Every state Stevenson won had been part of the Confederacy. Democrats won the Jim Crow South and lost the rest of the country.[73] Democrats won both the House and the Senate but, as a national party, the Democratic Party looked to be falling apart. White southern Democrats kept threatening to leave the party if it fought for civil rights, but without fighting for civil rights the Democratic Party could not win back the presidency.

Ed Greenfield, man of ideals, and man of ideas, decided that the Democrats needed a miracle, and he intended to provide it.[74] He needed more men. He intended to build a machine. Sometimes he called it an Issues UNIVAC. Sometimes he called it a Voting Behavior Machine. It would run the numbers, super fast, and tell a candidate the consequences of taking a position on any issue, anywhere, state by state, county by county, voter by voter, issue by issue. It would have to be ready in time for the election of 1960. To build that machine, he knew, he needed a very rare sort of man: a computer man.

CHAPTER 4
ARTIFICIAL INTELLIGENCE

The FORTRAN language is intended to be capable of expressing any problem of numerical computation. . . . However, for problems in which machine words have a logical rather than a numerical meaning it is less satisfactory, and it may fail entirely to express some such problems.

— Fortran: Automatic Coding System for the IBM 704, *1956*

"Sometimes, along about four in the morning," Alex Bernstein told *The New Yorker,* "I'd be bent over my chessboard, making encouraging noises at 704, and, glancing up, I'd see somebody staring in at me through the Madison Avenue windows. From the look on his face, I'd know he thought I was absolutely balmy."[1] Faces pressed to the glass, passersby would watch him, puzzled, bewildered, unable to believe their eyes: Was this funny little man really playing chess against a machine?

The IBM 704 took up "about an eighth of

an acre of brightly lighted floor space" in the gleaming ground-floor lobby of IBM's twenty-story world headquarters in New York, a showroom, on the corner of Madison Avenue and Fifty-Seventh Street. Anyone walking down the sidewalk could peer in through the giant plate-glass windows and see the 704, which, as *The New Yorker* observed, "consists of eighteen glass-fronted cabinets; an immense console studded with rows of push buttons and red, green, and yellow lights; and a printing machine."[2] It was like something out of a movie, extraterrestrial.

Bernstein had been born in 1931, in Milan. His parents, Russian Jews, had fled first to France and then to Italy.[3] His father died, but he and his grandfather, mother, and sister eventually made it to New York. He was an avid chess player. "I've been playing since I was nine, and I'm fairly good," he told *The New Yorker.* He went to the Bronx High School of Science and then to the City University of New York. During the Korean War, he served in the U.S. Army Signal Corps, where he worked on the electronic calculating machines used to break codes and to send signals. (The Signal Corps would retire its last pigeon in 1957.)[4]

Electronic computing machines capable of making calculations with unprecedented speed emerged out of the military priorities of the Allies during the Second World War, when engineers in the United States and the United Kingdom built them to compute missile trajectories and crack ciphers. These machines were top secret. The Mark I, developed by Harvard and IBM, was programmed by a navy lieutenant (later rear admiral) and Vassar professor of mathematics named Grace Murray Hopper. The ENIAC, the Electronic Numerical Integrator and Computer, had been invented by J. Presper Eckert and John Mauchly at the University of Pennsylvania. After the war, Eckert and Mauchly formed the Eckert-Mauchly Computer Corporation and hired Hopper as senior mathematician at their company, which was soon bought by Remington Rand, a typewriter manufacturer. In 1946, they unveiled the ENIAC to the public. "One of the war's top secrets, an amazing machine which applies electronic speeds for the first time to mathematical tasks hitherto too difficult and cumbersome for solution, was announced here tonight by the War Department," reported the *New York Times,* in a front-page story that included a photograph of a thirty-ton machine the size of a room.[5] A newsreel called it the world's first "giant electronic brain."[6]

It wasn't only reporters who likened the new machines to the human brain. In 1948, the MIT mathematician Norbert Wiener

118

published a book called *Cybernetics,* in which he compared the nervous systems of living things with the automatic control systems of machines.[7] At Remington Rand, Hopper devised the first "compiler," which allowed programmers to write code in something close to English.[8] In an essay called "The Education of a Computer," Hopper reported that "it is the current aim to replace, as far as possible, the human brain by an electronic digital computer."[9]

Remington Rand unveiled the UNIVAC in 1951, but the machine made its real public debut on Election Night 1952, when CBS News commissioned it to predict the outcome of the election. Greenfield got giddy at its promise. Still, the development of the first commercially available general purpose computers raised fears about unemployment, about machines replacing men. "Machines should be used instead of people whenever possible," advised a staffer for the National Office Management Association in 1952. At least at first, though, most machines replaced not men but women, taking over daily office tasks done by female clerks, typists, and filers, and "data-processing" tasks like handling payroll and inventory and generating paychecks.[10]

The commerce in computers grew slowly. By 1953, Remington Rand and IBM had, between them, installed a grand total of nine computers in the United States.[11] The practice of referring to these machines as "electronic brains" and "thinking machines" elicited fears of a robot invasion, a nightmare. In 1954, a journalist who visited an IBM factory in Endicott, New York, reported that, although signs on every wall urged him, with IBM's motto, to think, he left the factory with a melancholy vision of a future "of pale, silent people standing submissively before exorbitantly active metal boxes."[12] It was just this sort of thing that alarmed the passersby on Madison Avenue who spied Alex Bernstein, pale and exhausted, playing chess in the black of night with an IBM 704, metallic and untiring.

IBM had launched the 704 in 1954, the only computer that could handle complex mathematics.[13] (By the standards of the day, it was blindingly fast, handling twelve thousand floating-point operations per second. By the standards of the twenty-first century, it was prehistorically slow. The 2012 iPhone 5, for instance, handled 171 million floating-point operations per second.) Eugene Burdick had probably seen an IBM 704 at Stanford; it's exactly the sort of machine he describes in his 1956 novel, *The Ninth Wave*. At MIT, Ithiel de Sola Pool used one at MIT's new Computation Center, which housed a single IBM 704, meant to serve every research university in New England.[14]

But if these machines were to do more than make very fast calculations, they'd need subtler instructions. IBM charged those of its mathematicians who, like Bernstein, were known to be good chess players, with devising a new compiler, to better translate English-like language into instructions to be given to a machine. In 1956, they created FORTRAN, short for FORmula TRANslation.[15]

FORTRAN is usually called a language but it's not exactly a language, at least in the human sense; it's a set of instructions used in programming a computer, instructions like READ, FORMAT, GO TO, and IF. IF instructed the 704 to think a thought like this: If A is true, then X; else if B is true, then Y. This is the way a FORTRAN program instructed an IBM 704 to "think"; it isn't necessarily the way a person thinks. FORTRAN wasn't designed to teach a 704 to think like a person. It was designed to instruct a 704 to undertake mathematical operations. "The Fortran language is intended to be capable of expressing any problem of numerical computation," the manual explained. But then it took pains to warn: "However, for problems in which machine words have a logical rather than a numerical meaning, it is less satisfactory, and it may fail entirely to express some such problems."[16] This caveat notwithstanding, computer men began to call

FORTRAN a language, as if it could express everything that mattered. And they began, too, to try to collapse the distinction between an electronic brain and a human one.

The Simulmatics Corporation got its name from two words, mashed together: "simulation" and "automation." Ed Greenfield always figured "simulmatics" would one day be like the word "cybernetics," a catchphrase. Instead, "artificial intelligence" became that catchphrase. Still, "artificial intelligence" is pretty close to what Greenfield meant by "simulmatics."

The expression "artificial intelligence" was coined in 1955 when four men proposed a summer research seminar, to be held at Dartmouth College, in Hanover, New Hampshire: John McCarthy, a young mathematics professor at Dartmouth; Nathaniel Rochester of IBM, who'd designed the IBM 701; Claude Shannon, of Bell Labs; and Marvin Minsky, who'd just finished his PhD at Princeton. Before then, schemes to get a computer to think like a human had fallen under the broad heading of "automation." In the four men's proposal for the seminar, they also referred to simulation: they described their work as proceeding from "the conjecture that every aspect of learning or any other feature of intelligence can in principle be so precisely described that a machine can be made to simulate it." If the brain is like a machine — a very big *if* — the simulation of human intelligence is limitless, because "if a machine can do a job, then an automatic calculator can be programmed to simulate the machine." To distinguish what they wanted to do from mere automatic calculation and mere computer simulation, they elected to call what they hoped to study that summer "artificial intelligence."[17] And that's what it came to be called eventually, but in the 1950s, on the streets of America, outside of the Dartmouth conference, the words for this sort of thing were still "automation" and "simulation."

The word "automation," in the sense of machines that do the work of humans, was first used in 1948. By the mid-1950s, it was everywhere. (Between 1954 and 1955, its frequency of occurrence in major American newspapers multiplied sixfold.)[18] "Simulation," in the sense of computers using data to create a mathematical model for a real-world behavior — explained to the public as "testing without making actual models" — entered the vernacular at the same time.[19] Computer simulation came out of the war effort and the need to model objects in flight, including airplanes and missiles. After the war, the most important computer simulation project began at MIT in 1950, and led to the development of SAGE (for Semi-Automatic Ground Environment), a network of computers that monitored airspace in order to warn of and intercept Soviet airborne attacks.[20] Meanwhile, the aeronautics industry engaged in what newspapers called "the new field of Flight Simulation," too, because, as one aerospace industry executive explained, "airplanes simply blast through space faster than the human mind can think."[21]

A calculating machine, given enough information about an aircraft and enough data about factors like weather, can very usefully simulate a flight. Gravity is a law. Also, $F = ma$: the force that acts on a physical object is equal to its mass times its acceleration. But the computer simulation of human behavior or human cognition is much more difficult. Behavior is not a law, even if behavioral scientists wanted to make it one. Laws of human behavior, like Eugene Burdick's $f + h = p$, are something between whimsy and bunk. Fear plus hate does not equal power, not in any mathematical sense, anyway. Nor is cognition a law.

Lacking a theory of cognition, people who worked in the new field of "artificial intelligence" in the 1950s were less interested in getting a machine to think like a brain than in getting a machine to do things a brain can do, like playing chess. The man versus machine chess match has a long history. Beginning in the eighteenth century, a chess-playing automaton known as "the Mechanical Turk" — a wooden figure dressed as a Turkish man — had defeated players across Europe and the United States. Benjamin Franklin had tried to penetrate its secrets. So had Edgar Allan Poe, who in the end figured out that the Turk was really a very tiny man, confined in a box below the chessboard, moving the chess pieces by way of levers.[22] But the mechanical Turk lived on in the memory of mathematicians, as a kind of dare: whoever could first teach a machine to play chess would have broken through a wall. In 1950, the brilliant British mathematician Alan Turing tried to devise a chess-playing program. Alex Bernstein said that Turing's machine "played a very weak game, made stupid blunders and usually had to resign after a few moves."[23] Claude Shannon had tried, too.[24] Where these men had failed, Bernstein, working through the night at IBM, had succeeded.[25] He was one of only a handful of men invited to the Dartmouth conference on artificial intelligence in the summer of 1956.

The People Machine that Ed Greenfield came to think his company ought to build for the Democratic Party in time for the election of 1960 got its start on the campaign trail, with Adlai Stevenson. It got its start on Madison Avenue, in boardrooms with tiled harlequin floors. It got its start at the Center for Advanced Study in the Behavioral Sciences, in the hills of Palo Alto. It got its start in the Pentagon, behind guarded doors. It got its start on the ground floor of IBM's headquarters in Manhattan. And it got its start at Dartmouth, during the first ever conference on artificial intelligence. During all these years, on the streets of every Jim Crow state in the American South, the people protested, the people marched, the people shouted for freedom, the people cried for justice.

On one of the last days of 1955, in Montgomery, Alabama, a forty-two-year-old seamstress and longtime civil rights activist named Rosa Parks refused to give up her seat on a bus to a white man. Martin Luther King Jr., the elegant, twenty-six-year-old pastor of Montgomery's Dexter Avenue Baptist Church, spoke at a mass meeting four days later, at a Baptist church on Holt Street. Outside of preaching, he'd never before spoken in public. "There comes a time when people get tired of being trampled over by the iron feet of oppression," he told the crowd. That day, blacks in Montgomery had begun a boycott of the city's buses.[26] Every day of the summer of 1956, while Adlai Stevenson campaigned for reelection and prepared for the Democratic National Convention, every day in the summer of 1956, while the nation's leading computer men gathered at Dartmouth to found the field of artificial intelligence, the people of Montgomery boycotted the city's buses. "If we are arrested every day, if we are exploited every day, if we are trampled over every day," King said, after he himself was arrested, "don't ever let anyone pull you so low as to hate them."[27]

That same summer, Alex Bernstein left his office at IBM in New York and went to New Hampshire, for the summer seminar on artificial intelligence. He'd just gotten married to June Atlas, a schoolteacher who'd graduated from the University of Michigan. They lived in Brooklyn Heights.[28] It must have been hard to leave her during her summer vacation. But he went anyway. Maybe she came, too. Either way, he must have been nervous. He might have felt a bit over his head. These people were university professors, PhDs. He had a bachelor's degree, from the City University of New York, in medieval literature, a subject he'd also studied at Columbia as a graduate student. But, although they'd invited him, the leaders of the seminar had been dubious about his computer chess program. Claude Shannon had said he was "interested in chess" but wasn't "too enthused over its relevance." Marvin Minsky was "*very* much against chess," on the grounds that it was little more than a stunt.[29]

On August 8, 1956, in a room in Dartmouth's Mathematics Department, Bernstein presented an early version of his chess program. Getting a computer to play chess, he'd write, was a lot different from getting it to do "an ordinary job performed by a computer, say calculating John Doe's pay check." To the question of what to pay John Doe for forty hours at a particular wage, and given a particular tax rate, there is one and only one answer. "But in a chess game there are only two questions to which absolutely definite and unavoidable answers can be given: 'Is this move legal?' and 'Is the game over?' " Bernstein explained. "To all other questions there are various possible answers, though some may be more acceptable than others. The problem is to equip the machine with a system of evaluating the merits of the alternatives."[30] Bernstein had written a program that asked the computer to think through the consequences, over the next two moves, of every possible move.

It took the IBM 704 eight minutes to think through the consequences of any next two moves, a speed nevertheless known as "real time." Bernstein's instructions made the 704 into a pretty good chess player. "Once or twice it played so well that it rattled me," he said. "I'd actually find myself asking it, 'What the devil are you up to now?' "[31] Still, "anyone who can plot a three-move trap can beat it." Bernstein could have instructed the

704 to think through the next three moves, but then it would have taken eight hours, instead of eight minutes, between each move. A better improvement, he explained, would be to allow the 704 to learn from its mistakes, instead of starting each game as if it had never before played chess.[32] That would require teaching a machine how to learn.

The Montgomery bus boycott ended, on December 20, 1956, with the implementation of a Supreme Court ruling that declared the state and municipal laws that had segregated Montgomery's buses to be unconstitutional. The boycott had lasted 381 days. Eisenhower had not acted to stop the violence in the South, the John Birchers, the burnings and the lynchings, the White Citizens' Councils that fought against *Brown v. Board of Education*. Congress began debating a civil rights act, the first since Reconstruction. The bill passed the House in June, with more support from Republicans than from Democrats. And in August, despite the longest filibuster in American history, led by South Carolina Democrat Strom Thurmond, the Democratic Senate managed to pass it, by way of the political wrangling of Texan Lyndon B. Johnson, Senate majority leader. Johnson, a down-home New Dealer who'd come to Congress in 1937, on FDR's coattails, understood that it was time for the Democratic Party to renounce segregation, once and for all, and to take the side of civil rights.[33]

Despite Johnson's turnabout, segregationists dug in their heels. In September 1957, Orval Faubus, the governor of Arkansas and a Democrat, sent some 250 soldiers of the National Guard to Little Rock's Central High School, to bar any black children who tried to get inside the school's doors. Fifteen-year-old Elizabeth Eckford was turned away by soldiers stationed at the school's entrance and into a crowd of white students who cried, "Lynch her! Lynch her!"[34] Days later, Eisenhower signed the 1957 Civil Rights Act and then, in a nationally televised address, announced that he'd ordered a thousand paratroopers from the 101st Airborne Division to Little Rock. The federal government had at last intervened to guarantee the constitutional rights of schoolchildren.

Meanwhile, the nation was engaged in another battle, a battle of machines. Ten days after the 101st landed in Little Rock, the Soviet Union launched a satellite named *Sputnik* into orbit. The Russians had beaten the Americans to space. With *Sputnik,* the Soviets had far better surveillance capacity than the United States. *Sputnik* also raised the possibility that the Soviets would soon be able to launch nuclear missiles from space. Democrats were determined to put the ensuing panic to political use.

Kennedy vied with Johnson for the leadership of the party, each looking ahead to the election of 1960. Even before *Sputnik,* Kennedy had launched an assault on Eisenhower for failing to devote sufficient money to an American missile program. The United States, he said, had fallen behind the USSR in the arms race and suffered from what Kennedy dubbed a "missile gap" (it has since become clear that no such gap existed).[35] Johnson took much the same tack. "People will soon imagine some Russian sitting in *Sputnik* with a pair of binoculars and reading their mail over their shoulders," a party strategist wrote to Johnson in October. "The issue is one which, if properly handled, would blast the Republicans out of the water, unify the Democratic Party, and elect you as President." In November, Johnson convened hearings into why the United States was so far behind the USSR. He warned the American people, "Soon, the Russians will be dropping bombs on us from space like kids dropping rocks onto cars from freeway overpasses."[36] It was quite clear, to anyone paying attention, that federal government funding for military research would soon be growing at an even faster rate than it had grown since the Cold War began.

After the summer of 1956, Marvin Minsky and John McCarthy moved to MIT, where they founded its Artificial Intelligence Project. The next summer, Claude Shannon went to the Center for Advanced Study in the Behavioral Sciences, where he spent a year with Ithiel de Sola Pool. "We have a first rate group," Pool wrote from Palo Alto. Pool brought to the center another computer man, Manfred Kochen, a staff mathematician from IBM.[37] Kochen and Pool presented their work on the theory of social networks.[38]

But Pool had much else on his mind during his year at the center. "There are some complexities that arise from having two families" was how he put it.[39] His two sons from his first marriage lived in San Francisco with their mother. In November 1957, his new wife, Jean, gave birth; she and the baby nearly died; the doctor's forceps caused terrible hemorrhaging and also blinded the baby, Adam, in one eye. Their lives were saved by transfusions of blood donated by the center fellows and their wives.[40]

Pool and Kochen aimed to discover "through how many introductions would one have to go to get from person A to person B." Their justification was national defense: understanding social networks, Pool wrote in a funding proposal, would be useful for "decision-making, communication, morale, psychological warfare, and intelligence." Still, he had doubts about this line of research. He asked himself, in a note: "Is this a politically and diplomatically wise move?"[41] He seems to have decided that it wasn't.[42] He and

Kochen wrote up their findings in an essay that circulated, in manuscript form, for decades, but they didn't publish it.[43] Pool disciplined himself to pursue other lines of research instead. He'd seen that the federal government was changing course, and he changed course, too. He had lots of lines of research, ongoing research. But he decided to concentrate his new efforts on helping Ed Greenfield found the company that would become Simulmatics.

Two weeks after the Soviets launched *Sputnik,* Eisenhower held a meeting in which he asked the nation's leading scientists "to tell him where scientific research belonged in the structure of the federal government." He resolved to establish two new agencies: the National Aeronautics and Space Administration, or NASA, an independent, civilian agency, and the Advanced Research Projects Agency, or ARPA, as part of the Department of Defense.[44] Together, NASA and ARPA would pump staggering sums of money into the electronics and computer industries, as well as to universities, sums never before seen, in an ever accelerating space-and-arms race. The Soviets had been the first to space. With NASA, the Americans would be the first to the moon. The Soviets had been the first to build a satellite. With ARPA, Americans would build better weapons.

Leaders of the civil rights movement would

come to view the space race and the arms race as a flight of a different sort, a flight from justice, a flight from commitment, a white flight. "It will cost thirty-five billion dollars to put two men on the moon," the National Urban League's Whitney Young would complain. "It would take ten billion dollars to lift every poor person in this country above the official poverty standard this year."[45] At the launch of *Apollo 11,* headed for the moon, protesters would carry signs reading, "$12 a day to feed an Astronaut. We Could feed a Starving Child for $8." "We may go on from this day to Mars and to Jupiter and even to the heavens beyond," the head of the Southern Christian Leadership Conference would say, "but as long as racism, poverty, and hunger and war prevail on the earth, we as a civilized nation have failed."[46]

The unease ran deeper, too. The environmentalist Rachel Carson worried that, under the programs called for by Eisenhower, "man seems actually likely to take into his hands — ill-prepared as he is psychologically — many of the functions of 'God.' "[47] In 1958, the philosopher Hannah Arendt wondered at the sanity of a press that described the launching of a satellite as a "step toward escape from men's imprisonment on earth."[48] When, exactly, had the earth become a prison? The stars had been a dream since the dawn of humankind. Were there no dreams left but dreams of conquest, dreams of machines, dreams of artificial intelligences, dreams of simulated worlds? What about dreams of justice, dreams of equality, dreams of freedom?

Simulmatics began as a dream, Ed Greenfield's dream, a dream that had to do with his commitment to civil rights, his appetite for power, his envy of academics, and his fascination with the latest, best, fastest machines. A dream of perfect persuasion, of information extraction, of voter prediction, a dream of a world where Adlai Stevenson could win the presidency in the election of 1960 and even win the "negro vote" because he was the best man, even though he wasn't the most electable man. Alex Bernstein hadn't ever really thought about the possibility of trying to simulate an election, at least not until he was hired by Ed Greenfield to help write a program that would allow candidates to plot out their next moves on a computer, without actually making them, as if politics were a game of chess, played on magnetic tape on a spinning reel. But for that program, Ed Greenfield needed one last man: a mathematical genius from Columbia known as Wild Bill.

Chapter 5
Project Macroscope

I am skeptical of people whose God is testing.

— Sylvia Plath to her mother, 1960

Minnow Emery McPhee teaching preschool, c. 1950.

Minnow Emery McPhee was quiet and she was loving and she knew how to treat a fever and what finger paint is for. She didn't have a PhD, but, like most people who teach three-year-olds and four-year-olds, she was an expert on human behavior. In 1956, Ed Greenfield had tried to hire her husband, a dazzlingly brilliant mathematical sociologist, to do some work for the Adlai E. Stevenson campaign and McPhee had dipped in a toe or two, but, then, he was an Eisenhower man. He was also something of a crazy man.

Minnow's real name was Miriam. She'd been born in Colorado in 1923, one of five children; everyone called her Minnow, like the little fish.[1] She came from can-do women. Her mother had gone to Bryn Mawr. Her aunt Ruth Washburn had served in the Red Cross during the First World War, after which she'd studied at the London School of Economics before earning an MA from Radcliffe and a PhD from Yale; Ruth Washburn, a professor of child development, was a distinguished leader in the child education movement. She also lived with another woman, an archaeologist. Washburn was Minnow's hero.

In 1942, Minnow Emery, hoping to follow in her aunt's footsteps, enrolled in the Nursery Training School in Boston, a school whose philosophy was the radical proposition that children are people. Right about then, Minnow's brother Charles's best friend asked her to marry him.

Wild Bill McPhee was born in Colorado in

1921, to a family of loggers and cattle drivers. He went to Yale on a scholarship in 1940 but left during his sophomore year to serve in the war, flying a single-engine Piper Cub in the Himalayas. He crashed twice and was the only man in his unit to survive the war, a fact that shook him all of his days.[2]

She said yes to his proposal of marriage. That was probably a mistake. Minnow Emery was famously tender. Bill McPhee famously fierce. But there was no backing out. After they got married, Bill started a public opinion company in Colorado, later called Research Services, Inc.[3]

They had children straightaway. Wendy was born in 1946, and John, called Jock, in 1949. Bill had a terrible temper. He menaced everyone, and then later, when he calmed down, apologized. He drank and he drank and he drank and he smoked and he smoked and he smoked. You couldn't see across the living room. Jock got pneumonia four times, from the smoke.[4]

Research Services, Inc., caught the attention of Paul Lazarsfeld, who in 1951 recruited McPhee to come to the Bureau of Applied Social Research at Columbia. The McPhees packed their bags and drove across the country. Bill had never finished college, but he started graduate study with Lazarsfeld, completing the voting behavior work Lazarsfeld and Berelson had begun in 1948; he's

138

the third author of their landmark 1954 *Voting* study, the book that knocked the socks off Eugene Burdick, when he read it, during the year he spent with Lazarsfeld at the Center for Advanced Study in the Behavioral Sciences.[5] Naturally, when Ed Greenfield started trying to figure out how to program a computer to simulate voting behavior, he asked McPhee if he'd like to be part of a company he was planning to start. He didn't have a name for the company yet. But Greenfield came up with a code name for something McPhee was working on. He called it Project Macroscope.

The marriages of midcentury, white middle-class liberals were a bad bargain. Betty Friedan's *Feminine Mystique* began in 1957 as a set of interviews she conducted with her Smith College classmates, from the class of 1942, for their fifteenth reunion, women Minnow's age. Friedan wrote in the tradition of the social critic Vance Packard who in 1957 had published an indictment of the advertising industry, *The Hidden Persuaders.* Friedan was trying to trace an invisible force, too, "the problem that had no name," the quiet, lonely, miserable boredom of the American housewife. Wasn't there, possibly, something more? Friedan asked, "Who knows of the possibilities of love when men and women share not only children, home, and garden, not only the fulfillment of their biological roles, but the responsibilities and passions of the work that creates the human future and the full human knowledge of who they are?"[6] Minnow McPhee knew just what Betty Friedan was talking about.

But there was a bigger problem, too. And it didn't have a name, either. It had to do with knowledge itself. In the 1950s, when women's work was not work, women's knowledge was not knowledge. This had a disastrous effect on the two new fields of knowledge that this era produced: behavioral science and artificial intelligence.

When Bill was in graduate school, Minnow took care of Wendy and Jock and she also got a job at a nursery school. Like her aunt, she made a study of human behavior, not the behavior of voters but the behavior of babies and toddlers.[7] Computers learning language and following rules for behavior are a lot like infants acquiring intelligence and figuring out how to be a person, a similarity that appears to have been entirely lost on the men who studied behavioral science and artificial intelligence in the 1950s. Minnow probably thought about that, but she'd never have had the nerve to point it out to Bill or any of the men he worked with. She found them impossibly intimidating, "all Columbia people with P.H.D.'s," people she had to persuade to "chat on my level of information."[8] She was always doing that, undermining herself. And when she didn't, Bill did it for her. It does not seem to have occurred to him that she knew things about how people behaved. Instead, he toiled on his computer program, looking for wisdom in IF and THEN and ELSE.

Minnow McPhee was a prolific letter writer. She also wrote children's books, mainly books of poems composed during stolen moments at playgrounds. She'd carry her bundle of pages to editors, thrilled and terrified. "I go to the publishers next Thursday with my poems!" she wrote her mother. "Nothing will come of it — but it amuses me to think I'll even trespass the doorstep of a publisher."[9] She sent her poems to Simon & Schuster and then to Viking and Doubleday.[10] "All I want is to break the field so that I can have someway to rake in a penny now and then when we're low," she told her aunt.[11]

Minnow McPhee was no Sylvia Plath. "Look at me today / My dress is yellow / sticks out just right / My slip is lace / my hat is white / Daddy! Daddy! Boo! Look I'm new."[12] This poem of Minnow McPhee's, from a book of children's poetry she titled *Watch Me,* is a far cry from Plath's "Daddy": "Daddy, daddy, you bastard, I'm through."[13] Still, Minnow McPhee and Sylvia Plath lived in much the same stultifying world of intellectuals and faculty wives, of darning and sewing and diapering and writing and demurring, demurring, demurring, demurring.

McPhee's letters to her mother are full of much the same details as Plath's letters to her own mother, Plath with her yards of white-and-blue flowered fabric and her sewing patterns, stitching little dresses for her daughter, all the while worrying about the "occasional black Moods" of her husband, the English poet Ted Hughes.[14] "The Girl Scout meeting was boring as the devil" was the sort of thing Minnow was forever writing.[15] And McPhee's letters, like Plath's, are haunted by the constant threat of violence and humiliation. Once, McPhee came home wearing a new pair of high-heeled shoes, French, with an ankle strap.

"Those are the kind of shoes other ladies wear but not you!" said six-year-old Wendy, delighted.

"Yes," said Bill. "Whores."[16]

The McPhees' life revolved around the place Minnow only ever called "the Bureau": the Bureau of Applied Social Research, a grittier, East Coast version of the Center for Advanced Study in the Behavioral Sciences. She found it dreadful. "It's hard to find people we both enjoy," she wrote home. "Most of the Columbia men have rather miserable wives."[17] They went to bureau bowling games and bureau dinner parties and bureau base-

142

ball games and bureau cocktail parties. "I must say the Bureau picnics leave me cold," wrote Minnow.[18] The women looked dead-eyed.

"Last night we went to the Lazarsfelds after dinner," Minnow reported in 1951. "Half the time I need a dictionary to know what's going on but it's all very restful and uplifting to hear people talk about concepts, logic, psychology, sociology, politics, and what have you." Minnow liked Lazarsfeld, and about his wife — his third — who was herself trying to finish a PhD, she added, "She grows on you."[19]

There were two other couples Minnow liked: her brother Charles (Chuck) Emery, who was pursuing a PhD in philosophy at Columbia, and his wife, Jane (Janey) Aycrigg, who worked at the bureau, and Indiana-born James Coleman, a graduate student in sociology, and his wife, Lucille (Lu) Ritchey, who also worked as a secretary at the bureau. Chuck suffered from depression and was haunted by what he had seen in the Second World War. Jane wanted desperately to have a baby but could not. Lu Ritchey had met Jim Coleman while they were undergraduates together at Purdue. Like so many women, she'd dropped out of school when she got married; they'd moved to Rochester, where Coleman worked for the Eastman Kodak Company, and then to New York, where they lived in a walk-up apartment in Morningside Heights while Lu took care of Thomas, born in 1955, and John, born in 1957; they had one more son, in 1963.[20] Somewhere along about 1954, Jim Coleman and Jane Emery began an affair.[21] The bureau was that kind of place: swinging social scientists.

"Musical beds is the faculty sport around here," says George, the history professor, to the biology professor in Edward Albee's play *Who's Afraid of Virginia Woolf?*[22] Some sport. Minnow was bored, really and truly and painfully bored. "I don't know when I've wanted a baby so much," she wrote in 1953.[23] They moved to the Westchester town of Hastings-on-Hudson, where things looked up, because at least she got to meet people who weren't part of the bureau. "This neighborhood is a delight to live in," she wrote. "Predominantly jewish with a smattering of negroes and gentiles mixed in — I've never met such friendly people."[24] In Hastings-on-Hudson, the McPhees were also neighbors with Kenneth and Mamie Clark, researchers who provided the pivotal evidence in *Brown v. Board of Education*. "He is a negro and has just been down studying the segregation problem in the South," she told her mother later, reporting that Clark had "stayed with Reverend King who's [*sic*] home has been bombed by the White Citizens Committee."[25]

Minnow McPhee tried to put the pieces of

her world together. The bureau. *Brown v. Board.* The quantitative analysis of human behavior for purposes of prediction. So much political tragedy nationwide; so much pettiness at bureau picnics. The gossip, the affairs, the intellectual posing. She was tired. She felt helpless.

She had little luck with her poetry but better luck with the pregnancy. In April 1954, she gave birth to a girl named Sarah.[26] "The babby as black hair," Wendy reported to her grandmother, spelling as best as she knew how.[27] That summer, the McPhees, the Emerys, and the Colemans went to Hanover, New Hampshire, for bureau meetings in Dartmouth. Minnow sat in on one of their seminars. "A man spoke on philosophy of science and I didn't understand a word of it!"[28]

Was this her fault? Or his?

Ed Greenfield started wooing Bill McPhee in 1956. IBM was wooing him, too. Given the ingenuity and originality of his work, a lot of people were wooing Bill McPhee. The more people wooed him, the worse he treated his wife. "Please try to realize it's possible to be a great man without being a bastard," Jane Emery warned him.[29] He found this advice impossible to take. Everything in the culture told him that great men had to be bastards.

The bad bargains of the middle-class marriages of the 1950s were bargains that turned

145

out to be hard to keep. Minnow devotedly cared for the children and worked at the nursery school and also did all the housework. "It is almost bed time and I've just finished the dishes," she wrote her mother.[30] She was very often sick, suffering from exhaustion and also, possibly, from attacks from her husband, who at times weighed 250 pounds.[31] Bill would then insist that Wendy take over the housework. Once, when Bill ordered Wendy to clean up the kitchen, Minnow pointed out that Wendy, then nine, needed to go to bed. Bill yelled, "Minnow, either you run this house or I do."[32]

He did.

After they moved to Hastings-on-Hudson, it got harder for Minnow to leave the house. She couldn't drive ("I find New York traffic terrifying and wilt at the thought of learning to drive"), which led to still more battles.[33] "Oh how glad I'll be when Bill is through with his P.H.D. and I can cease to play this 'Momma-Poppa' game,' " she wrote her mother.[34] But him finishing his PhD would only make things worse.

McPhee said no to Greenfield in 1956 because he didn't want to work for Adlai Stevenson. And he said no to IBM because he didn't like the idea of working for a giant corporation. He stayed at the bureau and won grants from the Ford Foundation and the National Science Foundation. It was by now as easy to get money to do quantitative social science as it was hard to get money for non-quantitative scholarship. "The N.S.F. has turned down my research proposal," the distinguished sociologist C. Wright Mills complained to Lazarsfeld. "So has the Ford Foundation, the Health Department, and Columbia's own Council of Social Research." Mills asked Lazarsfeld if he knew where to find "two or three thousand dollars to hire a part-time secretary."[35] But for the kind of predictive analytic work McPhee wanted to do, you could get funding almost anywhere.

In the summer of 1956, McPhee and Coleman went to Dartmouth, for more seminars with Lazarsfeld. They met there at the same time as the artificial intelligence conference, where Alex Bernstein presented his chess program. There's no record that McPhee attended the meetings of Dartmouth's artificial intelligence seminar, but he was in Hanover that summer, and it's hard to imagine he didn't drop in. Minnow, eager to escape her cage, took the occasion of his absence to learn how to drive. "Daddy will be in Hanover and won't know his precious car is being misused," she wrote Janey.[36]

When Bill got back to New York, he threw himself into his dissertation. The deeper he got, decoding the mysteries of human behavior, the more viciously he treated his wife. He recruited her to help out — she coded his data — but the gulf between them widened. (She hated coding. "Never again," she wrote her mother.)[37] "He doesn't know how I'm feeling," Minnow had earlier written to Janey, despairing, "and every time I start to tell him he starts on this long harangue." It was impossible to talk to him. "If only I could talk to him without him blowing a fuse."[38]

He was drinking more, hostile, screaming, nasty, brutal, treating Minnow "as if a good beating or bawling out were the best approach." Minnow's brother Chuck Emery warned him to stop. "You as a scientist of the first rank can be so blind in your most intimate relationships," Emery wrote, and "this blindness is, as I've said before, destroying your wife."[39]

"I really don't understand why she doesn't shoot him," Janey wrote.[40] It was around this time that McPhee told Greenfield about one of his new computer simulation ideas: he wanted "to create a computer simulation of the U.S. television-viewing public."[41] It could be used to help television stations and networks sell advertising. Or it could be used to help out a political campaign. Eisenhower wouldn't be running again in 1960, and McPhee had no problem working against Ike's successor, Richard Nixon.

Greenfield asked McPhee to tell Ithiel de Sola Pool about this new project. Then the three men got together and talked it over.

Pool and Greenfield urged McPhee to pursue a slightly different idea: a simulation of the U.S. electorate.[42] Greenfield must have done more than urged. He would have promised McPhee the sky, the moon, and the stars. He would have wined him and dined him. He would have treated him like a messiah, about to deliver American politics from the age of ignorance to the age of knowledge. He would have told him he needed this political simulator now, yesterday, two presidential elections ago, but, damn it, he needed it now. By God, it could save the country, and it could make them all rich!

By early 1958, with *Sputnik II* now orbiting the earth, McPhee was working obsessively on this new project. He got even crazier, descending into a mania. Janey reported that Bill "is completely involved in his model of voting and yet so much in need of assurances he's worse than he ever was about monopolizing the conversation and boasting about his contributions to science."[43] He refused even to come to the dinner table. "Bill types all day long and even right through dinner which is the biggest lot of nonsense," Janey wrote. She'd about had it with her brother-in-law.

On Saturday, April 12, 1958, Janey took Wendy, Jock, and Sarah to the zoo so Minnow could rest. That night, Janey and Minnow went to see *Peyton Place*, an adaptation of Grace Metalious's 1956 blockbuster novel about the sexual repression and thwarted ambitions of women living in a small New England town. "Indian summer is like a woman," the novel famously begins. "Ripe, hotly passionate, but fickle, she comes and goes as she pleases so that one is never sure whether she will come at all, nor for how long she will stay." The plot involves adultery, rape, incest, and abortion. "Men were not necessary, for they were unreliable at best, and nothing but creators of trouble" is the view of one of the main characters.[44] The film starred Lana Turner. In its pivotal scene, a girl who has been raped by her stepfather bludgeons him to death. Earlier that month, in real life, Turner's daughter had killed Turner's abusive lover, Johnny Stompanato. Minnow told Janey that "there was something wrong with our society that so many men were crapped up."[45]

They really were crapped up: haunted by the war, deluded by Freud, trapped in terrible straits, raised to be strong but not brutal, astute but imperturbable, lines so easy to cross. "Bill ought to be shot," Janey wrote to Chuck.[46] The next week, when Janey told Minnow about a story she'd read in the newspaper, about a woman who'd shot her husband three times, Minnow said — and Janey described Minnow as uttering these words "in a voice mixed with weariness and hope" — "Why, I never thought of that."[47]

■ ■ ■ ■

The men often left their women and children to confer with other men about how to train machines to behave like humans, by which they meant: men. In the summer of 1958, Bill McPhee attended a Ford Foundation–funded conference on computer simulation held in Santa Monica and organized jointly by the Social Science Research Council and RAND.[48] That RAND summer of 1958 is legendary in the history of computers, almost as legendary as the artificial intelligence summer at Dartmouth in 1956. People who had never programmed a computer before learned how, mainly on RAND's computer, which was called JOHNNIAC. Attendance was by application. Most of the twenty participants stayed for three or four weeks.[49] Jim Coleman was there, too.[50] Minnow and Lu and the children stayed home, in Hastings-on-Hudson.

An elated Bill McPhee wrote a letter to nine-year-old Jock about the amazing machines. "Jock, the reason I'm having so much fun here is that the computers (Univac, Manniac + "Johnniac" — I work on Johnniac) are just like big electric trains for grown-up boys."[51] To Minnow, he started talking about all the money he expected to make off his new project, his voting prediction machine.

151

"Could he really make all the money he claims he could if he wanted to — or not?" Minnow's sister wondered.[52]

Sometime along about this time, apparently not long after Bill got back from Santa Monica, and during the months when he was refining the program upon which the Simulmatics Corporation would be founded, Minnow had him committed to an insane asylum. He was lost. He could not stop.

Bellevue Hospital, which started out as an almshouse, had a long-standing relationship with Columbia. It made sense that Minnow sent him there, and not someplace else. But its psychiatric ward was dire, bleak, and brutal, its inmates forlorn and broken and abandoned. It was the age of electric shock and lobotomies. Ken Kesey wrote *One Flew over the Cuckoo's Nest* in 1959, a story set in a different hospital but bearing witness to the same horrors. Minnow hadn't shot her husband. But, in commiting him to Bellevue, she hadn't done nothing, either.

Of the Simulmatics Corporation's many birthplaces — Madison Avenue, Palo Alto, the Pentagon, IBM — the bleakest was the psychiatric ward of Bellevue Hospital, with its white walls and linoleum floors and locked doors, smelling of antiseptic and urine, amid the mutterings of misery and screams of pain and cries of anguish and peals of deranged laughter and the quieter sounds of weeping. McPhee, using a typewriter, wrote to Coleman and told him Minnow had had him committed. "I've got a bunch of lawyers fighting it," he assured him. The lawyers sound like a delusion.

It was from inside Bellevue that McPhee toiled over his new theory, typing it on eight pages of yellow paper. He illustrated it with sketches. It described his plan to devise a voter prediction machine, as well as the theory behind it. He planned, he said, to create "literally a set of <u>rules of thinking</u>" to "reproduce or better <u>simulate</u> the thinking behavior of voters." In another letter, he told Coleman that "the yellow-sheet letter" had troubled his psychiatrists: "they seemed to doubt if it was a real letter to any real person."[53] But it was to a real person. And it was a real idea. Conceived in a mental hospital, it became the core intellectual property of a new company.

McPhee got out of Bellevue. Minnow took him back. He must have suffered terribly. Maybe he got help, one of the newer drugs to treat manic depression, drugs that had wretched, sedating, numbing side effects. If he'd taken them, he'd likely have found it difficult to work, the race car of his mind dragging.

He finished his dissertation. He presented his campaign simulator to his advisers at

Columbia, a program designed to make possible "A Fully Observable Electorate."[54] (It is not, at an elementary level, any different from what Cambridge Analytica sold as its services to the Trump and the Brexit "Leave" campaigns in 2015 and 2016.) The voting studies made it possible to know about voters as "discrete units at the 'microscopic' level," but this system would provide a "macroscopic picture of how, when it is all put together, the aggregate system works."[55] The earlier voting studies were mere child's play compared to what McPhee had invented: a voting prediction machine.

Greenfield wanted it, wanted it, wanted it. He'd been wanting it since 1952. Minnow overheard Bill talking to Ed: "mumbling over the phone on a big business deal."[56] It would make both their fortunes, he'd have promised. Greenfield took McPhee's paper, "A Model for Analyzing Macro-Dynamics in Voting Systems," and shaped it into a proposal, part intellectual justification, part business plan, for what Greenfield called "Project Macroscope." He imagined spinning off his Social Science Division into a new company, to run this project. Its objective was to direct Democratic strategy for the 1960 election.

Greenfield marked the Project Macroscope proposal CONFIDENTIAL and had copies made. He circulated it very selectively. It was a trade secret. He also knew it would be controversial.

With Project Macroscope, Greenfield proposed to build an "information bank" out of election returns and public opinion surveys. They'd sort voters into voter types, down to a microscopic level, classifying, for instance, "working-class Negroes in northern cities." For each voter type, information from the election returns and the public opinion surveys would be sorted by issue. "There is nothing mysterious about this on the surface, the input to the machine being the information about real individuals obtained in surveys," Greenfield wrote. "However, once this information is inside the high speed storage facilities of the machine, it is a different world." The machine, crammed with microscopic data about voters and issues, would act as a "macroscope": you could ask it any question about the kind of move a candidate might make, and it would be able to tell you how voters, down to the tiniest segment of the electorate, would respond.[57] It would be like sitting on Mount Olympus, with the gods, and looking down on the mortals.

Alex Bernstein's father-in-law was always asking him, whenever they talked about politics, "Yes, yes, but what does it mean for the Jews?" Bernstein, trying to explain to his father what Project Macroscope would do, said, "It will tell you what it means for the Jews!"[58] Or, and more to the point, it could

155

advise Democrats trying to figure out what to say about civil rights.

"Suppose that during the campaign, the question arises as to the possible consequences of making a strong civil rights speech in the deep South," Greenfield wrote. "We will, from our model, be able to predict what such a speech would mean to each of 1,000 sub-groups of the population, and how many individuals belonging to each sub-group there are in each state. We would therefore be able to predict the approximate small fraction of a percent difference that such a speech would make in each state and consequently to pinpoint the state where it could affect the electoral vote. We might thus advise, for example, that such a speech would lose 2 to 3% of the vote in several Southern states that we would carry anyhow, but might gain 1/2 of a percent of the vote in some crucial Northern state."[59]

Somewhere along the road, Ed Greenfield had lost hold of his ideals. Maybe he'd just gotten tired of Democrats losing elections. Maybe all that mattered anymore was winning. Or maybe he couldn't figure out any other way to push the Democratic Party to take a bolder stand on civil rights, something he'd been wanting it to do for years. Should a politician make a strong speech about civil rights in the South because it was the right thing to do? No. A politician should make a strong speech about civil rights in the South when and where the Macroscope advised it, based on the analytics.

Project Macroscope aimed to solve the problem of Adlai E. Stevenson. "The machine is no substitute for human beings," Greenfield wrote. "Rather it is designed to give human politicians the data so that for the first time they can make really intelligent decisions about how to make the democratic process work."[60] He was not unaware of how this whole thing looked. "I gather that questions have been raised about the political morality of using advanced computer programs in political research," he admitted. "There is also the notion that human beings are some-how going to be brainwashed." This was hogwash. "All that machines do is provide more data to more people more quickly than otherwise." Also: "It is not proposed to do anything to the voters. The brainwashing analogy is therefore completely irrelevant." And: "Machines can do nothing but speed up communication," which means that "they restore the possibility of ready discourse about important matters in large societies."[61]

Greenfield, in New York, sent his confidential proposal for Project Macroscope to Newton Minow, in Chicago.[62] Minow, thirty-three, and the father of three girls, was one of Adlai Stevenson's closest advisers, a partner at Stevenson's law firm; he'd also served as

Stevenson's counsel during his two presidential campaigns. Minow was a formidable and principled man. He became best known for serving as chairman of the Federal Communications Commission, under John F. Kennedy. In 1961, as FCC chairman, he would famously declare television a "vast wasteland," urge the development of educational television for children, and get Congress to pass legislation that established the public broadcasting systems that would become PBS and NPR. He'd known Ed Greenfield since Greenfield had worked on the Stevenson campaign in 1956.

Newton Minow read Greenfield's proposal and his jaw dropped. He was so alarmed that he made the rather fateful decision to send the proposal on to the Harvard historian and Stevenson speechwriter Arthur Schlesinger Jr., at his office in Cambridge, deep inside the stacks of Widener Library.

Schlesinger, forty-one, and seldom seen without his signature bow tie, was one of the most respected American historians in the United States. He wrote mostly about the presidency. His 1945 book, *The Age of Jackson,* had won a Pulitzer Prize. He'd win a second Pulitzer for a biography of John F. Kennedy. He was fascinated by political power. Like anyone who is fascinated by power, he was always in danger of falling into its thrall.

In 1952, Schlesinger, like Greenfield, had worked for Averell Harriman's campaign, but once Stevenson won the Democratic nomination, he'd given the Illinois governor his support. Schlesinger was also capable of great deviousness and betrayal; in 1956, he'd passed the Stevenson campaign inside information about how to defeat Harriman.[63] He'd become a full professor at Harvard in 1954, even while spending a great deal of his time writing speeches for Stevenson. He was close to the Kennedys, especially Jack; he'd known him since they were undergraduates together. And it happened that in Cambridge, on quiet, tree-lined Irving Street, he lived two doors down from Ithiel de Sola Pool. The Schlesingers lived at 109 Irving Street, the Pools at 105.

Schlesinger read Minow's letter behind the locked door of his study in the library. "Do you remember Ed Greenfield?" Minow asked. "He's the fellow who in '56 organized our so-called depth studies, which never really indicated the depth of our despair." Greenfield had given Minow permission to send the confidential proposal to Schlesinger. Minow wanted Schlesinger's advice: "Without prejudicing your judgment, my own opinion is that such a thing (a) cannot work, (b) is immoral, (c) should be declared illegal. Please advise."[64]

Schlesinger looked over the proposal. He

might well have already known about Project Macroscope, from Pool. He handled Minow's request for advice delicately. "I have pretty much your feelings about Project Macroscope," he began. "I shudder at the implication for public leadership of the notion . . . that a man shouldn't say something until it is cleared with the machine." But he wasn't willing to try to thwart it: "I do believe in science and don't like to be a party to choking off new ideas."[65]

Newton Minow had asked for help. He thought, as a citizen and as a lawyer, that what Ed Greenfield proposed to do was immoral and ought to be declared illegal. Arthur Schlesinger told him to do nothing.

Project Macroscope went ahead. It's going on still.

■ ■ ■ ■

Part Two:
The People
Machine

■ ■ ■ ■

"You know, Prestwick, when I look into the future I am certain we are entering upon an era in which religion and mass communications will put all distrust and antagonism behind them, and learn to co-operate to their mutual advantage."

"That's certainly a thought, R.V."

— *Michael Frayn,* The Tin Men, *1965*

Chapter 6
The IBM President

When a machine takes the job of ten men,
where do those ten men go?
— John F. Kennedy, 1960

John F. Kennedy campaign leaflet, 1960.

163

The Simulmatics Corporation opened for business on February 18, 1959, with offices at 501 Madison Avenue, on the corner of Fifty-Second Street, in a pale brick, thirty-story building five skip-and-a-hop blocks south of IBM's world headquarters, at 590 Madison Avenue, where Alex Bernstein had played chess with a 704 in the tiny hours of the morning, bishops taking queens, rooks taking pawns, until the final, clattering printout: ***THANK YOU FOR AN IN-TERESTING GAME***

Simulmatics was a decidedly smaller operation than IBM, occupying 1,625 square feet of office space (less than a third the size of IBM's lobby alone), in rooms on an upper floor sublet from another tenant for $375 a month, and no equipment more sophisticated than a typewriter. "The Company does not own, and does not intend to acquire, any computer facilities," Ed Greenfield explained. Instead, at least while it was starting up, it would rent time on the machines at IBM, Columbia, and MIT. Its offices were as modest as its ambition was grand: "The Company proposes to engage principally in estimating probable human behavior by the use of computer technology."[1] The company proposes to predict the future.

"This is going to be done," Greenfield promised Newton Minow in March. Minow had asked Arthur Schlesinger for help in trying to stop it; Schlesinger had urged Minow to back off. But, in fact, this thing had already been done: a month before Greenfield wrote to Minow, he'd filed papers of incorporation with the state of New York.[2]

The day a door opened with THE SIMUL-MATICS CORP. etched into its glass, there might have been some small ceremony. But if there was, Bill McPhee missed it. He'd been in Chicago, scheduled to fly back to New York on American Airlines Flight 320, but he'd canceled his reservation at the last minute and extended his stay in Chicago. Flight 320 crashed into the East River while trying to land at LaGuardia through ice and fog; all but eight passengers were killed. McPhee kept his unused ticket all his life, a good luck charm.[3]

If there was an opening day ceremony to christen the new offices of the Simulmatics Corporation, one more man was missing, too. Ed Greenfield had asked Eugene Burdick to join the company.[4] Burdick had just published a new novel, *The Ugly American,* set in a barely disguised Vietnam. It had gone through twenty printings in its first five months; it would spend a staggering seventy-six weeks on top of the bestseller list.[5] Eisenhower read it over a weekend at Camp David.[6] John F. Kennedy provided a copy to every member of the U.S. Senate. (Though when Kennedy met Burdick, he flattered him by mentioning his admiration not for *The Ugly American* but for his 1956 novel, *The Ninth Wave*.)[7] Burdick sold the film rights — the movie would star Marlon Brando — and was by now better known in Hollywood than at the annual meetings of the American Political Science Association.[8] He was still teaching political theory at Berkeley — and the girls hadn't stopped lining up outside his office door — but he'd also started offering a course called The Political Novel, teaching not Aristotle and Hume but *The Manchurian Candidate, The Secret Agent, All the King's Men,* and *The Naked and the Dead*.[9] The political thriller.

When Greenfield called Burdick to ask him to join Simulmatics' research board, headed by Ithiel de Sola Pool, Burdick declined. He said he wasn't sufficiently qualified in mathematics. He was busy. He was a literary celebrity. Also, he thought Simulmatics sounded dangerous. He soon became its fiercest critic. "This may or may not result in evil," Burdick would warn. "Certainly it will result in the end of politics as Americans have known it."[10]

Politics as Americans had known it would end in the 1960s with or without Simulmatics. The liberal consensus so widely if wrongly perceived in the 1950s would fall apart, and

166

fall to pieces. The long march of the civil rights movement would lead to the Civil Rights Act of 1964 and the Voting Rights Act of 1965. And yet injustice would endure, and protesters would riot on the streets, and police officers would attack them, with dogs and tear gas and bats and hoses. The United States would lose its way in the jungles of Vietnam, a war whose toll would include Lyndon B. Johnson's presidency and the Great Society itself. Universities would erupt in protest, students staging sit-ins and occupying buildings and, sometimes, threatening violence. A New Left would rise, and a New Right, both dedicated to a politics of confrontation and humiliation and evisceration. By 1967, Arthur Schlesinger would conclude that each of Johnson's ambitions and ideals, "the fight for equal opportunity for the Negro, the war against poverty, the struggle to save the cities, the improvement of our schools," had been "starved for the sake of Vietnam."[11] On the darkest of days, the political leaders of an entire generation would be assassinated, one after another: John F. Kennedy, Malcolm X, Martin Luther King Jr., and, in the last blow of a devastating decade, Robert F. Kennedy. And the decade would end, on an early May day in 1970, with soldiers of the Ohio National Guard — the government itself — firing on unarmed students at Kent State University.

The Simulmatics Corporation, founded in 1959 and bankrupted in 1970, played a role in most of these events, and its history offers a vantage on all of them, a shadow history of the 1960s. It's as if Simulmatics had left behind not a narrative of the decade but a box of punch cards waiting to be decoded, a cryptic chronicle of the unmaking of American politics.

It began with the effort to draft Adlai Stevenson to run for president, yet one more time. He'd lost twice before. Why would he agree to run again? To save the country from Richard Nixon. Eisenhower had served his two terms, and in 1960, Nixon would be topping the GOP ticket. Stevenson could not abide the idea of a Nixon presidency. "It seems to me unthinkable that a man with his background of slander, abuse, innuendo, expediency and resort to all the most devious political devices should ever occupy an office which we have tried for generations to exalt in the esteem of young people and the world," Stevenson wrote.[12]

Still, Stevenson had not announced his own intention to run. Instead, he weighed the decision, month after month after month. On one hand, a lot of people who'd voted for Eisenhower in 1956, including Bill McPhee, shared Stevenson's opinion of Nixon, had no intention of voting for him, and favored Stevenson over every other likely Democratic contender, including John F. Kennedy.[13] On the other hand, Stevenson's prospects in 1960 looked in many ways much grimmer than they had in 1956. "Not even his most adhesive admirers claim that Adlai Stevenson is a great candidate," the acerbic columnist Mary McGrory wrote in 1959. "They insist, nonetheless, that he would make a great President."[14] His liabilities hadn't gone away: the divorce, the bad jokes. Plenty of people who admired Stevenson wished he'd sat out the unwinnable election of 1956 and let Estes Kefauver lose to Eisenhower instead. They asked themselves: What if, what if, what if?

Ed Greenfield was selling something more than second guesses. He was selling a people machine, run by a team of scientists, the What-If Men, custom-built for the Stevenson campaign. In the spring of 1959, about the time he wrote to Newton Minow, Greenfield also wrote to Thomas K. Finletter, Stevenson's close friend and campaign aide, the man Greenfield had reported to when working for the Stevenson campaign in 1956. Since then, Finletter had started up a brain trust called the Democratic Advisory Council, somewhat in tension with the Democratic National Committee.[15] Greenfield told Finletter about Project Macroscope, informing him that it had now become possible "to develop a computer program which will predict the result of alternative campaign strategies."[16]

Finletter arranged for Greenfield to present his proposal to a group of wealthy donors that had established, for the DAC, a Committee for Special Projects.[17] The committee was led by Agnes E. Meyer, a philanthropist whose husband, Eugene Meyer, had bought the *Washington Post* in 1933. (In 1946, Eugene Meyer had stepped down as publisher of the *Post,* naming his son-in-law Philip Graham as his successor. In 1963, after Graham's suicide, the Meyers' daughter Katharine Graham would take over.) The committee had retained William Attwood, a former Stevenson speechwriter and editor at *Look* magazine, to conduct campaign research for Stevenson.[18] Greenfield asked the committee to hire Simulmatics, too, and to the same end.

Greenfield got a nod from the committee, and in May 1959, Greenfield, McPhee, and Pool went to Washington and presented their proposal to members of the Democratic Advisory Council and the Democratic National Committee.[19] The DAC and the DNC agreed to a four-month trial and the New York committee commissioned Simulmatics to conduct a single study, for $35,000.[20] In November, they met again, this time with a team of evaluators that included Harold Lasswell and Paul Lazarsfeld, resulting in a grant of another $30,000.[21] (The combined $65,000 is more than half a million dollars in

2020 money.) Pool called it "a kind of Manhattan Project gamble in politics."[22] It was as if they were building a bomb.

At the time, the Simulmatics' Manhattan Project was the largest political science research project in American history. Pool and McPhee began by collecting punch cards from one hundred thousand surveys conducted by Gallup and Roper in 1952, 1954, 1956, and 1958. They sorted the voters queried in those surveys into 480 voter types. "An example of what we mean by a type of voter is: Midwestern, rural, Protestant, lower income, female." Then they sorted the questions asked on the surveys into about fifty "issue attitudes." Finally, they recoded all of this data onto new punch cards and built a data set that included election returns from each of those years.

Only a tiny slice of Simulmatics' bank of voters were black: 6,564 in all, of whom 4,050 were in the North. But the company's ability to study African Americans as a voting type — and their *interest* in African Americans as voters — represented a major change in the measurement of public opinion. George Gallup had notoriously failed to include blacks in his surveys, both because he'd assumed that most of them were unable to vote, given the violence and repression in the Jim Crow South, and because his polls appeared in a nationally syndicated news-

171

paper column and southern newspapers had threatened to drop his column if he reported on the opinions of black Americans.[23] Knowing that blacks in the South could not vote — due to voting suppression measures that included poll taxes, literacy tests, and outright violence — most pollsters, following Gallup, had paid very little attention to black voters, anywhere. But Simulmatics wanted to convince the Democratic Party that black votes mattered, and to do that, they needed to count them.[24]

Meanwhile, Kennedy, in his bid for the Democratic nomination, began wooing the intellectuals who'd for so long supported Stevenson. He'd nearly become Stevenson's running mate in 1956; two years later he'd won reelection to the Senate with a whopping 73 percent of the vote.[25] Still, he had liabilities: he was Catholic, and the United States had never before elected a Catholic for president. He was young, only forty-two. He was weak on civil rights. And liberals didn't trust him. His family had close ties to Joseph McCarthy. His father had donated money to McCarthy's campaign. His sister had once dated him. His brother Bobby had worked for him. And when Congress voted to censure McCarthy, 67–22, Senator Kennedy, in the hospital recovering from near-fatal complications following surgery, did not appoint an aide to vote on his behalf: he was the only member of the Senate not to cast a vote.[26] Liberals had never forgiven him. Kennedy set about seeking that forgiveness.

"One morning in mid-July 1959, as I was sitting in the sun at Wellfleet, Kennedy called from Hyannis Port to invite me to dinner that night," Arthur Schlesinger recalled. "This was my first visit to the Kennedy compound." It would not be his last. Schlesinger began falling in love with Kennedy, and with his wife, too. "Underneath a thin veil of lovely inconsequence, she concealed tremendous awareness," Schlesinger observed.[27] The wealth, the glamour, the beauty. Kennedy impressed Schlesinger with his vigor and determination and intelligence and also with traits that Stevenson lacked, including decisiveness and ruthlessness. Schlesinger quietly switched sides. For months, he positioned himself as a middleman, calling Stevenson and visiting Kennedy, passing along messages and setting up meetings between the two men, and trying to get Stevenson to throw his support behind Kennedy and declare himself out of the race.[28] None of it worked.

On January 2, 1960, Kennedy announced his bid for the Democratic nomination. Square-shouldered Minnesota senator Hubert Humphrey entered the race, too, though he later said, tellingly, "I felt as competent as any man to be President with the exception of Stevenson."[29] And still Stevenson, Prince of Danes, refused to say he was running, or that he wasn't, instead only letting it be known that, if called to serve the country, he would not refuse.[30] This was hardly sufficient to inspire a restless electorate.

The election of John F. Kennedy to the White House in 1960 became the stuff of myth and legend. It carries an air of destiny. This is, in large part, the result of a single book, *The Making of the President 1960*, the defining, iconic account of the campaign, by the reporter Theodore H. White.[31] It became a bestseller. It won a Pulitzer Prize. It also helped create the myth of Kennedy as an inevitable president when, in fact, the story was rather more complicated than White allowed. White shadowed the Kennedy campaign, and favored the Kennedy campaign, and got far too close to the Kennedy campaign, and to Kennedy himself. In truth, Kennedy's bid for the nomination had been audacious from the start and it was contested to the last: his victory in the general election was one of the closest in American history.

In February 1960, Simulmatics was still at work on its study of black voters when, at a whites-only lunch counter in a Woolworth's in Greensboro, four black students from North Carolina A&T refused to give up their seats. Their resolve stirred the nation. The sit-ins spread across the South, to Tennessee and South Carolina, to Georgia and Virginia, to West Virginia, Texas, and Arkansas. "Who do you think you are?" white students carrying Confederate flags taunted them one day. The A&T football team, carrying tiny, hand-held American flags, leapt in and said, "We the Union Army."[32] And still Simulmatics sorted punch cards, dividing the electorate, voter by voter, issue by issue.

Meanwhile, on the campaign trail, Kennedy seemed to be speaking a new language, "talking in the Stevenson idiom," as Schlesinger put it, "stressing peril, uncertainty, sacrifice, purpose." Stevenson's influence on the party had been profound, but if Kennedy had begun sounding like Stevenson, this was in no small part due to all the time he'd been spending with Stevenson's former speechwriters, including Schlesinger himself. "Kennedy was emerging as the heir and executor of the Stevenson revolution," Schlesinger wrote.[33] But it was Schlesinger who crowned him as that heir.

Kennedy's path to winning the Democratic nomination was to win as many primaries as possible, by as wide a margin as possible, in order to convince the party that he could defeat Nixon, the inevitable Republican nominee. Kennedy built a campaign organization headed by his brother Bobby, with help from another brother, Teddy, much abetted by his long-term aides Theodore

Sorensen and Lawrence O'Brien, pressman Pierre Salinger, and pollster Lou Harris. And then he entered nearly every possible primary race, beginning with New Hampshire, which he won on March 8, with 85 percent of the vote. More significantly, he won the Wisconsin primary on April 5, with 56 percent of the vote, a painful defeat for the midwestern Humphrey, who, observing the efficiency of the Kennedy campaign, said, "I feel like an independent merchant bucking a chainstore operation!" On April 12, Kennedy won the primary in Stevenson's home state of Illinois. Finally, in May, after Kennedy won the West Virginia primary, which Humphrey had been expected to win, Humphrey withdrew from the race.[34] And still Stevenson had not said whether he would run, or not run.

On May 15, 1960, Simulmatics presented the results of its first study, "Negro Voters in Northern Cities," to the Democratic Advisory Council. It began with a bird's-eye view. In 1960, a year when 269 of a possible 537 electoral votes were needed to win the presidency, eight states with high African American voter turnout — New York, Pennsylvania, California, Illinois, Ohio, Michigan, New Jersey, and Missouri — would together account for 210 votes. African Americans had long voted Republican, supporting the party of Lincoln, the party of emancipation, but in the 1930s FDR had pulled them into his New

Deal coalition. Simulmatics reported that in the 1950s this coalition had begun to fall apart: in 1956 and 1958, Democratic candidates had lost black voters in the North, especially middle-class black voters, because the GOP had a stronger position on civil rights. "The shift was not just a swing to 'Ike,'" Simulmatics reported. "It was definitely a shift in party loyalty," as evidenced by the swing in the 1958 midterms, when Eisenhower was not on the ballot. "The Negro defectors . . . were party defectors, not 'Ike' enthusiasts," Simulmatics explained. "The thing that won them over was not the father-image of Eisenhower (who was, however, not disliked) but the image of what each party had done for the Negro people."[35] The GOP was the party of Lincoln, the Democrats the party of the Confederacy.

Simulmatics, in short, made two main points: (1) the Democratic Party could not win back the White House without black voters and (2) the party could succeed in winning back black voters who'd defected to the GOP only by taking a stronger position on civil rights. It might not have seemed to require a team of behavioral scientists, an IBM 704, and $65,000 to make this case, but, arguably, it had.

Stevenson's supporters might well have found Simulmatics' report edifying. But they refused to commission any further work. As

Pool explained, delicately, "for obvious reasons our original backers no longer felt justified in making commitments for the campaign the head of which was soon to be chosen."[36] By May 1960, Stevenson had not even entered the race, Humphrey had withdrawn, and Kennedy appeared poised to win it: no Democratic Party organization, no matter how loyal to Stevenson, was any longer in a position to continue to fund campaign research on his behalf.

For Pool, this was a particular blow: Simulmatics' work had been put on hold in mid-May, exactly when the academic semester ended, just when he and the rest of the academics working for Simulmatics were most available to get meaningful research done. But now, they could do very little except sit tight until the Democratic convention.

Kennedy's nomination remained uncertain. Even though Stevenson had neither entered the race nor in any other way indicated that he'd stand in Kennedy's way, plenty of Democrats tried to block and even tackle him. Texas senator Lyndon B. Johnson and other Democrats launched a "Stop Kennedy" campaign. Johnson, a bear and a bully who stood six foot three, knew how to make deals. Born into poverty in Stonewall, Texas, he'd hitched up his pants and clawed his way to power, fighting for Roosevelt's New Deal and

Truman's Fair Deal. He worked harder than any man in the Senate. But he strenuously disliked Jack Kennedy, the heir apparent, the pampered son of wealth. Johnson was fifty-one, Kennedy, forty-three. When Johnson called Kennedy "the boy," it wasn't about their difference in ages; it was about Kennedy's coddling.

Johnson's campaign didn't have much of an operation; he was waging more of a protest than a campaign. But the Stevenson campaign didn't even have a candidate. Nevertheless, Stevenson's supporters cobbled together a campaign staff. The *New Republic* and the *Nation* endorsed Stevenson and begged him to run. "Draft Stevenson" groups sprouted up all over the country. In May, it seemed not at all improbable that Stevenson would walk into the race and run away with it.

On May 21, the day after the Oregon primary, Kennedy went to visit Stevenson at his house in Libertyville. He asked Stevenson to deliver the nominating speech at the convention. He said, "Look, I have the votes for the nomination and if you don't give me your support, I'll have to shit all over you. I don't want to do that but I can, and I will if I have to." Stevenson refused.[37]

And then Schlesinger publicly defected from the Stevenson camp to the Kennedy camp. "Had Stevenson declared his candidacy, I suppose I would have stayed with him," he later said, "though I was among those who hoped he would not declare."[38] He believed Stevenson was the better man but Kennedy the better candidate.[39]

Once he switched sides, Schlesinger exerted his power to knock Stevenson out of the race. It was an odd move for a Harvard history professor, an overstepping from which he never really came back. In late May, Schlesinger arranged a meeting between Stevenson and Kennedy. Stevenson refused to endorse Kennedy, pledging only that he would equally refuse to join a "Stop Kennedy" movement. On June 5, Stevenson visited Cambridge and stayed at Schlesinger's house on Irving Street, two doors down from Pool's house, and just over the garden gate to John Kenneth Galbraith's house, at 30 Francis Avenue (the Schlesingers' backyard bordered the Galbraiths' backyard). Again Stevenson refused to endorse Kennedy.[40]

Then began the battle of the intellectuals. On June 13, some of the leading liberals in the United States, including Eleanor Roosevelt, Reinhold Niebuhr, Archibald MacLeish, John Hersey, Carl Sandburg, and John Steinbeck, sent a petition to the Democratic National Committee endorsing Stevenson. Four days later, on June 17, there appeared a counterpetition, likely written by Schlesinger and signed by a list of intellectuals headed by Schlesinger and Galbraith. "All of us supported Adlai Stevenson in 1952 and 1956 . . . but he insists he is not a candidate in 1960 and Senator Kennedy, a man of whom liberals can be proud, is an active candidate." The petition urged that "the liberals of America turn to Senator Kennedy for President" and contained Kennedy's assurance on civil rights "that he favors pledging the Democratic Party to Congressional and Executive action in support of the Supreme Court's desegregation decisions."[41]

Not everyone jumped ship. Schlesinger's wife, Marian, told newspapers she was still for Stevenson. "Can't you control your own wife — or are you like me?"[42] Robert Kennedy wrote in postscript scrawled at the bottom of a letter to Schlesinger. The unimpeachable Newton Minow was among the many former Stevenson supporters who did not defect.[43] Schlesinger later said that he regretted that the statement had come out so soon after Stevenson had been a guest in his house on Irving Street. "I felt badly that I hadn't warned him that the statement was coming out," he said.[44] But it was a terrible betrayal. One of Galbraith's friends accused him of the "worst personal betrayal in American history."[45]

Ithiel de Sola Pool jumped ship, too. He sent Kennedy's aide Ted Sorensen a copy of Simulmatics' STRICTLY CONFIDENTIAL report on black northern voters. "I would be most interested in any comments you might make about it, or in any ways in which future reports could be improved from the point of view of political usefulness," Pool wrote Sorensen, all but offering Simulmatics' services to the Kennedy campaign.[46] Sorensen, it seems, did not bite. At least not before the convention.

The 1960 Democratic National Convention was held in the Los Angeles Memorial Sports Arena, a brand-new venue just south of the University of Southern California. It was scheduled to begin on July 11. On July 5, Lyndon Johnson formally entered the race, not to win but to stop Kennedy. Three days later, on CBS News, Stevenson said that if drafted, he would run, and then he told the *New York Times* that he would do his "utmost to win."[47] The Kennedy campaign was rattled.

Stevenson's campaign staff, organizing a campaign that did not exist, headed to California. The phantom campaign's press secretary was a young writer named Thomas B. Morgan, who happened to be a very close friend of Ed Greenfield's. Early in July, Morgan flew from New York to Los Angeles, wearing a "Draft Stevenson" button on his lapel. He was supposed to set up the headquarters for the Stevenson campaign. The DNC, the Kennedy campaign, and the press all set up their headquarters at the Biltmore Hotel, "an enormous, tawny, old-fashioned hostelry, which stands eleven stories high, dominating the green rectangle of Pershing Square," as Theodore H. White wrote. The Kennedy campaign occupied a four-room suite, Suite 8315.[48] But Paul Butler, the head of the DNC, barred the Stevenson campaign from billeting in the Biltmore, on the grounds that Stevenson was a "noncandidate," so Morgan set up camp at the Paramount Building, across the square from the Biltmore, where Stevenson's campaign, all volunteers, slept in sleeping bags on the floor until, eventually, Butler agreed to give the Stevenson people two small rooms on the mezzanine, shabby quarters that the press took to calling "Butler's pantry."[49]

Kennedy arrived in Los Angeles on Saturday, July 9. So did Stevenson; he was greeted at the airport by ten thousand supporters, still, *still,* madly for Adlai.[50] At Logan airport in Boston, Schlesinger, waiting to board his own plane, miserable, wrote in his diary, "If AES had any chance, I would feel happier in Los Angeles if I were working for him, or at least I think I would; I think I would feel happier for myself."[51]

Kennedy's hope was to win on the first ballot, when delegates pledged to him in the primaries had no choice but to vote for him. Stevenson's plan — or, rather, the plan of his supporters — was to court enough unpledged delegates to keep Kennedy from winning on the first ballot in order to beat him on the second or third. Tom Morgan, trying to spin this, told every reporter he met the same thing: "If Stevenson were to receive the vote of every delegate who really prefers him but is voting for Johnson to stop Kennedy, or for Kennedy to stop Johnson, it would be Stevenson on the first ballot."[52] But that's not how a convention works.

Pool and Greenfield were there, too. Pool stayed at the Beverly Hilton. They'd have been talking up Simulmatics, taking names and passing notes scribbled on Simulmatics Corporation stationery. Simulmatics had no formal relationship with either campaign. But Greenfield and Pool wanted to make sure that Simulmatics' report on black voters in the North got into the hands of the platform committee, to urge the committee to take the strongest position possible on civil rights. Before the convention began, they'd gotten one copy of the report to Chester Bowles, a Kennedy supporter and former governor of Connecticut who served as chairman of the platform committee, and another to Harris Wofford, a Kennedy staffer and friend of Greenfield's who would draft the platform's civil rights plank.[53] ("Let me suggest that some time soon you try to talk with a good friend of mine, a very astute public relations man, Ed Greenfield," Wofford had written to Martin Luther King Jr. earlier that year.)[54] Bowles had appointed a twenty-man drafting panel, dominated by liberals. He'd included only four southerners. The platform committee met on Sunday, July 10, and endorsed a platform called "The Rights of Man."[55] Its boldest plank concerned civil rights, staking out the most liberal position on civil rights ever taken by either party.[56] *Time* described it as "a far-out liberal manifesto."[57]

Offstage, presidential nominating conventions involve a lot of hanky-panky and jiggery-pokery and more glitz and glamour and too-rich food and bad champagne than most people see in a lifetime. Gore Vidal hosted a party that, as Schlesinger reported, included "everyone from Max Lerner to Gina Lollobrigida" — that is, all sorts of celebrities, from serious journalists to sex symbols.[58] But politicking happens at parties, too. At a party at the Beverly Hilton, hosted by Agnes Meyer, Stevenson's supporters begged him to run. The next morning, Newt Minow went to see Stevenson in his bungalow at the Hilton, which was crowded with Draft Stevenson volunteers. Unlike most of Stevenson's most avid supporters, Minow had counted delegates, pledged and unpledged. He wanted to spare Stevenson defeat. To get away from the din, he pulled Stevenson into a bathroom, for a private word.

"Governor," he said, "you can listen to what you hear from those people or to me. Illinois is caucusing in fifteen minutes and it's almost one hundred percent for Kennedy." Stevenson's home state was about to set him aside.

"Really?" asked Stevenson.

"Really," said Minow.

"What do you suggest?"

"I suggest you not go out of here a defeated guy trying to get nominated a third time," said Minow. "I suggest you come out for Kennedy, be identified with his nomination, and unite the party."[59]

Stevenson hesitated. And hesitated. To run, or not to run. Hamlet to the end.

The convention proper began on Monday, July 11. In the Kennedy suite at the Biltmore that morning, Bobby Kennedy held a staff meeting. He took off his coat and loosened his tie and climbed onto a chair. "I want to say a few words about civil rights," he began. "We have the best civil rights plank the Democratic party has ever had. I want you fellows to make it clear to your delegations that the Kennedy forces are unequivocally in favor of this plank." In Schlesinger's account, it was the best, most sincere, and most inspiring speech of the convention.[60]

Outside the arena, thousands of Stevenson supporters gathered, chanting, "We want Stevenson!! We want Stevenson! We want

Stevenson! WE WANT STEVENSON!" They carried banners that read, "ADLAI IS A MORAL MAN"; "A THINKING MAN'S CHOICE — STEVENSON!" "WIN WITH ADLAI!" "STICK WITH STEVENSON"; "NOTHING LESS THAN THE BEST — STEVENSON!" "FACE THE MORAL CHALLENGE — STEVENSON"; "WE WANT STEVENSON!" Even Theodore H. White admitted, "This was more than a demonstration, it was an explosion."[61]

By Tuesday, the number of Stevenson supporters outside the arena had doubled. Stevenson still refused to declare his candidacy, but he began courting state delegations. When Stevenson went to visit the Minnesota delegation, Schlesinger, who had just finished speaking to them, stuck around while Stevenson spoke. In the back of the room, while the delegates applauded Stevenson, Schlesinger wept. He was moved to tears, he admitted in his diary, though he was also newly convinced that Stevenson could never win, but would instead bore voters with his "passionate and rather waspish historical lectures about which no one gave a damn."[62] Fair enough.

On Tuesday night, Stevenson entered the convention hall, not as a candidate but as a delegate for Illinois. He was greeted with seventeen minutes of applause.[63] Between Monday and Wednesday, White reported, the mood of the convention had changed. "On Monday the delegates had accepted as fact the rumor that Stevenson himself was about to nominate John F. Kennedy. On Wednesday as they woke, they accepted as fact that Stevenson was running and had decided to fight for it." By Wednesday afternoon, papers had printed new headlines: "KENNEDY BAND-WAGON FALTERS" and "KENNEDY TIDE EBBS."[64]

Stevenson asked Minnesota senator Eugene McCarthy to nominate him. Anticipating the nomination, Stevenson's campaign committee decided to pack the galleries. This required a certain underhandedness. They got Stevenson-friendly Kennedy ticket holders to give them their tickets; they collected used tickets from the gallery and balcony floors; they had tickets smuggled out and reused. They bribed a security guard.[65] On Wednesday evening, as thousands of Stevenson supporters streamed into the arena, the Los Angeles Police Department called for reinforcements. McCarthy rose to make his speech.

"Do not leave this prophet without honor in his own party," he pleaded. "Do not reject this man." The crowd swooned. The applause lasted for twenty-seven minutes, by Morgan's count. And still came the unending chant, WE WANT STEVENSON! WE WANT STEVENSON! They unfurled banners that read WE WANT STEVENSON! Golden balloons fell from the ceiling.

"The chairman pleaded for order," White reported. "McCarthy pleaded for order; the Convention band attempted to blare above their shouting; the lights were turned out; and still they chanted: WE WANT STEVENSON."[66] Morgan later wrote, "Then, as banners snaked along the aisles, waved from the balcony, fluttered from the rafters, a giant papier-maché 'snowball' — made of petitions bearing more than a million signatures calling on the convention to 'Draft Stevenson' — rolled out from behind the rostrum. It floated above the crowd as though lifted on an invisible wave of human enthusiasm." As the giant white ball floated overhead, someone called out, "Look, it's Sputnik!"[67]

H. L. Mencken once said that going to a convention is something between attending a revival and watching a hanging.[68] The snowball named *Sputnik*? That was old-fashioned politics, the politics of banners and trumpets and signatures, not the politics of punch cards and printouts. The galleries seemed to surge and throb. But, down on the floor, the delegates were strangely silent.

McCarthy asked the delegates to set themselves free from whatever pledge they'd made, no matter the caucuses and the primaries. Nothing, certainly nothing legal, barred them from voting whatever way they wanted to. But by now, in the history of the nominating convention, a set of expectations about the importance of the caucuses and the primaries, the voice of the people, had grown, and many delegates believed themselves bound. "Kennedy held honest title to these instructions from common voters in New Hampshire and Wisconsin, in West Virginia and in Michigan, in Oregon and Indiana," White wrote. "This the delegates knew; but not the galleries."[69]

In any event, Stevenson had already lost. Earlier that day, he'd tried to convince Chicago's Richard Daley to deliver the Illinois delegation to him, and Daley had refused. McCarthy knew that, given the numbers, Stevenson couldn't win: McCarthy's speech had been less a nomination than a tribute. That night, Morgan's assistant issued a press release from the Stevenson campaign committee, its last. It consisted of a single sentence: "A funny thing happened to us on the way to the nomination."[70]

Kennedy, unflinching pragmatist, was persuaded to offer the vice presidency to Lyndon Johnson, who was not expected to accept it. The offer, though, was thought to go a long way toward ensuring that Kennedy, if elected president, would have the full support of Johnson as Senate majority leader. Johnson, though, set Kennedy back on his heels. "You just won't believe it," Kennedy said, when he came back from meeting with Johnson. "He wants it!"[71]

Kennedy accepted the Democratic nomination late in the afternoon on Friday, July 15. He was exhausted. He delivered an acceptance speech so lackluster that it warmed Richard Nixon's heart. It also gave Nixon confidence that he would have no trouble handling Kennedy in a televised debate, if it came to that. (At the urging of Newton Minow, Adlai Stevenson had debated Estes Kefauver on television during the Democratic primary contest in 1956 — the first ever television debate between presidential candidates — and pressure was growing on Nixon and Kennedy to agree to a televised debate.)[72]

Kennedy's speech left Schlesinger troubled. Stevenson, unlike Schlesinger, found both politics and power distasteful: what a lot of people admired about him also made him unelectable. Kennedy was tougher, stronger, more willing to fight, hungrier for power, and less trustworthy. "I believe him to be a liberal," Schlesinger wrote in his private diary. "I also believe him to be a devious and, if necessary, a ruthless man." And he suspected that Kennedy would renege on what had been long assumed: that he would appoint Stevenson secretary of state. (He was right. Kennedy instead appointed Stevenson ambassador to the United Nations.) To "the inevitability of Kennedy," he conceded in his diary, with sorrow.[73]

"My admiration for Kennedy's strength and ability has increased during this convention," Schlesinger wrote in his diary, but "my affection for him and personal confidence in him have declined." Something died in Los Angeles that summer, Schlesinger thought: "Something I greatly value has gone out of national politics." It was the end of the age of Adlai Stevenson.

The next day, Ithiel de Sola Pool submitted a formal proposal to Robert F. Kennedy, at Kennedy for President Headquarters. He offered the Kennedy campaign the services of the Simulmatics Corporation.

Many voters found it difficult to tell Kennedy and Nixon apart. Both were perfectly packaged political candidates, made for cornflakes boxes and television screens, equally detached in their dispositions, with similar military experience in the Second World War and similar records in Congress and the Senate, not to mention the same square jaw, the same wavy hair, the same issuelessness. Schlesinger, hoping to distinguish them, raced to write a fifty-page book: *Kennedy or Nixon: Does It Make Any Difference?* It sold out in bookstores. "If both men appear at times cool in their attitude toward issues," he wrote, "this is, I would say, because Kennedy reasons about them, and Nixon doesn't much care."[74]

Schlesinger's little book so rattled Nixon's supporters that William F. Buckley shipped a donkey, an actual ass, to Schlesinger's door, at 109 Irving Street in Cambridge, with a note about *Kennedy or Nixon*. Schlesinger's wife, Marian, shipped the animal straight back to Buckley, at his house in Connecticut. Buckley kept it, and named it "Arthur."[75]

Two doors down, at 105 Irving Street, Ithiel de Sola Pool continued to court a commission for Simulmatics from the Kennedy campaign. The campaign already had a pollster, Lou Harris. Simulmatics was unknown and expensive and, quite possibly, too slow to be of any use. But Simulmatics, Pool promised, "has the facility for providing overnight answers to many questions of policy and strategy of the kind which are bound to arise between now and November."[76]

In a fifteen-page memo, Pool described "the Simulmatics Method" and promised that, although the initial data entry and checking and rechecking of all the earlier polls had taken thousands upon thousands of hours, the company's bank of data could now be consulted almost instantly, on any question. It had taken a very long time to set the thing up, but now that it was ready, it would be faster than anything else out there. "By use of the 704 Computer every one of these entries can be analyzed and re-combined in any desired pattern in forty minutes," he explained. "The printing of results takes an additional few hours. The preparation of an analytical report is a matter of a day or two at most." Voilà! Also, no one in the world was better qualified to do this work for the Kennedy campaign, Pool promised, because "the directors of the project are among the very few persons anywhere who combine an expert social science knowledge of voting behavior and an ability to use the IBM cards and computers."[77]

On July 16, the day Pool first wrote to Robert Kennedy, he also wrote to Lawrence O'Brien, who had worked on Kennedy's Senate campaigns in 1952 and 1958 and served as the director of his presidential campaign organization in 1960. Simulmatics had already lost precious time, Pool explained. But now the clock was ticking: "We are ready and raring to go."[78] This wasn't puffery; this was the truth.

The Kennedy campaign looked on the proposal with no small suspicion, given that it came from an outfit that had worked on Stevenson's campaign in 1956. Then, too, George Belknap, a polling consultant for the Democratic National Committee, had some doubts about the method, mainly because Simulmatics had cobbled together polls of uneven quality.[79] But Simulmatics pressed its case, promising "to test group responses to events that have not yet taken place. . . . With speed, accuracy, precision, and efficiency, simulation provides a unique test tube for the political strategist. <u>Before</u> he acts, it tells him what would be the effect of his handling of certain issues on the voting behavior of the electorate."[80]

It was a hard sell. But Schlesinger might well have vouched for Pool and, in any event, Pool and Greenfield and McPhee had come up with a smart way of talking about what they were doing. "The relationship between such current intelligence and a simulation model developed out of historical data is analogous to the relationship between current weather information and a climatological model," they'd say. "One can predict tomorrow's weather best if one has not only current information but also historical information about patterns into which current reports can be fitted."[81]

On August 11, the Kennedy campaign, persuaded, hired Simulmatics, commissioning three reports, which Simulmatics would have to put together very quickly. The first debate was scheduled for September 26. The election was to be held on November 8.

A mad scramble began, punch cards flying. Greenfield, McPhee, and Yale psychologist Robert Abelson, who had also signed onto Simulmatics, decamped to Pool's house in Cambridge. Ithiel's fifteen-year-old son,

Jeremy, who was spending the summer with his father, sat in on some of their meetings, watching them pore over endless reams of computer printouts.[82] Then they reconvened at Wading River, where Buckminster Fuller was just beginning work on the geodesic dome he was building for Frank Safford, next to Ed and Patty Greenfield's house.[83] They'd have had to send a staffer to take the train into New York, to feed punch cards into and collect printouts from the IBM 704 at Columbia.[84] On August 25, they headed to Washington, to submit Simulmatics' results to Bobby Kennedy and the top campaign staff during a briefing held in RFK's office.[85]

Simulmatics' three new studies were as shrewd as their initial study on black voters. In a report called "Kennedy Before Labor Day," they noted that Kennedy was behind Nixon in the polls, narrowly, but nearly a quarter of voters had not yet made a decision. "Democrats generally favor Kennedy, Republicans favor Nixon, and the only major source of deviation was the religious issue," they wrote. "The issue of anti-Catholicism and religious prejudice could become much more salient in the voters' minds. If that occurs, what will happen?" IF this, THEN that. They ran a simulation through the computer, analyzing the effect of further discussion of religion on each of 480 voter types concerning "(1) its past voting record; (2) its turnout at the polls; and (3) its attitude toward a Catholic candidate."[86] As a result of this analysis — the computer simulation of an election that had not yet happened — Simulmatics recommended that Kennedy confront the religion issue head-on, not to avert criticism but to incite it: "The simulation shows that Kennedy today has lost the bulk of the votes he would lose if the election campaign were to be embittered by the issue of anti-Catholicism," they wrote. "The net worst has been done."

IF/THEN: "If the campaign becomes embittered he will lose a few more reluctant Protestant votes to Nixon but will gain Catholic and minority group votes." Not only that, but if the contest became embittered, then Kennedy would be better off because "bitter anti-Catholicism in the campaign would bring about a reaction against prejudice and for Kennedy from Catholics and others who would resent overt prejudice."[87] IF/THEN. If Kennedy were to talk more about his Catholicism, then he would be attacked for it, and that attack would shore up support where Kennedy needed it most. Pointing out that "Negro voters are a danger point for the Kennedy campaign" and that Jewish support for Kennedy was fairly weak, Simulmatics suggested that "a straightforward attack on prejudice will appeal to these minorities since they are ideologically inclined to oppose such prejudices."[88]

If the Kennedy campaign wanted to avoid uncertainty, then it had only to read Simulmatics' reports. Two more reports, "Nixon Before Labor Day" and "Kennedy, Nixon, and Foreign Affairs," followed the same method and made similar recommendations. "What we have demonstrated is how data from past situations can be used to simulate a future situation."[89] They believed they had reinvented American politics.

Much of what Simulmatics recommended to the Kennedy campaign appears to have been fairly commonplace political wisdom among his close circle of advisers. There's a lot of bluster and nonsense in the archival trail left behind by flimflam men. There's really no objective way to measure the influence of the Simulmatics reports on the Kennedy campaign. As Pool admitted, "When Kennedy decided to confront the bigots head on, he himself could not say what part in his decision was played by any one piece of evidence."[90] That said, after the Simulmatics briefing, the Kennedy campaign quite exactly followed Simulmatics' advice about what Kennedy should do after Labor Day.[91]

Kennedy, who had trailed Nixon in the polls all summer, gained on him after Labor Day for three reasons: his advocacy for civil rights, his stance on religion, and his performance in four televised debates with Nixon. In each of these cases, the approach he took had been recommended by Simulmatics.

Simulmatics had specifically urged a stronger stance on civil rights in its first report, presented to the DNC in May, sent by Pool to the Kennedy campaign in June, and distributed to other Kennedy staffers at the convention. Immediately after the convention, Kennedy, the Democratic candidate least appealing to African Americans, set up a civil rights "division"; it was headed by Harris Wofford, the friend of Ed Greenfield's who'd drafted the civil rights plank on the party's platform.[92] In late October, Kennedy dramatically improved his standing with African American voters after he called Coretta Scott King to express concern and support when Martin Luther King Jr. was arrested while leading a protest in Atlanta. The suggestion that Kennedy call Mrs. King came from Harris Wofford.[93]

Kennedy began discussing his Catholicism frankly in early September. Simulmatics had specifically recommended this approach in its second report, delivered to the campaign on August 25. Instead of deflecting anti-Catholic opposition, the campaign drew attention to it, stirred it up, and sought opportunities for the candidate to respond by decrying religious prejudice. He accepted an invitation to speak before Protestant pastors at the Houston Ministers Conference on September 12, where he squarely condemned religious intolerance. "I may be the victim, but tomorrow it may be you," he said. And he warned, "I am not the Catholic candidate for President. I am the Democratic Party's candidate for President who happens also to be a Catholic. I do not speak for my church on public matters — and the church does not speak for me."[94]

The weekend after Labor Day, the last beach weekend of the summer, Ed Greenfield hosted twenty people at his house at Wading River. "Lordy — there were twenty-five mouths for his wife Pat to feed," Minnow McPhee wrote sympathetically. Minnow and Bill McPhee went, with Wendy, Jock, and little Sarah in tow.[95] McPhee, Greenfield, and Pool were likely working on a strategy for Kennedy during the upcoming debates, not least because Pool and McPhee were the nation's leading authorities on the image of politicians on television.

One of the Simulmatics reports given to Kennedy specifically addressed the upcoming debates, describing them as a risk for Nixon: "The danger to Nixon is that Kennedy can make use of his more personable traits — including a range of emotions such as fervor, humor, friendship, and spirituality beyond the expected seriousness and anger — and thus cause Nixon to 'lose the debates.' "[96]

Pool later claimed that the Simulmatics "machine" played a role in Kennedy's approach to the debates. "Some said the debates would hurt him; others said they would be decisive," Pool later told the press. "The machine was another voice that the candidate consulted."[97] The machine, the People Machine.

John F. Kennedy spent Election Night, November 8, at his family's compound in Hyannis, where a pink-and-white children's bedroom on the second floor of Robert Kennedy's house had been converted into a data center. The children's things had been cleared, White wrote, "the beds removed, the baby chairs thrust to the side," and in their stead had been placed a long table, where pollster Lou Harris pored over "reports received from the communications center downstairs and from the four teletypewriters of the wire agencies installed in the adjacent bedroom."[98]

All night, Kennedy walked back and forth across the lawn, from his own house, where the always elegant Jackie, eight months pregnant, was trying to rest, to Bobby's house, where the teletype keys were clattering. But ordinary Americans had their own data centers in their own living rooms, or their children's bedrooms, or their kitchens: wherever they watched television. The election of 1960 was "the fastest and best reported election in U.S. history." All three networks used computers. CBS News relied on IBM.[99]

The two machines that had upended American political culture in the 1950s, the television and the computer, held center stage in 1960. The first Election Night reported by computer had come in 1952, when CBS hired a UNIVAC to tally the votes and predict the outcome, in coverage that had turned out to be something of a fiasco. But by 1960, IBM had a much faster computer, the IBM 7090, which tallied the returns at nearly twice the rate of 1956's computations.

If the election of 1960 was the fastest ever calculated, it was also the closest presidential race since the 1880s. Sixty-eight million votes from 171,311 precincts were tallied by an IBM RAMAC 305 in Studio 65, the CBS News Election Headquarters on Twenty-Sixth Street in New York. Meanwhile, "in the IBM Datacenter, about a mile and a quarter uptown from Studio 65, the same IBM 7090 that was being used for the forecasting operation was put to work analyzing trends of historical significance as they developed."[100]

IBM, which had tried to hire Bill McPhee, had undertaken a project not unlike Project Macroscope, not for the purpose of devising a campaign strategy but for the purpose of reporting the voting. In advance of the election, IBM had compiled a data bank based on earlier elections by conducting a voting behavior study in 502 precincts in seventeen states: "Leading political scientists at universities in each of these states researched the precincts in that state to determine predominant voter characteristics. If a precinct had largely white-collar, rather than blue-collar voters, it was listed as a white-collar precinct. It was similarly classified for the rural or urban, racial, ethnic, and religious characteristics of its voters. Each precinct was identified not only by these characteristics, and by its state and region, but also by available voting records for presidential and House elections in 1956, 1952, 1948, and 1928." (IBM chose 1928 because it was the only other year that a Catholic, Al Smith, had run for president.) As the company explained, "All of this historical information on the special precincts was stored in the memory of the IBM 7090 computer to provide a representative sample of the national voting population."[101]

IBM, following the model pioneered by Simulmatics, used voting behavior research to compile a set of historical data on which it could make super-fast predictions. On Election Night, IBM's 7090 made its first prediction at 7:26 p.m., when hardly any polls had closed, and with less than 1 percent of all precincts reporting. CBS announced this prediction, cautiously.[102] (Only in later years would news networks agree to delay making predictions in any state until its polls had closed.)

The mood at the Kennedy compound turned gloomy. Not many minutes later, at 8:12 p.m., with less than 4 percent of all precincts reporting, the 7090 offered a new prediction: Kennedy would win with 51 percent of the popular vote to Nixon's 49 percent. CBS broadcast this prediction — absent any caveat — at 8:14 p.m., "the first forecast of a Kennedy victory to go on the air."[103]

Even more significant than its predictions, IBM asserted, was its on-the-spot election analysis. The very minute the polls closed in each of its 502 precincts in seventeen states, CBS News correspondents on the scene phoned the results to the data center, where these results were punched into cards, placed on a conveyor belt, and fed into the 7090. "Because of its ability to sort these returns and relate them to the voluminous data previously stored in its memory, the computer could then speedily answer all the thousands of questions which might be put to it on what the vote meant — in any one of those precincts, in any combination of them, or in the entire sample," IBM reported. Its 7090 could answer virtually any question in fifteen seconds.[104]

In the end, IBM's election prediction held.

With a razor-thin popular vote margin, Kennedy won with 34,226,731 votes, or 49.7 percent, to Nixon's 34,108,157, or 49.5 percent.[105] Kennedy's electoral margin of victory, 303 to 219, was wide. But the popular vote was close enough to lead to two recounts, efforts led by the Republican National Committee but not endorsed by Nixon, who said, privately, "Our country can't afford the agony of a constitutional crisis — and I damn well will not be a party to creating one just to become President or anything else."[106]

Nationally, if a single vote in every precinct had gone the other way, Nixon would have won the election. As Simulmatics predicted, "Negro voters in the North" turned out to be crucial to the Democrats' victory. Without them, Kennedy would have lost.[107]

John F. Kennedy became the thirty-fifth president of the United States. Computers came to rule Election Night. And the Simulmatics Corporation launched a publicity blitz.

CHAPTER 7
BILLION-DOLLAR BRAIN

"How important is that machinery? Is it as vital as they say?"

"Computers are like Scrabble games," I told her. "Unless you know how to use them, they're just a boxful of junk."

— Len Deighton, *The Billion-Dollar Brain,*
1966

The January 1961 issue of *Harper's* Magazine hit newsstands the week before Christmas and stayed there nearly until Kennedy's inauguration. It featured a shocking story about how a top secret computer called "the People-Machine," invented by the "What-If Men" of a magnificently mysterious organization known as the Simulmatics Corporation, had in effect elected Kennedy. Harold Lasswell announced, "This is the A-bomb of the social sciences."[1]

The *Harper's* story was picked up all over the country. It hung over the incoming administration like a storm cloud.

The *New York Herald Tribune* reported that

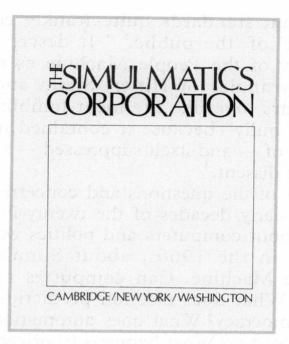

A Simulmatics Corporation brochure, c. 1961.

the People Machine, "a big, bulky monster called a 'Simulmatics,' " had been Kennedy's "secret weapon."[2] According to the *Chicago Sun-Times,* in the future, the directive of all politicians, before acting, would be to "Clear it with the P.-M."[3] An Oregon newspaper editorialized that, by way of Simulmatics, the Kennedy campaign had reduced "the voters — you, me, Mrs. Jones next door, and Professor Smith at the university — . . . to little holes in punch cards, or whatever device our new Ruler uses — and we were fed into the maw of the new Ruler and came out as the new standards to which we will ultimately

207

conform, standards quite frankly called a 'model' of 'the public.' " It described the tyranny of the People Machine as making "the tyrannies of Hitler, Stalin and their forebears look like the inept fumbling of a village bully" because it contained no possibility of — and itself suppressed — the very idea of dissent.[4]

Most of the questions and concerns raised in the early decades of the twenty-first century about computers and politics were first raised in the 1960s, about Simulmatics' People Machine. Can computers rig elections? What does election prediction mean for democracy? What does automation mean for humanity? What happens to privacy in an age of data? Most of these questions had been asked in that very first story, in *Harper's:* "If, in a free society, information is power, how do we prevent tampering with the data provided by the machine? As we approach a consensus of opinion, what happens to freedom and spontaneity? As we seek more and more data for the machines, can we maintain our traditions of privacy?" The *Harper's* story didn't offer answers to those questions. Nor did the Great Brain of Behavioral Science, Harold Lasswell, who was quoted in the piece offering this particle of wisdom: "You can't simulate the consequences of simulation."[5] The oracle had spoken.

The *Harper's* piece about Simulmatics had been written by Simulmatics' publicity guy, Thomas B. Morgan, identified in the magazine as a freelance writer. Morgan was sly that way. "I speak a disarming Corn Belt dialect," he once wrote. "This has seemed to put some people at ease."[6]

Morgan was what would come to be called a New Journalist, a features guy, and also a smart aleck, best known for his profiles of the likes of Sammy Davis Jr., Roy Cohn, and Gary Cooper. "Blithe and blue-eyed as I am, I have not gone around intending to think satiric thoughts about human folly, but looking back now, I see that I have," he'd later write.[7]

Born in Springfield, Illinois, in 1926, Morgan stood six foot four. He played football in high school. His father gave him an orange fountain pen for his bar mitzvah. He told him, "You want to be a writer, write with Eversharp." His family lost everything in the Depression, including the house, the car, and the family store. Morgan served in the Army Air Forces during the Second World War and studied to be a navigator, but by the time he was ready to go into combat, the war was over. He went to Carleton College on the GI Bill and graduated with a degree in English in 1949, the year after Joan Zuckerman. They married in 1951 and soon had two children, a girl and a boy, Kate and Nick. Settled in New York, Morgan became a prolific magazine writer (between 1950 and 1964, he published more than one hundred articles in *Look, Life,* and *Esquire*). Among the handful of assignments he turned down was a profile of Ayn Rand, which he abandoned after he read *Atlas Shrugged* and hated it. And among the assignments he accepted but never delivered was a profile of Arthur Schlesinger, who said he was happy to be written about but forbade Morgan from ever quoting him. "Nothing I could write on his terms would do him justice," Morgan admitted. "He is a lively man."

Morgan was not a dispassionate observer of either the election of 1960 or the People Machine. He'd been raised a Roosevelt Democrat and had first voted for Adlai Stevenson, for governor, in 1948. He'd written publicity materials for Stevenson's presidential campaigns in 1952 and 1956, and he'd covered him for *Look* magazine. Morgan loved Stevenson. He also pegged him, exactly: "He was a Henry James character in a Joseph Heller world."[8]

In 1960, Morgan had served as a press aide for the Draft Stevenson campaign. He was also one of Ed Greenfield's oldest and closest friends. The Morgans spent their summers in

Wading River, with the Greenfields. The Greenfields had that big Victorian house on the beach; the Morgans had a bungalow up on the bluff.[9] And when the Kennedy campaign hired Simulmatics to undertake those three reports, in the summer of 1960, Greenfield had asked Morgan to edit them. Morgan acknowledged none of these ties in his *Harper's* essay.

In "The People-Machine" story, Morgan, purporting to be a mere disinterested freelancer, explained the history of the company, the nature of simulation, and the work Simulmatics had done for the DNC and for the Kennedy campaign. As to its influence on Kennedy, Morgan offered a series of judicious caveats: "No one in Simulmatics was privy to the decisions of the Kennedy strategists, nor have I had any access to them," he admitted, suggesting that "at most, the simulation may have lent some psychological support to those Kennedy strategists who favored its conclusions anyway." Without Simulmatics, would the outcome of the election have been any different? Impossible to say, but it seems unlikely. Still, this much could not be disputed: what Simulmatics had predicted had been borne out by events: "In retrospect, the evaluations seem to have foreshadowed the direction of the campaign to an uncanny degree."[10]

The Kennedy campaign, now the Kennedy

transition team, was furious about Morgan's essay, which they accurately perceived as a very thinly veiled Simulmatics publicity stunt. If anyone were ever to use this thing, a Stevenson aide had warned Newton Minow about Project Macroscope the year before, "the operation should be conducted in a locked basement under maximum security to prevent the public from discovering the fact."[11] That's more or less how the Kennedy campaign felt, too. They'd used Simulmatics; they hadn't expected Simulmatics to use them.

The story that Kennedy had manipulated the election by way of a computer spread as fast as an IBM 7090. The Sunday before Christmas, it was picked up all over the country, including on CBS Radio, where a commentator said that one of the inventors of Simulmatics' People Machine "claims the machine 'thinks like voters.' "[12] What a mess. If a machine had gotten Kennedy elected, how was he to be trusted? "Did IBM Computer Draft Strategy Send Kennedy to White House?" was the headline of a wire service story. "A secretly designed robot campaign strategist nicknamed a 'people-machine' was said today to have been put to work by President-Elect John F. Kennedy's top advisers to suggest alternative methods of influencing voters," it reported.[13] There ensued a

212

national hullabaloo.

From the vantage of the Kennedy campaign, the Simulmatics story risked undermining Kennedy's presidency even before it began. It also must have looked like a continuation of the undermining of Kennedy by the Stevenson campaign. Even if people in the Kennedy camp didn't know that Morgan had helped Pool edit the reports submitted to the Kennedy campaign, they certainly knew that the Thomas B. Morgan who wrote "The People-Machine" was the same Thomas B. Morgan who had been Adlai Stevenson's press secretary in Los Angeles. Bobby Kennedy had at one point during the convention run into Morgan, who was wearing his "Draft Stevenson" button; Kennedy had given him an ice-cold stare.[14]

There was another problem, too. During the campaign, John F. Kennedy had made an issue of automation. "If Automation takes over your job . . . who will you want in the White House?" read one Kennedy brochure; it featured an illustration of a giant, spidery machine monster looming over a man in overalls carrying a lunch box. "The Republican administration has done nothing about the problem of automation. Nixon has said nothing. John Kennedy understands the *human* problems created by automation." Kennedy had called for retraining programs, extended unemployment compensation, and new employment services. In Charleston, West Virginia, in September, he'd delivered a speech about "the growing crisis of automation — the replacement of men by machines."[15] And here he'd gone and hired a People Machine.

The day the issue of *Harper's* hit newsstands, Pierre Salinger, Kennedy's press secretary, decided to put a stop to a national conversation about the computer that had supposedly gotten Kennedy elected. Salinger, a former reporter for the *San Francisco Chronicle* who'd been a navy captain in the Second World War, was witty and urbane, but he could also be a bulldog when bulldogging was called for. Before the end of the day, he'd gotten the wire services to send out what amounted to a public rebuke.

"An electronic brain designed to estimate voter reaction to campaign issues made strategy recommendations for Sen. Kennedy, it was reported Sunday," according to the wire service story that ran the next day, December 19. "But the top Kennedy aides denied receiving or following the recommendations." The piece quoted Salinger as saying (i.e., lying), "We did not use the machine. Nor were the machine studies made for us."[16] The denial ran in papers all over the country, in cities that included Albuquerque, San Francisco, Oklahoma City, Cincinnati, Muncie, Honolulu, Indianapolis, St.

Louis, Montgomery, and Salt Lake City, under headlines like "Kennedy Camp Denies Use of an Electronic 'Brain.' "[17]

But even though many newspapers reported the position taken by the Kennedy team, noting that Salinger had denied the story, they still ran it.[18] If anything, the denial only generated more coverage.

Simulmatics' publicity blitz had started even before Kennedy was elected. "Ed is roiling in the juices of optimism," Robert Abelson had written to Pool days before the election, when Greenfield was already taking credit for the Kennedy win with an eye toward lining up commercial clients for Simulmatics' services.[19] Pool, keen to burnish his academic credentials, had begun talking to the press. As early as November 13, less than a week after the election, the *Boston Globe* had run a feature story about Simulmatics — "Political Weather Map Put Kennedy Up There" — a description of "this latest egghead impact on American politics." Simulmatics, the *Globe* reported, had told Kennedy "why and what he must do." It credited Simulmatics with his victory in the debates and his position on civil rights ("after Kennedy and other leading Northern figures read the Simulmatics report, they decided not to compromise on civil rights"). Pool told the *Globe,* "The Kennedy brothers read our reports the day they got them."[20]

Pool also intended to write a book about Simulmatics' role in the 1960 election. He'd begun drafting an outline for the story of "how this starry-eyed notion on the frontiers of science became a reality." He wanted it to explain "what simulation can do that polls can't." And he intended to raise the question "Is computer simulation (and polling) the end of the democratic tradition?" so that he could answer, emphatically, "No." "The computer does not make the politician an automaton," he wrote in a book proposal.[21]

But before Pool really got started writing, word got out that Theodore H. White was writing a book about the campaign, too. "It will be a narrative history telling of the key decisions and will describe the basic forces and background of the primary and election campaigns," the *New York Times* reported about *The Making of the President 1960*.[22] Pool planned a different sort of book. He had a few ideas for what to call it, including the tidy *1960* and the aspirational *The People Decide*.[23] Morgan told him he ought to call it, simply, *Simulmatics,* an echo of Norbert Wiener's *Cybernetics*.[24]

By January, between Pool's trumpeting of Simulmatics to reporters, Morgan's essay in *Harper's,* the radio coverage, and the wire service reports, news that a top secret robot had rigged the 1960 election was eliciting editorials, too. "Mouth more or less agape and breath more or less bated, we have been reading how by 'simulmatics' this marvelous contrivance gave young Mr. Robert Kennedy the 'advance dope' on problems as opaque as the religious issue," wrote the editors of the *St. Louis Post-Dispatch.* "Hinkle-pinkle! If Mr. Kennedy took serious stock in the machine — 'a model of the American people' better than the original since it knew what the people would do even before they were sure — that would be enough to explain how he came so close to losing the election."[25]

The distance between these years and the present is the measure of how entirely the American voting public now takes these kinds of political shenanigans for granted. The machinery of every American political campaign above the level of dogcatcher is now run by companies like the Simulmatics Corporation using algorithms that are the great-grandchildren of the code written by Bill McPhee for the People Machine in 1959. *Mouth agape and breath more or less bated* is not how Americans view the ordinary undertakings of the thousands of data and analytics firms that advise American political campaigns in the early decades of the twenty-first century. Hardly anyone even notices them.

But in 1961, editorial writers who followed this story found it scandalous. Like Newton

Minow, who considered Project Macroscope certainly immoral and possibly illegal, they did not believe that using a computer to predict the outcome of an election, and even to direct it, was consistent with democratic self-government. Had Kennedy, somehow, cheated? Possibly. But, more broadly, democracy itself had been corrupted. Most informed observers had little use for Pierre Salinger's denial. "The fact remains," remarked the editors of the *Cincinnati Enquirer,* "that the machine knew whereof it spoke." The real story, they said, "is not so much what the machine did in the campaign that ended last month but rather in what its apparent canniness may mean for the nation's political future."[26] Could the republic survive this challenge to political self-determination?

The conservative columnist Victor Lasky, who had earlier coauthored an adamantly pro-Nixon book with Ralph de Toledano, the *Newsweek* editor who had written to Nixon in an effort to help Ithiel de Sola Pool get a security clearance, took this story the farthest. He phoned Salinger and pressed him, confronting him with the evidence.

"Pierre's recollection was somewhat refreshed," Lasky later wrote, "after it was disclosed that Bobby Kennedy had helped finance the People Machine by a sum of $20,000 and that reports based on its findings went directly to Bobby." (Lasky got

218

Salinger to admit to hiring Simulmatics, but Salinger lied to him about how much the Kennedy campaign had paid the company.)

In *JFK: The Man and the Myth,* a fierce and partisan attack on Kennedy published in September 1963, Lasky would all but argue that Kennedy had stolen the election from Nixon by using a robot.[27] Two months later, after Kennedy was assassinated, Lasky's publishers stopped printing the best-selling book, and Lasky, acting with honor, scrapped all of his scheduled lectures and TV appearances. "I've cancelled everything," he said the day after the president's death. "As far as I am concerned Kennedy is no longer subject to criticism on my part."[28] Thus disappeared, into oblivion, the one substantive investigation of Simulmatics' role in Kennedy's election.

But Nixon's defenders weren't the only people convinced that Kennedy had somehow cheated by hiring Simulmatics. And Kennedy's defenders weren't the only people frustrated by the *Harper's* story. Bill McPhee believed that Morgan "had overplayed the case."[29] Robert Abelson was so upset by the "nonsense on CBS radio" that he sent Greenfield a press release that he wanted issued by the company "in case our story becomes hot and absurd again." He titled the release: "CLARIFYING THE ROLE OF THE SIMULMATICS CORP. PROJECT IN THE PRESIDENTIAL

CAMPAIGN." Abelson insisted that the company be prepared to issue a statement that "it would be conceited of us to claim that our information and ideas played any sort of decisive role in the campaign." But telling the press that Simulmatics had helped elect Kennedy was part of Greenfield's business plan. Once Abelson understood this, he began to extricate himself from the corporation. He'd show up for board meetings, but he never did any meaningful work for Simulmatics again.[30]

Ed Greenfield didn't believe in bad publicity. This ruckus was just what he'd been hoping for. He put Morgan on Simulmatics' payroll as its head of public relations. He intended to make an initial public offering of Simulmatics' stock.

On January 9, 1961, John F. Kennedy traveled to Cambridge. He'd agreed to address the Massachusetts legislature, but he wanted to hold private meetings in the afternoon. He was putting together a brain trust, his best and the brightest. Lyndon Johnson witheringly called them "the Harvards." Kennedy didn't want to meet in a room on campus, so his staff arranged for him to hold those meetings at Arthur Schlesinger's house. Secret Service agents surrounded the property; Harvard and Cambridge police shut down Irving Street. Balding, bespectacled McGeorge

Bundy, a forty-one-year-old Harvard dean, rode his bike over; he'd be joining the Kennedy administration as national security adviser. Schlesinger, too, would be coming on board; he'd agreed to go to the White House as special assistant to the president.[31]

Pool, two doors down, might possibly have been invited to come by, but there's no record of that. Whether anyone, that day, had anything to say about the flap over the People Machine taking place in the pages of the nation's newspapers is doubtful. They were planning a presidency.

On January 17, Dwight D. Eisenhower, then the oldest man to hold the office of president, delivered a televised farewell address from the White House. Sitting at a desk in a three-piece suit, part grandfather, part commander, he issued an extraordinary warning about what he perceived to be a grave threat to American democracy: the influence of the arms race on American government and on American values. Citing the scale and influence of the arms industry, the three and a half million men and women working for the "defense establishment," and the unprecedented scale of military spending, he told Americans, "we must guard against the acquisition of unwarranted influence, whether sought or unsought, by the military-industrial complex." Even the "free university" had become beholden to the defense industry. "A government contract becomes virtually a substitute for intellectual curiosity," he observed, noting, too, that "for every old blackboard there are now hundreds of new electronic computers." What would be the consequences of the intellectuals of the nation being directed at nothing but defense? Noting the growth of both basic and applied research, he warned that "in holding scientific discovery in respect, as we should, we must also be alert to the equal and opposite danger that public policy could itself become the captive of a scientific-technological elite."[32]

Three days later, John F. Kennedy, the youngest man ever to be inaugurated, was sworn in as president before the Capitol by Supreme Court Chief Justice Earl Warren. Behind him stood his wife, who'd only just given birth to their second child, a son. Kennedy had met with Truman that morning — the first time Truman had been in the White House since the end of his presidency, in 1953, on the day of Eisenhower's inauguration.[33]

It had snowed all night, then stopped at dawn. At noon, the eighty-seven-year-old poet Robert Frost, a man born during the presidency of Ulysses S. Grant, stepped up to the podium and struggled to shield his eyes from the glare of the sun, which made it impossible for him read the poem he'd written for the occasion, celebrating "A golden age of poetry and power / Of which this noonday's the beginning hour." (Instead, Frost recited another poem, from memory.)[34] And then Kennedy, beneath a bright blue sky, wearing no overcoat, his breath billowing in front of him in the cold, took the oath of office and began his address. He declared that "the torch has been passed to a new generation of Americans — born in this century, tempered by war, disciplined by a hard and bitter peace, proud of our ancient heritage." He spoke of the length of the struggle for justice and for peace. "All this will not be finished in the first hundred days. Nor will it be finished in the first thousand days, nor in the life of this administration, nor even perhaps in our lifetime on this planet." He depicted the Cold War world as one in which "both sides" were "overburdened by the cost of modern weapons, both rightly alarmed by the steady spread of the deadly atom." But, in the spirit of his "New Frontier" agenda, he celebrated research: "Let both sides seek to invoke the wonders of science instead of its terrors."

All that winter and spring, Simulmatics' staff and scientists busily lined up clients, in anticipation of the corporation's planned public stock offering. They were exploring their own new frontier. They razzled and they dazzled. Tom Morgan, at the Simulmatics Corporation's Madison Avenue offices, began drafting a series of brochures, including one titled *Human Behavior and the Electronic Computer.* "Without exception, the men on the Simulmatics team are at the top of their professions," Morgan assured potential clients. "No comparable group exists in American business."[35]

Bill McPhee's mood, always up and then again down, improved. "He's feeling much cheered that this Simulmatics Corp. is really going to move," Minnow McPhee wrote to her mother. Her husband seems to have told her that the stock was already for sale (it wasn't). She wrote her mother, "I don't want to count my chickens before they hatch but people are buying the stock like mad and he's going to be allowed $50,000 worth of stock since it was his idea."[36] Minnow dreamed about all the things she might do "if that Simulmatics Corp. really does make us millionaires."[37] It seemed perfectly likely.

In early February, Greenfield convened a two-day Simulmatics conference at New York's Hotel Barclay.[38] The company decided to pitch projects to three types of clients: media companies, government departments, and advertising agencies. Pitching to media companies, Alex Bernstein and Tom Morgan led a Simulmatics team that met with executives from the Motion Picture Association of America, MGM, and Columbia Records, with the idea of setting up the sorts of data analytics that would, one day, lead to Netflix and Spotify. They also proposed a "mass culture model" to collect consumer data from companies across all media — publishing houses, record labels, magazine publishers, television networks, and moviemakers — in order to devise a model that could be used to direct advertising and sales by way of a meta-media-and-data corporation that sounds rather a lot like Amazon.[39]

The problem, Bernstein reported, is that it's not possible to build a model without data, and these media companies had surprisingly little data about who read the books and magazines they published or listened to or watched the records and movies they produced (television, because of Nielsen ratings, was an exception). "Not only do they lack data on current attitudes and practices in movie-going but what is even more surprising is the lack of data on movie attendance itself," Bernstein reported. "They have no way of determining how many people have seen a particular movie. Any model, therefore, that would be applicable to the industry would involve a stage of data collection."[40]

To government agencies, which generally boasted sizable collections of data, Simulmatics proposed custom-made simulations of particular problems, from air pollution to juvenile delinquency. They reached out to people in the new administration. Pool might have tried to call in favors from Schlesinger, now special assistant to the president.

Schlesinger had found, having taken a leave from his faculty position at Harvard, that he didn't have all that much to do. His office was in the East Wing, not the West Wing, and although the president occasionally consulted him, he was mostly there to bear witness, a court historian. "We'd better make sure we have a record over here," Kennedy would say during meetings, and Schlesinger would dutifully take out the stack of eight-by-four-inch index cards he carried around in his pocket and begin taking notes. "He didn't do a helluva lot," Bobby Kennedy would say, "but he was good to have around."[41]

Greenfield had already reached out to Newton Minow, the new chairman of the Federal Communications Commission. Two weeks after the election, Greenfield had sent Minow a list of "Possible Government Uses of Simulation" — tasks the Kennedy administration might hire Simulmatics to take on with departments of the federal government, from the Post Office Department ("model of mail flows") and the Department of Health, Education, and Welfare ("model of epidemiology of addiction and delinquency") to the Department of Labor ("model of future wage rate changes for different jobs") and the Department of State ("model of voting in the United Nations to project likely blocs under different conditions"). "The boys in Cambridge are responsible for it," Greenfield told Minow, "and I hope it gives some feeling of what potentially could be done in Washington."[42]

Contracts began to come in. Simulmatics' early commissions included a study of the effectiveness of government communication about the fluoridation of drinking water. Jim Coleman, McPhee's graduate school friend, had also joined Simulmatics. He secured an account with Bristol Laboratories to simulate physicians' acceptance of new drugs.[43] Simulmatics' other early government proposals included, for the USDA, studies of rural communications systems and rural agricultural practices and, for the Bureau of Public Roads, the simulation of automobile traffic.[44] The company also initially considered, but did not then pursue, projects with the Department of Defense, headed by Kennedy appointee Robert McNamara, the former head of the Ford Motor Company. Greenfield clipped from the newspaper and sent to Pool a *Washington Post* story, dated March 25, 1961, describing McNamara's quantitative, systems-analyst bent.[45] Pool tucked the clipping away, to be taken out another day.

For advertising agencies, Simulmatics devised a product called Media-Mix, a simulation of an advertising campaign that could predict its effectiveness. Advertising agencies, unlike media companies, already both collected consumer data and made these sorts of predictions for their clients; Simulmatics proposed to make those predictions more accurate, and offered to prove the effectiveness of its simulations by evaluating past media strategies. Media-Mix was a translation of McPhee's voting model into the world of commerce: voters were just like consumers, and brand switching was like party switching.

Much of that work of translation was done by Alex Bernstein. "Bernstein's interpretation of McPhee's model is that it will raise an agency's ability to forecast media strategies from the current level of 60% to 95% of perfect," Morgan wrote to Greenfield. "Our model is essentially a measure of ad exposure given an advertising campaign and could tell a client which people and how many people would see an ad or a set of ads in a campaign, how many times over a particular length of time."[46]

Simulmatics pitched its Media-Mix to most if not all of the major Madison Avenue agencies, and also peddled it to the in-house advertising agencies of major consumer goods manufacturers. Greenfield would woo executives with a day at Wading River, where they'd tour the geodesic dome, a vision of the future.[47] Pool, McPhee, Bernstein, and Greenfield made phone calls or sat down for meetings with a virtual who's who of American corporate giants: Philip Morris, Procter & Gamble, Quaker Oats, Anheuser-Busch, Colgate-Palmolive, Jergens, Goodyear, Lever Brothers, Bristol Laboratories, General Foods, Nestlé, Nescafé, and Ralston Purina. Most, if not all, of these companies became Simulmatics clients.[48]

Simulmatics claimed to have created a simulated population, a miniature United States, consisting of three thousand perfectly representative but entirely imaginary people, living in one thousand separate households. They could test anything on this population, ads for dog food or instant coffee or breakfast cereal or chocolate syrup, merely by tinkering with the program. They'd divide these three thousand imaginary people into 150 different types, by sex, education, region, favorite television shows, and more.[49]

Simulmatics boasted, accurately, that it was the only company in the United States capable of providing this service, with "the only social scientists capable of building models of the sort we propose." But as a warning in an in-house memo pointed out, that edge would not last forever, or even for very long.[50] Still, for a time, Simulmatics thrived in the dog-eat-dog world of Madison Avenue. In 1962, a market research firm, Home Testing Institute, proposed a merger, which Simulmatics rebuffed, believing the merger would deflate its own very high reputation as the unrivaled technical leader in the field, and because of doubts about the quality of Home Testing Institute's data.[51]

When Simulmatics' sales pitches failed, they failed for two reasons: the high cost of the Simulmatics model (many clients "seemed shocked by our price scale")[52] and the insufficiency of data on which to train the model ("calls have all bogged down on the 'past data' question").[53] Even as the company readied for its public stock offering, Pool raised the question that Simulmatics would never really answer: "What is the data we would need for this model?"[54]

Of Simulmatics' vaunted simulated population — three thousand imaginary Americans segmented into 150 different types — surely this was too few, observed the British magazine *Punch*. *Punch* proposed that Simulmatics' Media-Mix add a few more categories, including "dog-lover, flat-earther, doughnut dunker, milk-in-firster," so as to be able to determine, for instance, whether a new, seventy-five-cent pink ballpoint pen ink refill would be desired by consumers who shop on Wednesdays, have Republican sympathies, "fruit-juice breakfasts, bouts of depression, slight astigmatism, fitted carpets, thick eyebrows, and one or more cousins in the armed forces." Wrote *Punch:* "You need all kinds."[55]

Simulmatics' stock went public on May 15,

1961, offered at two dollars a share. Greenfield served as president; Bernstein, Coleman, and McPhee, vice presidents; and Pool and Abelson, directors. Greenfield was the largest shareholder, with 132,250 of the 280,400 shares of common stock issued to the officers, Pool the second largest, with 50,650. In the end, at the end, all of this stock would be worthless.

Simulmatics' stock circular made a series of claims and promises, as these things do. The company will collect data. "From the raw data so obtained, the Company proposes to design and construct, in mathematical or logical terms, representations of behavioral processes incorporating hypotheses as to the behavior pattern of the group." It will make mathematical models of this behavior and, by way of a computer, will "simulate probable group behavior under alternative hypothetical situations." IF/THEN. "The Company proposes to submit to its clients reports based upon these simulations." It had so far constructed one model, a model of the 1960 election. It intended to construct many more. Buying stock, then, was speculative, because "the Company's success will depend on its ability to obtain orders for, and profitably to construct and sell useful mathematical models." There was hardly any competition.[56]

At first, the stock was hot. On the first day of trade, it was quoted at nine dollars a share.

"That blasted Simulmatics Corp. is supposed to come through with some very substantial dough on the first of June," Minnow McPhee wrote, with mixed feelings. Her husband had been given 25,000 shares — half what he believed he'd been initially promised — in exchange for which he had signed a contract that forbade him from doing any other outside consulting, and he had agreed not to cash in his stock for at least three years. "With my luck it'll probably be worth a penny a share by then," Minnow wrote.[57] They held on to it for longer than that but she was right: in the end, a penny would have been generous.

Meanwhile, the Book-of-the-Month Club announced its selection of White's *The Making of the President* for the month of July.[58] Published on Independence Day, White's riveting account of the Kennedy campaign became a critical and commercial sensation. It climbed the bestseller lists. It was taken as the definitive account both of the election and of the president himself, despite White's deeply felt admiration for Kennedy and his unwillingness to acknowledge Kennedy's flaws.[59] It is quite difficult, for instance, to reconcile White's portrait of Kennedy with the president who in April 1961 approved a disastrous plan to invade Cuba at the Bay of Pigs, a botched invasion crushed less than two days after it began. Cuba, just over one hundred miles off the coast of Florida, had become a Communist state in 1959, after the revolution led by Fidel Castro in 1953. The invasion at the Bay of Pigs had been planned by the CIA during Eisenhower's administration, but Kennedy had not stopped it. Its failure proved an enduring embarrassment.

Theodore H. White's John F. Kennedy is more informed and more decisive than any other figure in American politics or history, a political wunderkind. White's omissions were many: he made no mention of Kennedy's womanizing (no one did) or his grave health problems (nor, of course, did other reporters). He also failed to even touch on Simulmatics and the advice it had provided to the campaign. Instead, White described Kennedy as a flawless political strategist who had "mastered all the power brokers and power forms that stir internal American politics" and had surrounded himself with the best men, so wise "that when crisis happens all necessary information is instantly available, all alternate courses already plotted" — which might be taken, I suppose, as an oblique reference to Simulmatics.[60] But White's Kennedy didn't need a computer to give him advice about politics; he was his own What-If Man.

Simulmatics' stock soon began to fall.[61] It likely didn't help that Morgan was listed on the stock circular as a shareholder and

233

information manager, at a salary of $1,000 a month, leading at least one reader of *Harper's* to demand a retraction for "The People-Machine." (Instead, Morgan clarified that he had not been employed by Simulmatics at the time he wrote the article, which, while technically true, was something of a nicety.)[62] Also, Simulmatics' stock offering had the whiff of a scam. Nationally syndicated financial affairs columnist Sylvia Porter, a bulwark of middle-class investment advice, indicted Simulmatics in a column about how unknown companies defraud investors by running PR campaigns to jack up prices. The fact that Simulmatics, offered at two dollars a share on its opening day, was quoted at nine dollars just a few hours later, made it the kind of stock that made "a paper fortune overnight" for the company insiders who got blocks of stock at the initial price. Still, Porter attributed Simulmatics' stock rise and fall not to fraud but to error. "Many of these companies are so new, so offbeat, that it's surprising firms can price the stocks at all," she wrote. "This was a 2-year-old company with a net worth of minus — minus! — $21,000. It's a wonder the underwriter settled on any price."[63]

If it was a risky venture, Simulmatics was not a scam. Having raised nearly $200,000 of capital by selling stock (about $1.75 million in 2020 dollars), the company didn't shut down and cash out: it tailored its business plan. In June, everyone involved with Simulmatics met in New York for a three-day conference at the swank, seven-hundred-room New Weston Hotel, at Fiftieth and Madison, just down the street from the Simulmatics offices. All the principals were there, along with a number of guests and prominent stockholders, including Harold Lasswell.

The chief problem, they seemed to agree, was the scarcity of data. There were three choices: they could rely on clients, who had little data, and what they had was usually old; they could collect new data; or they could buy data from data collection agencies.[64]

"Most models will require new data," McPhee said.

Pool disagreed. He thought old data could work.

They also debated what to do in the political field in the upcoming 1962 midterms. Robert Kennedy was a man known for bearing grudges, and he was furious with Simulmatics. Democrats wouldn't hire them, out of loyalty to the Kennedys, and if Simulmatics took on a project for the GOP, Jim Coleman said, "we would commit suicide in Washington." McPhee suggested more or less skipping the 1962 election season because Democrats were likely to do poorly. It would be better, they mainly agreed, to sit this one out and work for a news organization instead. Coleman was particularly worried about the wrath of Robert Kennedy. Still, he said, he didn't think Simulmatics "should shy away from public activity because of possible disfavor of Bobby Kennedy."[65] It was a pickle.

In the end, they agreed to Pool's proposal, which was to make a priority of improving Media-Mix.[66] This would require working as quickly as possible because, as Bernstein pointed out, "In two years, our secret will be out."[67] The leak came sooner.

The Simulmatics scientists and businessmen "all treat their wives like dirt," Minnow McPhee wrote to her mother. "You wouldn't believe it unless you saw it." Still, at the end of the 1961 school year, Minnow McPhee began readying the family for a summer on the seashore with those very men, and their much-demeaned wives. She felt she had no choice.[68] She packed the children's tiny suitcases and her own. Towels, toys, swimsuits, pens and stationery. She'd written a children's book called *Fun in Bed* and had sent it to Dell. ("I should do one for children + one for adults!" she wrote her mother, laughing at the book's title.)[69] She was pretty sure this would be the family's last summer together. Bill was a wreck again. She'd resolved to leave him.[70] Before the family headed for the beach, she told fourteen-year-old Wendy, who burst into tears.[71]

Patty Greenfield, Ed's sweet, artistic, and much-neglected wife, set about readying her family's beach house for a retreat, and readying, too, the house next door: Buckminster Fuller had finished work on the geodesic dome. It had a plywood floor and triangular windows, overlooking Long Island Sound. Inside, it was like a honeycomb.[72]

Patty and Ed arranged for all of the Simulmatics families to rent houses nearby, along the beach, up on the bluff, or near the creek. (Pool had invited Paul Lazarsfeld, but Lazarsfeld spent his summers in Vermont. He sent his regrets: "The world is full of tantalizing rumors as to what you are up to," Lazarsfeld wrote Pool.)[73] The work of hosting everyone would fall to Patty, who was also busy minding Michael, nine; Annie, seven; and Susie, six. Ed, an enormously affectionate father, would pack the kids up and drive them to the station whenever anyone new came to town. The kids would lay pennies on the tracks, for good luck.[74]

The families came, carting beach chairs and flippers, family dogs and transistor radios. They settled into the life of villagers, a hamlet of behavioral scientists. The children swam and they sailed and they built sand castles on the beach. They dug for clams and mussels, more for their mothers to steam and boil, pulling flesh from shells. Nearly naked bodies were everywhere; the shore smelled of sex. Patty Greenfield's sister was pregnant that summer, and unmarried. She'd lie on the sand and make a hole in the sand for her belly, while the children stared.[75] Jock McPhee took swimming lessons at the public beach. His sister Wendy got bored. She entertained herself by developing a crush on Jim Coleman. When Wendy and Jock and the Greenfield kids went to a fair, and Sarah, seven and home sick, couldn't go, they brought her back a blue-and-white panda bear, won at a shooting gallery; Sarah named the bear Michael: he was dreamy.

Minnow found the wives, at first, difficult to get to know. "Lou is the only one I find easy to chat with," Minnow wrote about Lu Coleman, early on, and though she thought Jean Pool seemed promising, the Pools' cottage was a bit too far away.[76] The Pools came along with four-year-old Adam and Ithiel's son Jeremy, who was sixteen and miserable. Ed Greenfield hired Jeremy to tutor Michael in math. Ithiel, noticing that Jeremy seemed bored, handed him a FORTRAN manual. "Here, Jerry, you might be interested in this," he said. When the scientists met, they'd let Jeremy sit in a corner. He listened, and didn't make a peep.[77] The men sent one another memos, delivered by the children, from house to house. "What is the most daring claim you can make for our theory?" Ed memoed Ithiel one long summer day in July.[78] He dared, and they dared.

On rainy days, the Simulmatics kids all played at the Greenfields' house, which had fourteen rooms. Minnow's tiny, four-room house slowly filled with sand, like an hourglass. "By the end of the summer I expect to have more sand in than out," she wrote her mother. "Then I shall borrow Jean Pool's vacuum cleaner and clean out." Minnow spent her scant quiet moments reading; she tried to borrow a book from Joan Morgan, but no matter where she looked, *Peyton Place* seemed to be the only book on any of the Simulmatics' wives' bookshelves. *Indian summer is like a woman. Ripe, hotly passionate, but fickle, she comes and goes as she pleases so that one is never sure whether she will come at all, nor for how long she will stay.* Minnow had already read it, more than once. And she'd already seen the film. Really, she'd already lived that story. And here she was, living it all over again, in Wading River.

The Morgans had a house right nearby. Sarah McPhee and Katie Morgan were the same age. The McPhee and Morgan kids would walk together, along the creek, with the McPhees' black poodle Sputnik wading in the water, all the way to Annie Greenfield's house, on the beach. The kids considered the Greenfields' house the center of the universe.

Katie was convinced Ithiel was a rabbi: "There was something important about him." After a while, the wives settled into a sorority. "I must say there are a nice bunch of wives in this group," Minnow told her mother. Patty, Lu, Jean, Joan, and Minnow.[79] Peyton Place, Long Island.

Bill McPhee, on his better days, taught Sarah how to swim. He'd head out into water deeper than she could stand in and shout, "Swim!"[80] There was no television in their house, but McPhee charged his children with collecting all the *TV Guide*s from all the other houses and marking them up: coding.[81] Simulmatics needed data for Media-Mix and, in particular, data about television schedules. This assignment — preparing a report called "Preliminary Codes for Television Program File (Media Mix)" — fell to McPhee, who passed it on to his children. But as one of the company's clients complained to Greenfield, "there is a serious question as to whether your coding equipment could be sensitive to a sufficient degree to have any real value."[82] That coding equipment? That coding equipment was Wendy and Jock.

Coleman had come to the conclusion that collecting enough data to train the model wasn't the only problem. Anticipating by a half century the commodification of attention that would drive the early twenty-first century's age of social media, Coleman observed in a memo that August that the problem was how to measure a consumer's attention: "The basic notion is <u>attention</u>, and the fact that he has a limited amount of it (thinking of attention as a kind of storehouse which is of limited size) and the fact that new things are coming in all the time, depending on the array of stimuli to which he is exposed."[83] What model could keep track of that?

At night, at the Greenfields', the grown-ups would drink and put on records, or else Patty would play the piano, and everyone would dance. The children played bartender, serving whiskey sours and Manhattans and gin and tonics. The grown-ups staggered and swayed. In the dark, the children played Kick the Can.[84] On warm nights, they gathered for bonfires on the beach. When the children ran out of marshmallows to toast, Patty and Minnow would dash out to the store, to get more.[85]

At the beach that summer, it would not have been unusual if there had been affairs, and fights, and recriminations; it would only have been unusual if there had *not* been affairs, and fights, and recriminations. "I always feel I'm at a play or something," Minnow wrote, the year before the debut of Edward Albee's devastating play about faculty marriages, *Who's Afraid of Virginia Woolf?* In the play, Albee's Martha says, "We both cry all the time, and then, what we do, we cry, and we take our tears, and we put 'em in the ice box, in the goddamn ice trays until they're all frozen and then . . . we put them . . . in our . . . drinks." Minnow McPhee might have said the same. At Wading River, they drank and they drank. And if they did not play musical beds, neither did they sleep soundly. The McPhees, the Colemans, the Greenfields, the Morgans: all their marriages were falling apart.

"I make lemonade in my sleep," Minnow wrote to her mother at the end of that summer. She'd decided to move back to Colorado.[86] The day Minnow wrote to her mother describing her plans, Ithiel Pool wrote a memo about how Bill, working on a proposal for a simulation of how people switch cigarette brands, had suggested that a brand-switching model ought to "take account of such things as marriage, change of jobs, moving from city to city."[87] Wives, it seemed, switched brands, too.

Minnow packed up Wendy and Jock and Sarah, and Sputnik. She sold the house in Hastings-on-Hudson and they flew to Colorado. She'd finally made her escape from the orbit of her husband.

Their marriage wasn't yet over. McPhee was leaving New York, too. At the end of that summer, he headed to California to accept a fellowship at the Center for Advanced Study in the Behavioral Sciences. His distance would prove a problem for Simulmatics, but his nearness hadn't been easy, either. He was erratic. He was difficult. For months, he held up work on Media-Mix, until Pool phoned him and they decided he'd step away from that project, so that it could go on without him. "The thing he wants to do more than anything else is a simulation of the physiology of alcoholism," Pool reported after the call.[88] McPhee, lonely and miserable, tried desperately to stop drinking. "I have quit you know what," he'd write from Palo Alto to Wendy in Denver. "I promised Mommy I would before I came back (or I wouldn't come back!) and I'm getting the agony over now." But he never really quit drinking, and Minnow never really took him back.[89]

In August, writing from beneath the geodesic dome, Greenfield issued a report to stockholders, announcing a new government contract with the Department of Health, Education, and Welfare, a study simulating the results of public health initiatives.[90] Morgan issued a press release, proclaiming that "a 'far-reaching revolution in all phases of business, economic and social research is likely in the immediate future' as the result of a new technique of electronic computer simulation."[91] That led to a story about the company in the *New York Times,* a puff piece illustrated with a photograph of Greenfield and Pool, in shirtsleeves, standing by a chalkboard, inside the dome.[92]

It would be left for Pool to undo the damage done by the story Morgan had published in *Harper's.* "The so-called People Machine," Pool told *Newsweek,* "does not eat people. Indeed, it is not even a machine. It is an activity that tries to predict human behavior."[93] The price of the stock began to recover. "For anyone interested in an offbeat RANK speculation where the risks are tremendous but the future exciting," one trader advised at the end of that summer, "we suggest you look closely at SIMULMATICS CORP."[94]

The children would always remember the summer, and the sand. The Greenfields would sometimes head up to Cambridge, to visit the Pools on Irving Street, by the Schlesingers and the Galbraiths. Jean Pool loved Patty Greenfield; they'd catch up in the kitchen. America's most beloved chef, Julia Child, lived next door, at 103 Irving Street; she was close with Jean. Upstairs, from the window of the guest room, Susie Greenfield could hear Child call out to her husband, in her inimitable, croaking, high-pitched voice.[95]

In the Pools' backyard, Adam and his father set up a playhouse they'd built, in the shape of a miniature geodesic dome.[96] The sun streaked through the honeycomb frame, and into their tiny world.

CHAPTER 8
FAIL-SAFE

Computers are too important to be left to the mathematicians.
— Eugene Burdick and Harvey Wheeler,
Fail-Safe, 1962

The fifteen-story New York Times building at West Forty-Third Street looked like a bloated French château, with its arches and spires, except for the no-nonsense, sans serif, brightly lit "TIMES" sign bolted to its topmost turret. The building, a block from Times Square, was known as "the Factory," and, especially out back, it looked like one: its skidded truck bays and greased lifts had for decades hoisted printing presses and line-casting machines and press plates and received shipments of gigantic rolls and reams of paper. The *Times* reported, investigated, wrote, and edited news; it also manufactured newspapers. Every day, men on those bays had delivered onto awaiting trucks bales upon bales of bundled newspapers that, if stacked, would have towered over every skyscraper from the East River to the Hudson.[1]

Then came the colossus. In the fall of 1962, the Times Building's bays and lifts had to be reinforced in order to bear the weight of an expected delivery: a four-ton IBM 1401, the Model T of the midcentury computer industry. It came boxed in wooden crates and wrapped in a brand-new packing material, the space age's answer to sawdust: bubble wrap.[2] It had been ordered for the *Times* by the Simulmatics Corporation.

The *Times* had hired Simulmatics to help run its coverage of the election of 1962. "Data journalism" wouldn't be called by that name for decades, but at the building at West Forty-Third Street, it started the day the *Times* signed a contract with the Simulmatics Corporation.[3]

The *New York Times* was founded in 1851, and it had been, from the start, an early adopter of the latest technologies available for covering elections. It reported the results of the election of 1852 by way of news carried by railroads and the pony express and that year announced its ambition to rely on transmission by telegraph "to enable the Press of the entire country to announce the result of the national election on the morning after the closing of the polls." So quick a turnaround would not be possible for decades, at least for the presidential election. As late as the election of 1896, *Times* reporters were still tallying returns carried by homing pigeons. But by 1904 the *Times,* like other big-city papers, had all sorts of ways of telling its readers about the outcomes, as soon as the numbers were in. On Election Night, it broadcast the results from its building in New York by way of searchlights that could be seen for thirty miles, as if the building itself had become a lighthouse. Steady light to the west meant a Republican victory in the presidential race, steady light to the east a Democratic one; flashing lights in different combinations broadcast the winners of congressional and gubernatorial races. This is what's meant by a news "flash."[4]

The emergence of radio in the 1920s hadn't put an end to the *Times*' Election Night searchlight. But television and the computer had. The *Times* had broadcast election returns by searchlight for the last, magical time in 1952, the year the UNIVAC calculated the returns on television for CBS News. And then, in 1960, CBS had brought in the IBM 7090. After Election Night 1960, a *Times* in-house postmortem concluded that no newspaper, no matter how many editions it printed over the course of the day and night, could beat the speed of television news aided by computers.

It wasn't only Election Night: newspapers struggled with the acceleration of reporting, by way of television, every single day of the year. By 1960, anyone who wanted to find out what had happened over the course of the day could watch the evening news on television. It was easy to perceive this as a race, in which case, the best strategy is: speed up. But the best newspapers realized that it really wasn't a race, and that the newspapers' best strategy would be: go deeper.[5] This realization would make newspaper journalism in the 1960s better than it had ever been before. No longer merely chronicling events but instead investigating and analyzing them, the newspaper press would become fiercer, more hard-hitting, and more adversarial, reporters questioning the very course of history itself.

This turn had begun in the 1950s, when, as one radio commentator pointed out, "The rise of McCarthy has compelled newspapers of integrity to develop a form of reporting which puts into context what men like McCarthy have to say." In 1958, the *Times* added "News Analysis" as a story category, a category that allowed the paper to print, even on the front page, stories with a decided point of view. "Once upon a time news stories were like tape recorders," the *Bulletin of the American Society of Newspaper Editors* commented in 1963. "No more. A whole generation of events has taught us better — Hitler and Goebbels, Stalin and McCarthy, automation and analog computers and missiles."[6] During what came to be seen as a golden age for American newspaper reporting, political reporters at the leading national newspapers also stopped deferring to the government and became more critical of it. In 1960, nine out of every ten articles in the *Times* were descriptive; by 1976, more than half were interpretive.[7]

When higher-ups at the *Times* conducted their postmortem after the election of 1960, they knew they couldn't win a race against television in announcing the results, but they understood that they had an advantage in another kind of reporting: analysis. The *Times'* decision to install a mainframe computer on its editorial floor came out of this broad commitment to news analysis. It needed, too, to find a company that could help it cover the midterm elections of 1962, by setting up a system to analyze election results, on the spot, so that *Times* reporters, working all night from the best possible information, could write the most informed articles, which would appear in the early edition the next morning, where readers could find out not only who'd been elected but why they'd been elected, and by whom.[8]

The *Times* had had a series of discussions with IBM, and then they'd gotten a proposal from Simulmatics. Going after this account made perfect sense for Simulmatics, given its decision to sit out the election of 1962, and not to try to sell its services to any Democratic campaigns, because of Bobby Kennedy's hostility toward the company. In October 1961, Simulmatics submitted a ten-page proposal to the *Times,* pledging to provide a newspaper equivalent to the services IBM had so stunningly provided to CBS News in 1960, services that Simulmatics claimed would allow the *Times* to evaluate the election in "real time" — a term new enough that Simulmatics scientists took pains to explain what it meant.

Real-time computing collapses the distance between something happening and a machine conducting an analysis of it. Covering an election, Simulmatics said, is like watching out for incoming missiles: both involve evaluating data instantly, with no lag for processing, because delay spells failure, and even disaster. "An example of a computer operation in real time is the SAGE system," the Simulmatics proposal explained. "In an early warning system, the computers must respond promptly and accurately at the moment that the blips appear on the radar screen." The problem of reporting on an election on Election Night was "clearly one of operation in real time."[9]

Simulmatics promised to provide the *New York Times* with an Election Night early warning system, a computer system that could watch the returns come in like so many missiles. Inside the *Times* building, at editorial meetings held at long wooden tables, not everyone agreed that hiring this Simulmatics outfit made sense. Ed Greenfield had given them a glitzy pitch. He'd brought in his leading scientists: Ithiel de Sola Pool had come down from MIT. James Coleman had come up from Johns Hopkins. Alex Bernstein had come over from the Simulmatics office on Madison Avenue. They seemed to be very impressive men. Still, the *Times* was slow in deciding whether or not to proceed. And, in the end, it came to regret that it had.

"The Times to Speed Fall Election Data," ran the eventual headline.[10] Everything seemed to be speeding up in 1962. The United States military was engaged in an arms race. NASA was engaged in a space race. ARPA was undertaking to build a network that would allow computers to communicate with one another across vast distances; soon, not only would computers process data in real time, they'd also communicate with one another in real time. Historians call this era the Great Acceleration.

Simulmatics, meanwhile, was engaged in a race of its own. The advertising industry was very quickly catching up with the kind of simulation Simulmatics had pioneered. "Time is of the essence," Pool pressed, begging a survey company for media data.[11]

251

Early in 1961, Pool met with a director of the leading advertising agency, BBDO, only to find out that BBDO was so interested in Simulmatics that it had begun building its own, rival model.[12] BBDO announced its plan to develop a "media-mix" that fall.[13] Within a week, Greenfield rushed out a draft of a Simulmatics promotional brochure, describing its Media-Mix.[14] It was a media simulation arms race.

The debut of Simulmatics' Media-Mix got a lot of attention and brought in a lot of clients, including big accounts with *McCall's* magazine, the Nestlé company, RCA, Ralston Purina, Colgate-Palmolive, and General Foods.[15] Pool peddled it at trade shows. At the annual meeting of the American Marketing Association, held in the Waldorf-Astoria, his talk titled "Simulation: How It Can Help the Marketer" drew the largest crowd of the conference.[16] Simulmatics' February 1962 brochure announced the launch of "a computer simulation of the activities of the American population to help solve the problem of allocating advertising dollars," designed to allow "advertisers, advertising agencies, or the media themselves to predict in advance, with considerable reliability, the audience that will be reached by each part of an advertising schedule for TV, radio, magazines, newspapers and Sunday supplements." It described a reel of magnetic tape containing "detailed information about 2,944 imaginary individuals," a perfect cross section of the population. "Where did all this data come from? The answer is that it comes from many sources and took the better part of a year to assemble."[17]

Simulmatics had a better computer model than BBDO, but its edge didn't even last the two years that Alex Bernstein had predicted.[18] BBDO and another Madison Avenue shop, Young & Rubicam, both unleashed their own media simulation models in 1962, one week after yet another swish agency, J. Walter Thompson, promised to "have a mechanical brain by 1963."[19] IBMs were being unboxed from monster-sized crates, their bubble wrap unfurled, up and down Madison Avenue. Each of these advertising agencies had better data than Simulmatics because, unlike Simulmatics, they had in-house data about markets and media and products and consumers. They'd been collecting all sorts of data for decades. J. Walter Thompson traces its origins to 1877. BBDO was founded in 1891. Young & Rubicam opened its doors in 1923. Simulmatics cobbled together publicly available data, like censuses and consumer price indices, with whatever it could get from its clients, but for larger data sets, it was left to beg. When Simulmatics asked Dr. Charles Ramond, director of the Advertising Research Foundation, for access to the foundation's data, Ramond said he wanted to help but was able to share data only with member advertising agencies.[20] Also, and even more importantly, big companies like BBDO and Young & Rubicam and J. Walter Thompson, unlike Simulmatics, were able and willing to invest in and install computers in their own offices.[21]

Whether advertising agencies' computers were providing much of any real value to their clients is harder to say. An advertising industry newsletter decried "The Great Computer Hoax" of 1962. Advertising agencies were making millions of dollars by charging for computer use but, said this critic, "computers are only as useful as the data fed into them." And the data, everywhere, was lousy.[22] Still, in the go-go years of the 1960s, no one appeared to mind. Big manufacturers were willing to pay advertising agencies for all the latest, best, new gee-whiz devices for chasing consumers. Everyone wanted a piece of a people machine.

That included, apparently, the *New York Times.* In April 1962, at the height of "the Great Computer Hoax," the *Times* finally made its decision and signed a contract for Simulmatics to provide its services for the upcoming midterm election and, if that went well, for the presidential election of 1964, too.[23] The *Times* had commissioned a new searchlight.

■ ■ ■ ■

Simulmatics rented more rooms, nearer to the *Times,* on Forty-Fourth Street. This time, there was plenty of money in the budget for the rent: the *Times* had agreed to Simulmatics' bid for a flat rate of $34,000 for its services in 1962, with fees for the 1964 election to be determined.[24]

Jim Coleman and Alex Bernstein headed the project. Their work for the *Times* built on Simulmatics' 1960 election study, and it built on Media-Mix, too. With executives at the *Times,* Coleman and Bernstein decided to focus on a small number of elections across the country: the New York gubernatorial and Senate races, governor's races in eight other states, Senate races in seven other states, and forty-two House races. Using Simulmatics' earlier models, they wrote a program designed to measure deviations from past votes, conduct a regression analysis to see how different groups voted, and make direct projections, allowing the *Times* to provide instant Election Night explanations for why certain groups voted the way they did.[25]

Coleman and Bernstein reported to the *Times*' Harold Faber, the newspaper's daily assignment editor. Faber, a City College graduate and World War II veteran, was a hard-bitten former war correspondent who'd lost a leg while covering the Korean War. He did not have a great deal of patience for Ivy League professors and IBM whiz kids who, he thought, had overstated their talents, overpromised and underdelivered.[26] The preparations went much more slowly than Coleman and Bernstein had expected, or Faber could tolerate. Down to the wire, it was unclear whether Simulmatics would be ready by the day of the election, November 6.

The operation proved tricky. The program Coleman and Bernstein were writing was buggy and difficult to test. Then there were the mechanical issues. Information had to be routed through a series of different connections, any one of which could fail along the way. The plan was to install two IBM 1401s on the editorial floor of the Times Building on Forty-Third Street, to be connected by telephone lines to an IBM 7090 at the IBM Datacenter on Fiftieth Street.[27] Each machine included some twenty thousand mechanical and fifty thousand electrical components; a failure in any one of them could crash the entire system.[28] IBM's 1400 series was designed for commercial applications, its 7000s for scientific applications. Both used magnetic tape for storage; a standard reel held the equivalent of four hundred thousand punch cards.[29] The data would be collected at the *Times;* it would be analyzed at IBM.

Collecting the data at the *Times* required a lot of steps, too. Returns from early reporting precincts would be received, by telephone and teletype, from the Associated Press. The people who answered the phones and read the teletype — an operation headed by a Mrs. Arbuckle — would write that information down on sheets of paper and then hand those sheets of paper over to women who would enter that data onto a punch-card-punching machine. The punch cards were to be read onto magnetic tape at one of the 1401s, then submitted, by modem, to an IBM 7701 at the IBM Datacenter, five blocks away, where the actual analysis would be conducted on yet another machine, a 7090.[30] A problem anywhere along this chain would crash the whole system.

By the beginning of October, things weren't going well. When the *Times* prepared to announce "the first usage by a newspaper of electronic computers in the preparation of news coverage," the paper's executives hesitated. The computers hadn't been delivered yet; the *Times* considered calling the operation off.[31] The announcement could hardly be coming at a worse time. Typographers at New York City's seven daily newspapers, whose contract would expire in early December, were ready to go out on strike, mainly because the city's newspaper publishers had proved unwilling to meet the demands of the typographers' union, which centered on compensation for automation. They were threatening to strike because computers were putting them out of work.[32] Under these circumstances, delivering giant crates full of computer equipment to the Times Building seemed, at best, ill-considered.

There's no good time for a newspaper strike, but weeks before an election is not among the better moments for newspapers to shut down. Also, at the time, although as yet unknown to the public, the United States stood on the brink of nuclear war.

On October 8, Adlai Stevenson, U.S. ambassador to the United Nations, speaking at U.N. headquarters in New York, warned that while the United States would not commit aggression against Cuba, it would not tolerate aggression from Cuba. The Soviets had been building up military forces in Cuba ever since the Bay of Pigs invasion, the year before. Americans feared that the Soviets were using Cuba as a base for a missile attack. Stevenson, echoing warnings issued by the Kennedy administration, said that the United States would consider the establishment of a missile site as an act of aggression. On October 14, Stevenson met with Kennedy in New York. That same day, a CIA-run U-2 flying a secret surveillance mission over Cuba took photographs that revealed the existence of a launching pad and at least one nuclear missile in San Cristóbal.[33]

The Cuban Missile Crisis had begun. Over the next thirteen days, the United States — and the world — would come closer to nuclear war than at any other point during the Cold War. On Tuesday, October 16, Kennedy called for the first of a series of secret meetings in the White House of a group that came to be called "ExComm" — the Executive Committee of the National Security Council. Chaired by Robert Kennedy, ExComm included Secretary of State Dean Rusk and Secretary of Defense Robert McNamara. National security adviser McGeorge Bundy presented additional photographs that identified the first of three SS-5 IRBM sites. ExComm debated possible courses of action, from conducting air strikes to take out the missiles (urged by Bundy) to seeking a resolution from the U.N. (urged by Stevenson) and offering concessions, including abandoning the U.S. military base at Guantánamo (also urged by Stevenson). On Thursday, October 18, at the White House, Kennedy confronted Soviet Foreign Minister Andrei Gromyko, who lied and denied the existence of the missiles. On Friday, Kennedy left Washington to campaign in Ohio and Illinois, stumping for Democratic candidates running in the upcoming midterm elections. But on Saturday, feigning a cold so as to not alarm the public, he canceled his remaining campaign stops and returned to Washington.[34] The public remained in the dark.

The next day, Sunday, October 21, Eugene Burdick and Harvey Wheeler's new novel, *Fail-Safe,* appeared in bookstores: the story of a nuclear standoff between the American president and the Soviet premier. *Fail-Safe,* a bestseller, a Book-of-the-Month Club selection, a *Saturday Evening Post* serial, became a critical sensation. "The most exciting novel I have read in at least ten years," said the influential critic and literary tastemaker Clifton Fadiman.[35] Linus Pauling, chemist and peace activist, blurbed that *Fail-Safe* had "made an emotional impact on me greater than any other novel that I have ever read."[36] A *Times* critic called it "a slam-bang thriller."[37] According to *Times* political reporter Jack Raymond, the novel had also "attracted considerable notice at the Pentagon."[38]

It was topical, of course. But in 1962, any new book by Eugene Burdick commanded extraordinary attention because of the staggering success and lasting influence of the blockbuster novel he had published in 1958, *The Ugly American,* coauthored with Captain William Lederer. *The Ugly American* had captured the attention of the nation and the world. It had been read by presidents and members of Congress. It had been attacked in *Dissent.*[39] It had been criticized on the floor of the Senate.[40] It had been published all over the world. A pirated edition had even been published in Arabic.[41]

Set in the fictitious Southeast Asia country of Sarkhan, *The Ugly American* aimed to expose incompetence, graft, corruption, and ignorance in the U.S. diplomatic corps and the mismanagement of U.S. aid programs. Lederer and Burdick had met at Bread Loaf, the renowned writers' colony, in 1948. After the war, Lederer had worked as a reporter, but he'd remained in military service, as a naval intelligence officer in Southeast Asia, until his retirement in 1958, when he became the Far East correspondent for *Reader's Digest.* He spoke six Asian languages.[42] Burdick met him again in the summer of 1957, while reporting a magazine story in Hawaii, and they decided to write a book together, about "the steady decline of American prestige and loss of power in the Far East."[43] They adapted their title from Graham Greene's *Quiet American,* but unlike Greene, they didn't advocate pulling out of Southeast Asia; they advocated engaging there more effectively.[44] The novel's many characters include U Maung Swe, a Burmese journalist who had spent a great deal of time in the United States — where he loved the Americans he met — who makes this observation:

261

"A mysterious change seems to come over Americans when they go to a foreign land. They isolate themselves socially. They live pretentiously. They're loud and ostentatious."[45] The real Sarkhan was unmistakable. Burdick and Lederer had first written *The Ugly American* as nonfiction, as a study of Vietnam.[46] Their original outline includes, for instance, a chapter on "How no one in Saigon spoke Vietnamese (I mean no American)."[47] Their editor at Norton, though, persuaded them to rewrite the book as fiction.[48] Still, in a "Factual Epilogue," they insisted that everything in *The Ugly American* was true.[49]

Burdick's 1962 blockbuster, *Fail-Safe,* claimed to be true in this sense, too. In the book's preface, Burdick and Wheeler — who, like Burdick, was a professor of political science — asserted that "there is substantial agreement among experts that an accidental war is possible and that its probability increases with the increasing complexity of the machine-made components which make up our defense system."[50] By "fail-safe" they meant a point after which planes sent to drop nuclear bombs can no longer be recalled. In the novel, a mechanical failure — due to a single burnt fuse in a computer — leads to a failure of communication between Strategic Air Command and a formation of B-52s traveling at fifteen hundred miles per hour toward Russia, each carrying two twenty-megaton bombs. Unable to recall two of the planes, mainly because of Soviet radio jamming, the American president first orders fighter planes after them, to shoot them down, and, when that fails, attempts to convince the Soviet premier, Khrushchev, that their proceeding to target Moscow is an accident.

Much of the novel consists of transcripts of telephone conversations, by way of a dedicated phone line on the president's desk.

"Premier Khrushchev," the president says, the first time he calls Moscow, "I am using the telephone line which your government and mine agreed should always be kept open. This is the first time it has been used."[51]

Unable to convince Khrushchev that the planes were not sent on purpose, and knowing that Khrushchev has no choice but to retaliate, the president instead orders the bombing of New York, as the only way to avert Armageddon. The American bombers, set on the course because of a computer glitch, cannot be stopped. Moscow will be annihilated. To prove to the world that the destruction of Moscow was an accident, the United States must sacrifice New York. In the end, the president and the premier commiserate. "We have become prisoners of our own machines," says Khrushchev from the Kremlin, where he is awaiting his own death.

The President paused. A calm fell on the line.

"Premier Khrushchev?" There was a tentative note to the President's voice.

"Yes, Mr. President?"

"This crisis of ours — this accident, as you say. . . . In one way it's no man's fault. No human being made any mistake, and there's no point in trying to place the blame on anyone." The President paused.

"I agree, Mr. President." . . .

The President continued, in part, thinking aloud: ". . . It's as if human beings had evaporated, and their places were taken by computers. And all day you and I have sat here, fighting, not each other, but rather this big rebellious computerized system, struggling to keep it from blowing up the world."

The American ambassador, seated with Khrushchev in the Kremlin, awaiting his own death, breaks in. "Mr. President, I can hear the sound of explosions coming from the northeast," he says. "The sky is very bright, like a long row of very big sky rockets. It is almost beautiful, like a Fourth of July." Then a screech, and silence. In an instant, Moscow is gone. And then, to save the world, the president of the United States orders the destruction of New York City.[52]

On Monday, October 22, 1962, the day after

the publication of *Fail-Safe,* the American military raised its state of alert to DEFCON 3. Kennedy called Truman and Eisenhower, as well as the prime minister of Britain, to brief them on the state of the crisis. Then he decided to alert the public. That evening at seven o'clock, grim and determined, Kennedy delivered a televised address from the Oval Office. Earlier in the day, he'd sent a copy of the statement he intended to make to Khrushchev, along with a letter. "I have not assumed that you or any other sane man would in this nuclear age, deliberately plunge the world into war which it is crystal clear no country could win and which could only result in catastrophic consequences to the whole world, including the aggressor," Kennedy wrote to his Soviet counterpart. On television that night, he spoke with a gravity that Americans had never before seen in him. He revealed the existence of the Soviet missile bases in Cuba. "The purpose of these bases can be none other than to provide a nuclear strike capability against the Western Hemisphere," Kennedy said. He detailed his conversation with Gromyko, flatly accusing the Soviets of outright deception. He had determined on a course of action: a complete blockade of Cuba. And, in dialogue that could easily have appeared in *Fail-Safe,* he issued Premier Khrushchev an unambiguous threat: "It shall be the policy of this nation to regard any nuclear missile launched from Cuba against any nation in the Western Hemisphere as an attack by the Soviet Union on the United States requiring a full retaliatory response on the Soviet Union." Finally, he appealed to Soviet premier Khrushchev "to abandon this course of world domination, and to join in an historic effort to end the perilous arms race and to transform the history of man."[53] All across the country, Americans sat in front of their televisions in hushed silence, struck dumb by the shock, a stillness that lasted long after the president spoke his final words.

The next day, parents kept their children home from school. Stores closed their doors. Kennedy and Johnson canceled all of their campaign stops. The *Times* ran a terrifying three-line headline:

**U.S. IMPOSES ARMS BLOCKADE
ON CUBA
ON FINDING OFFENSIVE-MISSILE
SITES;
KENNEDY READY FOR SOVIET
SHOWDOWN[54]**

Americans braced for war.

The Soviets did not back down. Instead, the crisis escalated. On Wednesday, October 24, Soviet ships were in sight of the quarantine and, as McNamara reported to Kennedy, they were being shadowed by a Soviet submarine. Stevenson, meanwhile, was working with Arthur Schlesinger on a speech he was to deliver to the U.N. Security Council. "There is a road to peace," he told the council on October 23, urging that this "be remembered not as the day when the world came to the edge of a nuclear war but as the day when men resolved to let nothing thereafter stop them in their quest for peace." Two days later, confronting the Soviet ambassador with photographic evidence of the missiles, Stevenson grew uncharacteristically heated. "ADLAI BLOWS TOP," ran the headline in the *New York Journal-American.* "I never knew Adlai had it in him," Kennedy said, watching the speech on television from the Oval Office.[55] In Washington, the American military raised its state of alert to DEFCON 2, the highest it has ever been.[56]

In the midst of this calamitous, gripping story — reporting on the possible end of the world itself — the *New York Times* could wait no longer to make a decision about its upcoming election coverage, under threat of a strike by newspaper unions. On October 23, the *Times* decided to go ahead with the installation of the computers ordered by Simulmatics, "strike or no strike," though they had to hire armed guards to protect the machines. The first IBM 1401 was unloaded at the Times Building at eleven-thirty a.m.

on Friday, October 26. That day, the White House received a secret communication from Khrushchev, sent by telegram to the State Department. It might have appeared in the pages of *Fail-Safe*. The premier wrote,

Mr. President, we and you ought not now to pull on the ends of the rope in which you have tied the knot of war, because the more the two of us pull, the tighter that knot will be tied. And a moment may come when that knot will be tied so tight that even he who tied it will not have the strength to untie it, and then it will be necessary to cut that knot, and what that would mean is not for me to explain to you, because you yourself understand perfectly of what terrible forces our countries dispose. Consequently, if there is no intention to tighten that knot and thereby to doom the world to the catastrophe of thermonuclear war, then let us not only relax the forces pulling on the ends of the rope, let us take measures to untie that knot. We are ready for this.

Khrushchev's letter seemed to offer a way out of the crisis. He offered to remove the missiles in exchange for an end to the blockade and a promise that the United States would not invade Cuba.[57]

But the next day, Saturday, Khrushchev sent, by formal, diplomatic means, an altogether different letter, offering to remove the missiles from Cuba only if the United States agreed to remove its own missiles from a base in Turkey. And still the situation worsened: over Cuba, a Soviet ground-to-air missile shot down an American U-2 plane, killing the pilot. At ExComm, some of Kennedy's advisers urged an immediate air attack on the missile bases in Cuba. But Robert Kennedy counseled the president to ignore Khrushchev's second letter and announce an agreement based on the first letter. Secretly, the United States nevertheless pledged to remove its missiles from Turkey — a course of action that had been urged by Adlai Stevenson all along.[58] On Sunday, October 28, Khrushchev agreed to this arrangement.[59]

As suddenly as it began, the crisis was over. Kennedy had steered the world clear of nuclear war.

Some things had changed forever over those harrowing thirteen days. It had taken nearly twelve hours to receive, decode, and translate the secret three-thousand-word message Khrushchev had sent on October 26. By the time it had been read in the White House, Khrushchev, not having received any reply, had already sent a second, harder-line message. The delay might have been fatal. In *Fail-Safe,* Burdick and Wheeler had argued that a failure of communications, even just a small computer glitch, could lead to Armageddon. After the Cuban Missile Crisis ended, representatives of the United States and the USSR met in Geneva to sign a "Memorandum of Understanding . . . Regarding the Establishment of a Direct Communications Link": they set up a hotline between the Kremlin and the Pentagon, a link by teletype. Eugene Burdick took credit for that idea, since this very hotline links Moscow and Washington in *Fail-Safe.* "We needed it as a dramatic device," Burdick explained. He said that when he was writing the novel, he'd asked Kennedy if such a hotline actually existed. Burdick recalled, "And he replied no but admitted that it might not be a bad idea."[60]

In a broader sense, too, the Cuban Missile Crisis changed the course of the Cold War. Coming so close to a nuclear war made it clear to anyone who had not yet understood as much that any strike would lead to another strike, and before too long, the whole world might be gone. The arms race didn't end; the two nations didn't disarm. But they had each come to see that the Manichaean battle between capitalism and communism would have to be fought not with nuclear missiles but on the ground. This would lead the United States to greater involvement in Vietnam. In 1961, Kennedy had named George Ball, Adlai Stevenson's former law partner and longtime campaign manager, as undersecretary of state. When Kennedy asked his closest aides about sending more "advisers" into Vietnam, Ball was the only one to object. He warned that the situation could escalate, and before too long, the United States might have three hundred thousand men wandering the rice paddies of Vietnam. Kennedy laughed. "George," he said, "you're supposed to be one of the smartest guys in town, but you're crazier than hell."[61]

The two giant IBM computers delivered to the Times Building on Forty-Third Street at the very height of the Cuban Missile Crisis were "up and running" by noon on Monday, October 29, the day after the crisis came to a close. They had arrived just in time for a rehearsal to be held that evening. Harold Faber kept an hour-by-hour record of the dry run. It started off badly, it went badly, and it ended badly. "Mrs. Arbuckle reported at 6 P.M. and immediately reported trouble reading her cards into 1402," Faber wrote. Also, it appeared to Faber that this operation would be a disaster, mainly because, as he said at a postmortem meeting that night, "It's obvious Simulmatics doesn't know how to operate machines."[62]

Another dry run was scheduled for Wednesday, October 31, Halloween, city doorways and porches and balconies decorated with dangling orange plastic jack-o'-lanterns, even

271

as rumors spread that a picket line would form outside the building at midnight. This time, things went so badly that the *Times,* which was renting all of this equipment from IBM, asked IBM to take it all back. Bernstein could not get the 1401 in the Times Building to communicate with the 7701 in the IBM building. Moscow and Washington needed a hotline. But so, apparently, did the *Times* and IBM. IBM decided to leave its equipment in place, come what may, arguing that much would be learned by a trial of the equipment on Election Night. Coleman, by now, was signing his memos "From Jim Coleman, desperate as ever."[63] Faber called Greenfield in a fury and told him "his men were running an amateur night." Simulmatics, Faber said, "frankly, not only underestimated, but completely misjudged the importance of competent technicians." IBM stepped in, promising to deliver a number of "girls" on Election Night, since they, unlike the men of Simulmatics, actually knew how to operate all the equipment.[64]

Greenfield and Pool came to watch the next rehearsals. By now, Faber had all but given up on the Simulmatics scientists, who, in his view, were useless. "Everyone who watched the dry-runs was convinced of this except Simulmatics, until Dr. Pool and Mr. Greenfield saw for themselves on Nov. 1 and 2," Faber wrote in his notes.[65] On the editorial floor, the machines hummed and rumbled and the chad of punched cards fluttered through the air, like ashes after a fire.

On Election Night, November 6, 1962, IBM's "girls" arrived at the Times Building at six o'clock, after dusk settled. At six-thirty p.m., even before the first polls closed, one of the data phones broke down. The 1401 kept finding it difficult to reach the 7701. Boxes of punch cards got kicked over, the cards spilling out like water. Everyone kept tripping over cables. To make room for the giant computers at one end of the room, all the editors' and reporters' desks had been crammed together. "Boys" charged with bringing notes taken by phone to the "girls" punching the data had to race across the room, but it was like crossing a snake pit.[66]

The results began to come in. Kennedy's handling of the Cuban Missile Crisis had strengthened his presidency and bolstered his party. Richard Nixon, running for governor of California, suffered a bruising loss. Democrats held on to both the House and the Senate, in victories that would make possible landmark civil rights legislation in 1964 and 1965. Little of this was as clear as it ought to have been in the Times Building on Election Night, where reporters were tripping over cords. It was a madhouse. Also, the AP reports of early returns came in much later than expected, holding everyone up.[67] Then, when they did come in, they came at a bad time. Reporters complained that they were "hit by a flood of projections just we were struggling to knock out a first edition lead that said something, but the projections were not only meaningless at this stage of the night but, in addition, cluttered up desk space at a key point in the evening."[68] A clutter, a chaos, a mess.

Ithiel Pool had come down from Cambridge. The Simulmatics people, he said, "worked as I have not seen anybody work since the war."[69] People at the *Times* thought differently.

"The robot age moved into the Times on election night," the *Times* would later report in its chipper in-house newsletter, trying to put the best face on it. "The 135 men and women on the election schedule — grinding out copy, compiling tables and estimating trends at top speed — had as their hand-maiden a cold but efficient electronic computer that gobbled up fresh election returns and coughed out projections and analyses with remarkable accuracy."[70]

Internally, though, *Times* staffers reported that the operation run by Simulmatics was "a completely disorganized shambles." The Simulmatics scientists walked around with instruction manuals, trying to read them on the fly. (One staffer wrote dryly: "One does not learn to operate a 1401 Computer system or a 7090 Computer system in three easy lessons.") They mishandled funds, charging the *Times* $1,045 for sixteen boxes of printer paper when they needed only five. IBM was bad, as one *Times* memo had it, but Simulmatics was "absolutely horrible": "That these people are political scientists and sociologists I do not deny, but I question whether they have ever seen a piece of business machinery other than an electric typewriter."[71]

With the exception of Alex Bernstein, the Simulmatics scientists, like most scientists, had very little experience actually running an IBM machine. Wrote Faber: "We were assured that they were experts in the field of computer technology, but it turned out that they were experts in programming, not in operations."[72] Historically, the work of operating these machines had always been done by women. That the Simulmatics scientists were expected to do this work — women's work — drove Coleman to distraction. He thought the scientists ought to be conducting only higher-order thinking. "To have four university professors in attendance at the Times on election night, and make the use of them that was made last night is not to be excused," he wrote in a furious letter the next morning.[73]

But to some veteran reporters, Simulmatics' higher-order thinking looked shoddy, too, especially the projections Coleman, Bernstein, and Pool made for the African American vote. In 1960, days before the election, the *Times'* Layhmond Robinson Jr., a pioneering black journalist and a savvy political reporter, had, without resorting to a computer simulation, rightly predicted that the black vote in New York would go to Kennedy. (Kennedy also won 68 percent of the black vote, nationally.) Robinson had also noticed the importance for the Democratic ticket of "the telephone call Mr. Kennedy made to the wife of the Rev. Dr. Martin Luther King," which, he said, had "eased some of the resentment stirred by the decision of Mr. Kennedy and Senator Lyndon B. Johnson to kill two civil rights bills submitted by President Eisenhower in the recent short session of Congress."[74] Robinson, born in Louisiana in 1925, had served in the navy as a photographer during the Second World War before earning degrees from Syracuse University and the Columbia School of Journalism. At the *Times* since 1950, he was the nation's best-informed reporter on the African American vote.[75] The Simulmatics team, in preparing its Election Night coverage, never consulted with him.

"I was surprised to learn, on Election night, from Mr. Pool and another of the Simulmatics men, that they were using only one Negro checkpoint in New York city, where about 350,000 of the state's 400,000 Negro voters live," Robinson later told Faber. "This checkpoint was on 109th Street, which is not an all-Negro area but one populated by a mixture of whites, Negroes and Puerto Ricans." It was ridiculous. "Unless we can do a better job on the Negro vote next time around I think we ought to forget it and stop confusing our readers," Robinson told Faber. "This time it was a flop."[76] Faber agreed.

The morning after the election, "in the harsh political dawn of Wednesday, Nov. 7," *Times* columnist Tom Wicker assessed the damage from the computer-cluttered newsroom, where "haggard knights of the typewriter" punched out their last reports.[77] The Democrats had won the day. But real-time reporting did not work, at least not yet.

In the weeks and months after the election, Simulmatics and the *Times* fell into a feud. Simulmatics had contracted the job at a projected cost of $32,500 but billed the *Times* for $89,400. Greenfield insisted that the overrun was an understandable and even inevitable error: "due to the experimental nature of our project (without precedent in computer programming history) any estimate we gave at the beginning was in danger of being off the mark."[78] The *Times* expected Simulmatics to absorb the $56,900 loss. Greenfield begged for an adjustment, and even "some financial assistance to carry his corporation over until after July 1." The *Times,* agreeing that one source of the overrun was that the news department had asked for much more analysis than had been outlined in the original contract, decided to give Greenfield $7,500, as a supplemental fee, along with a loan of $17,500, at a 6 percent interest rate.[79] This decision had been controversial, with Harold Faber complaining, "I don't think we have any moral obligation to help out a company that has so misestimated its costs. We know that it underbid deliberately to get the contract."[80] Faber was not wrong. Simulmatics must have deliberately underbid, hoping to knock out IBM. A $32,500 estimate for the entire job looked absurd, given that in the end Simulmatics billed the *Times* $29,545 for computer time alone.[81]

Meanwhile, the long-threatened newspaper strike finally began. On December 8, printers at four New York papers, the *Daily News,* the *New York Times,* the *World-Telegram and Sun,* and the *Journal-American,* walked off the job. Three hours later, the publishers of all seven New York dailies, acting according to a plan of solidarity, closed their plants, plants that had between them printed 5.7 million newspapers every day. The closure left some seventeen thousand New Yorkers out of work. Christmas shopping suffered. On January 10, 1963, members of the Newspaper and Mail Deliverers' Union joined the strike; more trade unions joined, too. The city had no news, and no newspaper ads. Hotels had fewer guests. Car sales and home sales fell. Near the end of February, the solidarity of the publishers began to crack when the publisher of the *New York Post* decided to negotiate with the unions.

The Great Newspaper Strike didn't come to an end until March 31, 1963, 114 days after it began. The unions and the papers reached an agreement; its provisions included a restraint on robots: "Automation is allowed to proceed to a certain extent but only in so far as it does not render any compositors redundant."[82]

That spring, the *Times* and Simulmatics came to an agreement, too, by renegotiating their original contract. The *Times* said Simulmatics' work in 1962 had "quite clearly showed that the computer operation has great value to our reporters and editors." Still, the newspaper was worried about "the financial ability of Simulmatics to continue in business."[83] In a revised contract, signed in June 1963, the *Times* set conditions on Simulmatics' continuance for the 1964 election, including an insistence that, at its own cost, Simulmatics "recruit trained personnel who will, well in advance of the 1964 election, operate and perform several trial runs."[84] On this promise, Simulmatics apparently failed to deliver, because on November 5, 1963, Election Day, the assistant managing editor of the *Times* notified Greenfield that he was terminating the agreement. "We don't believe that we have gotten or can get out of your organization the type of thing we feel we need for our election coverage," he wrote.[85] This setback nearly sank the company, except that, by then, Pool had begun edging Simulmatics into a new field: psychological warfare. The People Machine could earn votes and woo customers. Why not win hearts and minds?

On December 12, 1962, Pool flew to Washington to deliver, by invitation, the first talk on the subject of social science ever given before the Advanced Research Projects Agency of the Department of Defense. Before an audience of about a hundred people, deep inside the Pentagon, he clicked through his slides, Simulmatics charts, Simulmatics tables, Simulmatics graphs.

"The events of the Cuban crisis last fall brought home to all of us the central role communication can play in the evolution of international controversies and in the prevention of armed conflict," he explained.[86] A hotline between the Kremlin and the Pentagon was one thing. Pool believed that another kind of line of communication was needed: "in some kinds of crisis the important thing might be not only to communicate to the Kremlin but also to the Russian people."[87]

But how? Pool presented to ARPA the Simulmatics Corporation's Media-Mix simulation, detailing the company's findings about exposure, frequency, and attention, across television, magazines, and newspapers, tallied by market segment: male, female, urban, rural, suburban.[88] The research had been done for the sake of selling shampoo and dog food. But out of it came a new, Cold War endeavor: Project ComCom, short for Communist Communications.

Project ComCom would be run out of MIT's Center for International Studies. It would be funded by ARPA. "In order to achieve deterrence and not provocation, the government needs research on how information comes to the attention of and is interpreted by Soviet decision makers," Pool explained. "We need to know how leaks, rumors, and intentional disclosures spread." He had mathematical models ready to hand over: his own, earlier study of social networks, as well as Simulmatics' Media-Mix. "We intend to study the Soviet and Chinese communication systems by various research approaches, including the use of computer simulation," he explained. Pool had, in fact, already begun studying how news from the United States was picked up by the Soviet press; this involved "an hour by hour analysis of American news tickers, press releases, and radio," which had "revealed patterns of dialogue in half a dozen crises." But there was more to Project ComCom. "We are planning to replicate certain psychological experiments," he wrote. He requested a budget of $225,000, for eighteen months.[89]

Project ComCom, with Pool as director, would lie at the center of Pool's work for years. It would require a massive effort at data collection. "For example, among peasants, what is the distribution of newspaper reading by paper, by minutes per day, by type of story, etc.," he asked. Finding answers to questions like that would require great sums of money, grant after grant after grant. The work was never-ending. It also greatly intensified Pool's use of MIT's computers. By 1962, it was easier than ever before to use the machine at MIT's Computation Center, now an IBM 7090: MIT's John McCarthy, cofounder of its Artificial Intelligence Project, had pioneered what came to be called "time-sharing," using a single computer to run multiple jobs at the same time.[90]

Project ComCom was not a Simulmatics project. But it borrowed, heavily, from Simulmatics research. And the MIT students Pool hired to work on ComCom tended, also, to work on Simulmatics projects. After all, they were tinkering with what was, elementally, the same computer program. Pool's staff included three undergraduates: Tom Van Vleck, a math major; Noel Morris (the

282

brother of Errol Morris, the filmmaker), an electrical engineering major; and Sam Popkin, a mathematics and political science major. Pool assigned Van Vleck and Morris to a windowless office in Building 14, in the library stacks, with a single IBM 1050 terminal. The sign on the door said, "T. LEHRER, N. MORRIS." (T. Lehrer never showed up. He was Tom Lehrer, the Harvard mathematician who became a writer and singer of satirical songs.)[91] Popkin was assigned to a room nearby.

Pool hired Van Vleck and Morris to work on Project ComCom, and under its budget. He hired Popkin, at first, to work for Simulmatics, preparing an update of the 1960 election study, for the 1964 election. Popkin reprogrammed all of the original study and added updates from 1962. But Popkin worked on ComCom, too. Like Van Vleck, he billed expenses to ComCom. And he, Van Vleck, and Morris traveled together, for conferences on how to handle data archives.[92]

For Project ComCom, Van Vleck and Morris spent their time trying to analyze whatever data Pool gave them, in an attempt to build a simulated population of Communists. Their data consisted chiefly of censuses of Eastern European countries, economic surveys, and the results of interviews with defectors and political exiles, people who'd crossed the Iron Curtain to the West; they'd been interviewed about how they got their news. All of this was data entered into punch cards that Van Vleck and Morris were supposed to feed into a program that would be used to generate tabulations and compute statistics intended to suggest how this imaginary population of Communists would respond to different political messages. "Look, just tabulate every variable against every other and only print out the tables where the chi-square is large," they were often told. There was a joke at MIT at the time, that you could get a master's degree if you could find three significant chi-squares, but a PhD took seven.

Van Vleck did not consider ComCom sound. "There wasn't enough data," he said later. Also, Pool believed that he could organize people's perceptions of their world "into chains of If this/then that," and Van Vleck disagreed. He was worried that, at bottom, ComCom was a model without a theory. "MIT didn't teach political philosophy at the time," Van Vleck recalled. "You got, instead, one week of Plato." He couldn't see how ComCom could produce any meaningful predictions when it did not rest on any known theory of human behavior, not to mention any proven theory of human behavior. But Van Vleck and Morris were only undergraduates, and they needed the money, and it wasn't their job to argue with Professor Pool.[93]

Pool's student researchers worked for him for years. Popkin entered MIT's PhD program, with Pool as his adviser. Morris dropped out of school but kept working for Pool. Van Vleck graduated but stayed on as part of the full-time staff. Van Vleck and Morris, working so closely together for so long, sharing an office and a single terminal, linked up to MIT's time-sharing computer, got creative about how to better collaborate. They would leave disk files for each other, labeled "To Tom" or "Read Me Now," since they weren't usually in the office on the same shift. Other people at MIT's computing center had the same idea that Van Vleck and Morris had: to devise a command that would send mail. "That's great, where is it?" said Van Vleck and Morris. "Everybody's too busy to write it," came the answer. "Okay, we'll do it," they said. And they did. In 1965, Van Vleck and Morris, working for Ithiel de Sola Pool, invented e-mail.[94]

What is the proper public role of behavioral scientists? Eugene Burdick and Ithiel de Sola Pool had taken divergent paths by the end of 1962. Pool spent much of his time in the Pentagon. Burdick spent much of his time in Hollywood. Each man thought the other a danger: a danger to American political science, and maybe even a danger to the country. Burdick didn't trust putting information of importance to national security on computers. Pool didn't approve of scholars becoming celebrities.

Quite where the line ought to be, between the university and the public, was hard to say. Burdick and Pool, each in their own way, had crossed it. So had Arthur Schlesinger. He'd initially taken a six-month leave from his position in the History Department at Harvard. When that had expired, he'd asked for his leave to be extended for the longest time possible, two years, in order to work at the White House. "No professional historian in all our history has ever been privileged to see events from this vantage point, and I know I would always regret it if I declined the opportunity," he'd written to the university's president. By January 1962, his two-year leave had expired, and the university decided it had no interest in extending it. Schlesinger asked Kennedy for advice. "I think you would be more useful down here than teaching those sons of privilege up there," the Harvard-educated president told him. Schlesinger resigned his professorship.[95]

Burdick and Pool grew distant from their own universities in different ways, and Berkeley and MIT set different conditions than had Harvard, and made more allowances. Burdick's star kept rising. A film adaptation of *The Ugly American,* starring Marlon Brando, premiered in April 1963; the Writers

Guild of America nominated it for Best Written Drama.[96] Burdick himself had become such a celebrity that he served as the spokesman for Ballantine Ale. "WHO IS THE ALE MAN?" its ads asked, and answered: "A MAN WITH A THIRST FOR A MANLIER BREW." In magazines, Ballantine Ale ads pictured Burdick in scuba gear, just coming out of the water, his scuba mask pushed rakishly back on his head, and also at his desk, at his typewriter, about to drink a tall glass of beer. Eugene Burdick: "Underseas explorer . . . literary man . . . Ale man." Ballantine television ads followed Burdick, in his scuba gear, diving into the depths and then rising out of the water; cut to Burdick at his typewriter; and then cut to Burdick at a cocktail party, glass in hand, while the voice-over purred, "Let it make an Ale man out of you."[97]

But both Burdick and Pool wrestled, each in his own way, with the problem of nuclear brinkmanship. Burdick wrote *Fail-Safe;* Pool created ComCom. Burdick expected the novel to be made into a film, directed by the much-admired filmmaker Sidney Lumet, starring Henry Fonda as the American president, and produced by a new company, Entertainment Corporation of America, or ECA. But in February 1963, the month Pool launched ComCom, Burdick faced a lawsuit filed by filmmaker Stanley Kubrick.

Kubrick was in London shooting a different movie about an inadvertent nuclear war, *Dr. Strangelove; or, How I Learned to Stop Worrying and Love the Bomb,* starring Peter Sellers, produced by Columbia Pictures. Columbia had budgeted $1,700,000 for *Strangelove,* a huge investment and an expression of its faith in Kubrick, following his directorial turn at *Spartacus,* in 1960, and his 1962 film, *Lolita,* an adaptation of the 1955 novel by Vladimir Nabokov. In 1961, as he was finishing *Lolita,* Kubrick had read a 1958 novel by Peter George, published in the United States as *Red Alert* and in the United Kingdom as *Two Hours to Doom,* about a nuclear catastrophe. He'd decided to buy the rights, and although the film he actually ended up making bore little relation to the novel, it was on the basis of those rights that he sued Eugene Burdick. Kubrick, George, and Columbia Pictures filed a suit in New York federal court against Burdick and Wheeler, their book's publisher, and ECA, alleging that the novel *Fail-Safe* had been "largely copied" from *Red Alert* and calling on ECA to abandon the project.[98]

There is no evidence that Burdick and Wheeler plagiarized *Fail-Safe,* and much evidence that the Kubrick action was nothing more than a nuisance suit: Kubrick's main objective was to get *Dr. Strangelove* out in theaters before Lumet's *Fail Safe.* Burdick and Wheeler's *Fail-Safe* began as a short story called "Abraham '58: A Nuclear Fantasy," written by Wheeler in 1956 or 1957, rejected by major literary magazines, and finally published by *Dissent* in 1959. Wheeler discussed his story at a lunch in New York for the Fund for the Republic. Burdick was at that lunch and, sometime later, wrote to Wheeler to see if he would be interested in adapting the story into a movie; in the end, they decided to write a novel together.[99] Burdick insisted that he had heard about *Red Alert* only later, sometime in 1961, because the physicist–turned–peace activist Leo Szilard had given a copy to everyone he could think of, and friends of Burdick's who'd read it warned him that it sounded a lot like the plot of the book they knew Burdick and Wheeler to be writing.[100]

Burdick charged that it was Kubrick who was the real thief. In December 1961, Burdick wrote to Kubrick, enclosing a partial draft and outline of *Fail-Safe* and offering it "for your consideration in connection with the potential purchase by you of the worldwide motion picture and allied rights." By then, though, Kubrick had already bought the rights to *Red Alert.* Realizing that Burdick was likely to sell the rights to someone else, and might then charge Kubrick's film with being, itself, a theft of *Fail-Safe,* Kubrick sent Burdick and Wheeler a copy of *Red*

Alert, by way of registered mail, informing them of his ownership of "the exclusive world-wide motion picture and allied rights" to the story. This also conveniently established, as would become relevant in court, that Burdick and Wheeler could not defend themselves from charges of plagiarism by saying they had never heard of George's novel. This tangled legal battle ended in April 1963 when, in an out-of-court settlement, Columbia Pictures took over *Fail Safe* from ECA and guaranteed that it would not release the film until at least six months after the release of *Dr. Strangelove.*[101]

Kubrick's *Dr. Strangelove* appeared in theaters in January 1964, Lumet's haunting adaptation of *Fail-Safe* eight months later.[102] Critics adored Lumet's film, which was shot as a documentary, in black and white, and released without any musical scoring. Unlike Kubrick's brilliantly devastating satire, Lumet's film is dead serious and terrifying, all the more so because it depicts global thermonuclear war as triggered not by human error, as in *Strangelove,* but by computer error.[103]

Released in October 1964, *Fail Safe* failed at the box office. No matter, Eugene Burdick might have thought. Because, that same month, he published his most ambitious novel yet, *The 480:* a page-turning thriller about . . . the Simulmatics Corporation.

CHAPTER 9
THE FOUR-EIGHTY

It was all gone now.
— Arthur Schlesinger, *A Thousand Days,*
1965

Eugene Burdick and his wife, Carol, were at the beach on Friday, November 22, 1963, when they heard news of a shooting in Dallas. They'd have huddled around the nearest radio, antenna to the sky, waiting to learn whether President John F. Kennedy would live or die. The NBC Radio Network broke into its midday broadcast with a cryptic, frantic emergency announcement:

— several shots were fired near the president in Dallas a hotline bulletin coming to you in fifteen seconds, stand by all NBC stations, a hotline bulletin in TEN seconds, a hotline bulletin in FIVE seconds —

They stood by.

291

— Shots were fired near President Kennedy in Dallas —

NBC News' Robert MacNeil, reporting from the scene, where he'd been following Kennedy's campaign stop in Texas, came in by phone, his voice uneven, distant, electric:

> Several shots were fired as President Kennedy's motorcade passed through downtown Dallas. Crowds screamed and lay down on the grass as the motorcade went by. Police broke away and began chasing an unknown gunman across some railroad tracks. It is not known if the shots were aimed at the president. Repeat. It is not known if the shots were aimed at the president.

NBC's announcer passed on a report from the Associated Press:

> The president was seriously wounded. This is not confirmed. This is a flash. Repeat. This is a flash.

The Burdicks on the beach, like everyone all over the country, felt as if they were sinking, falling, losing their very bearings. The ocean was still there, the mighty Pacific, and the sand between their toes. But the world seemed, suddenly, to reel.

NBC played orchestra music for twenty seconds; then came a strange thrum for another ten, until:

— stand by all NBC stations. . . . In TEN . . . in FIVE seconds —

NBC reporter Bob Thornton on a crackly phone line from Dallas, talking fast, as if he was gasping for air.

— Information is still sketchy at this time. We have just talked with two eyewitnesses, a man and his wife who were standing near the president's motorcade. They said that a shot rang out from behind them; they thought at first it was at first a firecracker. Then according to the eyewitnesses, one woman, who was in a hysterical condition, told us that the president was hit in the side of the head and fell into his wife, Jacqueline's arms —

Then swing music. Then again the thrum.

— stand by all NBC stations —

Should they wait at the beach? Or go home, listening to the car radio on the way, so that they could sit by the television? Helplessness overcame them.

— A small Negro boy and a white man said

293

they had seen a man with a gun in the window of a building, overlooking the roof.

This is Robert MacNeil —

Burdick, the Ballantine Ale man, the man with a manlier thirst, hadn't been well. In 1959, he'd had a heart attack while playing squash with a graduate student. He'd spent two weeks in a hospital in Oakland. His heart was weak, and now his heart was breaking. He was supposed to be mindful of exertion, and stress, and shock. He could feel his heart thumping, racing.[1]

The Burdicks that night drank two quarts of vodka between them, trying to bear up, to steady their nerves, to soothe their sorrow and their fury.[2]

— The President, his limp body, cradled in the arms of his wife, was rushed to Parkland hospital —

Two quarts weren't quite enough.

— The president remains in the emergency room —

And the waves lapped the shore, and the sun shone, and everything was the same, except it wasn't, and it felt as though the world would never be the same again.

— two Roman Catholic priests have been

294

summoned to the emergency room —

Burdick's heart raced, as if the motor of his Jaguar were inside his chest, at full throttle.

NBC Radio and NBC Television joined up, in a single, simultaneous broadcast.

— He has received blood transfusions. He is in critical condition, but alive —

Burdick couldn't help but think about the book he was writing, a novel called *The Candidate*.[3] It was supposed to reach its climax at the 1964 Republican National Convention in San Francisco; Burdick and his publisher planned to release the book days before the opening of the convention. The whole plot turned on the futility of any Republican candidate running against Kennedy, wildly popular, undefeatable, the man who'd saved the world from Armageddon.[4] But not, apparently, invincible.

In Dallas, police arrested a slender white man, the suspected assassin, Lee Harvey Oswald. A rumor spread that Vice President Lyndon B. Johnson had been shot, too. All kinds of rumors spread. Minutes passed. Was the president alive? Was the president dead? Doctors called for a neurosurgeon. The priests performed last rites. *Through this holy anointing may the Lord in his love and mercy help you with the grace of the Holy Spirit.*

Thirty minutes after the president was shot, on NBC Radio —

— Here is a flash from Dallas. Two priests who were with President Kennedy say he is dead. . . . I repeat. Two priests who were with President Kennedy say he is dead. . . . I will repeat, with the greatest regret, this flash. Two priests —[5]

Burdick, reeling, a Jaguar trapped in his chest, drove to his doctor — or maybe Carol drove — and submitted to an EKG. He was okay. Except that nothing felt okay.

Ten days later, still unsteady, Burdick wrote to a longtime friend of his, the prominent international lawyer Donald Rivkin, a former Rhodes scholar who served as the vice president of Simulmatics. "Right now I am trying by dint of hard work and exercise to get the whole thing out of my head," Burdick explained. "One of the things I have to do is grapple with the assassination in the new book I am doing." He wondered: Could Simulmatics send him its confidential reports, and Tom Morgan's 1961 *Harper's* article, too? "I can assure you that Simulmatics will get fair treatment and excellent publicity in the book," he promised. "Could you pass the word to Ed Greenfield or whoever else is in charge?"[6] He knew that Greenfield could not say no, could never say no, to publicity.

The flag-draped coffin bearing John F. Kennedy arrived at the White House on November 23, in the last bleak hours before dawn. In the East Room, an unsteady Bobby Kennedy asked Arthur Schlesinger to open the coffin and look at the corpse, and decide whether the casket should be open or closed for the state funeral, to be held at the U.S. Capitol. Half of the president's head was gone. Schlesinger came back and said, "It's appalling." Secretary of Defense Robert McNamara, who was there, too, wondered whether it might not be wrong for a head of state to be hidden from view; Schlesinger as official presidential historian cited a precedent: FDR's casket had been closed. Bobby Kennedy decided. "Close it," he said.[7] Two days later, Kennedy was laid to rest in Arlington National Cemetery. And the photograph of Kennedy's three-year-old son, in a tiny jacket, knee pants, and little shoes, saluting his father's casket in the cathedral, was seared in the nation's memory.[8]

Lyndon B. Johnson, having been sworn in as the new president within hours of Kennedy's death, addressed a joint session of Congress on November 27. "All I have I would have given gladly not to be standing here today," he began. He spoke slowly, quietly, as if each word were the end of a sentence. "The greatest leader of our time has been struck down by the foulest deed of our time." He pledged to continue the work Kennedy had begun. He would honor the nation's commitments in Vietnam. He would see Congress pass, at last, a new civil rights law. "We have talked long enough in this country about equal rights for all Americans, whatever their race or color," he said, his voice rising. Congress interrupted him with applause. "We have talked for one hundred years or more. It is time now to write the next chapter, and to write it in the books of law."[9]

Johnson would pledge to create a Great Society. He would announce a war on poverty. He sought federal aid for public education, and for the care of the sick and the poor and the old, with the establishment of Medicare and Medicaid. Johnson was a legislative mastermind. "Kennedy couldn't have gotten the Ten Commandments through Congress," Johnson later scoffed, privately. Johnson's program was not Kennedy's program; compared to Johnson, Kennedy was a conservative. What cause had Kennedy died for? That wasn't clear, Johnson said. But he made use of it: "I had to take the dead man's program and turn it into a martyr's cause."[10] With staggering ambitions for domestic programs, he had no interest in Vietnam, none of the fascination with and dedication to foreign affairs that had so driven his predecessor. He didn't want the United States to fight Communist North Vietnam; he wanted the South Vietnamese to do it, said he wanted them "to get off their butts and get out into those jungles and whip the hell out of some Communists, and then I want them to leave me alone, because I've got some bigger things to do right here at home."[11]

Unsure of foreign affairs, Johnson kept most of Kennedy's key foreign policy advisers, including McNamara as secretary of defense and Bundy as national security adviser. Arthur Schlesinger he let go.

Schlesinger had begun work on a history of the Kennedy administration the day after the president's death. Johnson accepted his resignation, and Schlesinger set to work on a book to be called *A Thousand Days*. Part memoir, part history, it would serve as Schlesinger's memorial to Kennedy.[12]

Johnson said the best memorial to Kennedy would be to pass a civil rights act. "No memorial oration or eulogy could more eloquently honor President Kennedy's memory than the earliest possible passage," he said.[13] Congress finally obliged, and Johnson signed the Civil Rights Act into law on July 2, 1964.

It's not altogether clear exactly when Eugene Burdick first got the idea to write a novel about Simulmatics. It was plainly no memorial to John F. Kennedy. And he'd been planning it for years. In 1961, *Reader's Digest* had asked William Lederer, Burdick's coauthor on *The Ugly American,* to take a stab at an article about the company. "Simulmatics seems to be going great guns," Lederer's editor wrote him, pitching the assignment. "I don't know anybody who could do it better than you."[14] Lederer had written to Ithiel de Sola Pool. "Is there enough material and evidence available to do a piece on Simulmatics?" he asked. "Do you want to write it, if there is? Do you want me to write it? Or shall we write it together?" Pool had replied, "I think there may be a good story there."[15]

Lederer never turned in that piece. Maybe he passed the story idea on to Burdick. Right about that time, Pool hired Sam Popkin to help him write his long-delayed account of the 1960 campaign. It *was* a good story; he didn't want someone else telling it; he wanted to tell it himself.

Burdick always said his book about Simulmatics was a product of his year at the Center for Advanced Study in the Behavioral Sciences.[16] The story had started out as both a novel and a film script, to be developed and released in tandem. In the summer of 1963, Burdick signed a contract with Hope Pictures, Inc., a television and film production company started by the comedian Bob Hope, for production of a made-for-television film called *The Candidate,* written by Burdick, about a candidate whose campaign is controlled by a computer.[17] In October 1963, when Burdick read a rewrite of his script, he was appalled.[18] "The heroes become villains, villains become heroes," he wrote to his lawyer. "Also, the underlying theme of the control of politics by cybernetics is merely mentioned in the teleplay and never really described." Not only that, but the computer equipment itself had been left out of every scene, despite the fact that Burdick "had emphasized how important it was to depict the electronic equipment used visually so that the public would have some notion of what it was all about."[19]

Burdick was in a bind. In negotiating and signing the contract with Hope Pictures, Inc., he'd consulted neither his lawyer nor his literary agent.[20] Under the terms of that contract, he had no real remedies. He couldn't stop the made-for-TV film *The Candidate.* And while he could change the name of his novel, so that the book and the TV movie would no longer be connected, he couldn't sell the film rights to the novel to anyone else, no matter what he called it, at least not before 1970. In 1963, when both Universal and Paramount expressed interest in optioning the novel, which Burdick eventually titled *The 480,* Burdick was left to confess that he had signed away the rights.[21] That's why the film version of *The 480* was never made, and a good part of why the story of the Simulmatics Corporation vanished so entirely.

Not being able to sell the film rights hurt Burdick, not least because he was looking to cash in, straightaway, with a story that, as one reviewer aptly put it, "is as timely as a newspaper."[22] He left his publisher to accept a $200,000 advance for *The 480* from McGraw-Hill (something upwards of $1.6 million in 2020 dollars).[23] In 1960, McGraw-Hill had solicited a nonfiction book about Simulmatics from Ithiel de Sola Pool, but by 1963, Pool had yet to deliver a manuscript and McGraw-Hill, having contracted with Burdick, was no longer interested in Pool's book; Pool instead signed a contract with the MIT Press. In the end, Pool's book appeared within weeks of Burdick's: it sank without a trace.

The 480, the novel that Burdick had for a long time called *The Candidate,* tells the story of John Thatch, a ruggedly handsome American engineer who becomes a national hero after building a bridge between India and Pakistan. In 1964, Republican Party insiders, needing a disposable presidential candidate, since "Kennedy will beat whoever we put up," recruit Thatch to seek the GOP nomination. They just want someone to run, and to lose, so that their favorite candidate, Dr.

Bryant Clark, can sit out 1964 and run in 1968 without the hindrance of a defeat trailing behind him.[24]

After Kennedy's assassination, Burdick nearly gave up the book. The plot didn't make any sense anymore. But then he got a better idea: he'd incorporate the assassination into the story of the election, which would make the book particularly current, and would also make John Thatch's bid for the presidency less of an unwinnable gambit — against Lyndon B. Johnson, this guy just might win — which made for a more suspenseful story. Burdick decided that the GOP, with the stakes raised, would want to make sure Thatch did their bidding. In the novel, to ensure that their candidate says exactly what they want him to say, exactly when they want him to say it, they hire a company called Simulations Enterprises. To tell this story, though, Burdick needed to know more about the Simulmatics Corporation.

For a very long time, people at Simulmatics cooperated with Burdick, complying with his many requests for material — material that often came to him in the form of reports marked CONFIDENTIAL.[25] Burdick, after all, was a Simulmatics insider, a former staffer for Edward L. Greenfield & Co. Greenfield might not have trusted Burdick, but he would have appreciated his stature, the scale of his celebrity, the brightness of the limelight he could cast on Simulmatics. No one said no.

"Two of the election reports from '60' are confidential but I am sending them to give you a clearer understanding of our function in reporting to the Kennedy organization," a Simulmatics executive wrote to him, in answer to his first letter to Rivkin.[26] Alex Bernstein sent Burdick unpublished reports, printouts, and even piles of punch cards. "If you have any problems reconstructing the types from the paper, call Bob Abelson in New Haven," Bernstein scribbled in a cover note. "If there is anything else I can do, call me."[27]

In the novel, Burdick introduced his fictional company, Simulations Enterprises, as a rival to the Simulmatics Corporation. (Burdick sometimes confused the two names, and because the book was badly copyedited, that error recurs.)

"Anybody ever use this Simulmatics outfit in an election?" one GOP operator asks a Simulations Enterprises behavioral scientist, a character Burdick burdens with the name Madison Curver: a Madison Avenue ad man, mad man, quantifier, a curve plotter: a mad curver.

"Kennedy did in 1960," Curver answers. "He was interested to see how much of Ike's popularity would rub off on Nixon. All the pros were saying 'Don't worry about the

Negro vote. It will come back to the Democratic Party once Eisenhower is off the scene.' The Simulmatics people got out their tapes and fed them into the computer. They found that the Negroes had shifted not just to Eisenhower, but many had transferred their allegiance to the Republicans."[28]

The novel, in other words, is highly didactic. *The Ugly American,* for all its influence, was less a novel than a collection of crude and inelegant short stories that are more accurately understood as parables. Its characters are very grossly drawn. Critics compared it to a comic strip and said it was written with the "subtlety of a sledge hammer."[29] The same is true for *The 480.* Only more so. Burdick lifted whole passages straight out of Simulmatics reports and memos and even interoffice correspondence he managed to wrangle out of Rivkin and Greenfield and Bernstein. He even included, as an appendix, a list of Simulmatics' 480 voter types. He'd gotten the list from the New York office. "Simulmatics is coughing up the 480 groups themselves," his editor wrote him.[30] True to his word, Burdick also advanced Simulmatics' PR argument: that its People Machine had gotten John F. Kennedy elected president.

"Did the Simulmatics people tell Kennedy how to act during the TV debates in 1960?"

"They did," Curver said.

In January 1964, nearly done writing, Burdick settled on a new title, *The 480,* after Simulmatics' 480 voter types.[31] (Other editions used titles closer to *The Candidate:* an Italian translation was published as *Il Successore.*)[32] Burdick was, at the time, at the very top of his game. In February, the *New York Times* ran a photo essay called "Writers on Campus," about those of the nation's most distinguished writers who also had teaching appointments of one kind or another, the nation's leading academic lights. Burdick, ever the "Ale man," was pictured playing squash in shorts and a tank top — wishing himself well — in a story that also featured Saul Bellow (University of Chicago), Ralph Ellison (Rutgers), Bernard Malamud (Bennington), and Philip Roth (Princeton).[33]

In April, McGraw-Hill alerted Burdick to an announcement in *Publishers' Weekly:* "the M.I.T. Press is coming out with something called SIMULATION OF THE 1960 PRESIDENTIAL ELECTION, coming August 19th: 'The fascinating account of the first computer simulation ever used in a national presidential campaign.' " *Publishers' Weekly* called it "a must for political scientists."[34] Burdick wrote to the MIT press, asking for galleys of the book about Simulmatics, written by Pool, Abelson, and Popkin, with the unwieldy title *Candidates, Issues,*

Strategies.[35] It was scheduled to be released two months after Burdick's *480,* a book *Publishers' Weekly* had called "a tense, frightening story of what might happen if a presidential nominating campaign were maneuvered by computer results."[36] Worried, Pool, Abelson, and Greenfield decided to meet with Burdick's publisher. But there was, needless to say, very little they could do.

In the spring of 1964, Simulmatics needed a new trick. It had lost its edge in the advertising field; those clients had dried up. The termination of its contract with the *New York Times,* with whom the company had expected to conduct election coverage in 1964, had hurt badly. Simulmatics had made its reputation on election analysis, but it had no political contract for the upcoming election. Conservative Arizona senator Barry Goldwater was seeking the Republican nomination, running against the New York moderate, Nelson Rockefeller. Goldwater represented the hard right: he opposed *Brown v. Board,* voted against the 1964 Civil Rights Bill, and appeared willing and even eager to engage in nuclear war. His campaign approached Simulmatics in April 1964, at least according to Greenfield, and Simulmatics refused to work for him. "Democratic loyalty spoke louder than money," Greenfield proudly told the press.[37]

Greenfield was beginning to break under the pressure. The assassination of the president had knocked him down. It had knocked everyone down. He'd begun drinking more. He'd also begun an affair with a young and glamorous woman, Naomi Spatz. He left his wife, Patty, then went back to her, long enough for her to become pregnant with their fourth child, and then he went back to Spatz. When Patty, near her due date, was out at the house at Wading River, Greenfield zoomed up to the house with Spatz, in Spatz's red convertible.[38] Greenfield had gotten by on charm for only so long.

Greenfield thought Burdick's book would be good for business, but wiser heads at Simulmatics had begun to worry about the effect it might have on the company's efforts to get hired to work on Democratic political campaigns. Whatever had or hadn't happened with the Goldwater campaign, Simulmatics was desperately trying to secure a contract with the Johnson campaign. In New York on June 10, 1964, Simulmatics executives met with executives at McGraw-Hill, including its director of publicity, who was preparing the final press release for *The 480.* Some people at Simulmatics had been given the chance to read the manuscript, and they were alarmed. They regretted their cooperation. Simulmatics insisted that the press release for the book include a disclaimer: "A spokesman for The Simulmatics Corporation points out that the firm disagrees with the conclusions Mr. Burdick expresses in 'The 480.' "[39]

A third book about Simulmatics appeared in 1964, set in the year 2033. The science fiction writer Daniel F. Galouye doesn't appear to have gotten any secret reports from Simulmatics, but he described the company's practices with precision in his dystopian novel *Simulacron-3*. "The simulator is an electro-mathematical model of an average community," one of the company's scientists says in the novel. "It permits long-range behavior forecasts." *Simulacron-3* was Galouye's third novel, and although it was also published in England, under the title *Counterfeit World*, it didn't get a lot of attention outside the pages of sci-fi pulp magazines.[40] It's not clear that anyone from Simulmatics ever even heard about it.

The day after people from Simulmatics met with McGraw-Hill, Pool put together a Simulmatics proposal for the Johnson campaign. The Republican nominee was as yet undecided, but Johnson sought the Democratic nomination, uncontested, and was expected to easily defeat whatever candidate the Republicans settled on. Pool knew that Johnson was very likely to win but insisted that "how he wins may affect the future of American politics profoundly." And that, in turn, depended, in part, on his choice of vice president: "We propose a Simulmatics study that will provide objective estimates of the number of votes likely to be gained or lost in each key state by President Johnson's potential respective choices among running mates." The project would build on the work Simulmatics had done for Kennedy in 1960 — "our method combines behavioral science and computer technology" — using its 480 voter types and the "Simulmatics data bank." Pool proposed that the company submit a first report right before the Republican convention and a final report right after the Democratic one. He wrote, "These reports, research data, and every other aspect of our study will remain absolutely confidential" — a promise that looked hard to keep given that Simulmatics had just leaked its Kennedy reports to Burdick. Estimated cost: $73,000 (upwards of $600,000 in 2020 money).[41]

The week Pool sent Simulmatics' proposal to the Johnson campaign, the first installment of *The 480* appeared in the *Saturday Evening Post*. "There is a benign underworld in American politics," the excerpt, taken from the novel's preface, began. "The new underworld is made up of innocent and well-intentioned people who work with slide rules and calculating machines and computers. Most of these people are highly educated, and none that I have met have malignant political designs on the American public. They may, however, radically reconstruct the American political system."[42] A week and a half later, the Johnson campaign turned down Simulmatics' proposal.[43]

When Simulmatics' own book came out, that August, the MIT Press included a disclaimer of its own. "The research that the Simulmatics Corp. did for the Kennedy campaign in 1960 has become the subject of many controversial articles and even of a sensationalized novel," that disclaimer read. "Here at last is the complete and accurate report of what a computer simulation actually showed and how it was used in the campaign. . . . No one who reads this book can come away with lurid pictures of a political underworld, as some have portrayed it."[44] The problem was, of course, that hardly anyone read Simulmatics' own book. Instead, they read Burdick's. By July, days before the Republican National Convention, *The 480* was No. 10 on the *New York Times* bestseller list.[45]

"Simulations Enterprises," says one character in *The 480,* "is a dreadful name." But Simulations Enterprises' top behavioral scientist, Dr. Devlin, is a beautiful, alluring woman who makes even a mainframe seem sexy, in her cold, dark, creepy way.

"This is the IBM seven-oh-nine-four Computer," Dr. Devlin said. "Isn't it beautiful?" She moved her hand to indicate six gray boxes, each about the size of an upright coffin.[46]

By the time *The 480*'s fictional Republican convention is about to open, the Goldfinger-like Devlin and the simulations she runs on her IBM 7094 have all but assured John Thatch the nomination. This alarms the backers of Dr. Bryant Clark (a university president who was a conscientious objector during the Second World War and who appears to have been modeled on Burdick's friend and boss, University of California president Clark Kerr). Given that the Republican candidate will face not Kennedy but Johnson, Clark's backers decide to contest Thatch's nomination. They put Clark's name forward and threaten to leak to the press that Thatch's campaign has been driven by Simulations Enterprises.

"You go ahead and do that," dares Curver. "You'll walk right into a buzz saw. For every canned story on Thatch the Computerized Candidate there will be five on Clark as the conscientious objector who fell into clover while the rest of the boys were over there getting shot up."[47]

And so it goes, ruthlessly, down to the Republican convention, held in San Francisco's Cow Palace, a twelve-thousand-seat arena where the real-life 1964 Republican convention was to be held, too. At the novel's climax, Clark's campaign threatens to leak to the press that during the Second World War, Thatch's Filipina wife had prostituted herself to a Japanese prison guard to save her life. Distraught, Thatch's wife goes to the Golden Gate Bridge to kill herself, only to be saved, at the last minute, by Dr. Devlin, who, in the end, cares more about doing the right thing than about winning an election. Love beats hate, man beats machine.

On the eve of the real Republican National Convention in the actual Cow Palace, the last installment of *The 480* appeared in the *Saturday Evening Post,* and copies rested on the desks of politicians all over the country. The publisher's press release, playing on Orwell's *1984,* dubbed it "a novel of 1964."[48] McGraw-Hill sent a copy of the book to every governor of every state and to every member of the Senate, with a publishing-industry pitch letter from the editor in chief, calling it "one of the most important books we have ever published."[49] Burdick had asked McGraw-Hill to try to place the novel in the hands of television reporters and anchors, too. He had also specifically requested that copies be sent, courtesy of the author, to key Kennedy people, including Pierre Salinger and Arthur Schlesinger.[50] (Burdick and Schlesinger were longtime correspondents, readers of each other's work, and admirers of each other's productivity. "It is depressing to have my backlog of Schlesinger increase rather than decrease," Burdick had once written Schlesinger, congratulating him on another book. "Either my literary metabolic rate must speed up or yours must slow down.")[51]

The reviews took Burdick's book in stride, after *The Ugly American* and *Fail-Safe,* as the latest in his trilogy of nightmares, and the likeliest to end up as a blockbuster film. The *New York Times,* whose review of *The 480* ran alongside a half-page ad for the book, deemed it "a whale of a story," even while pointing out that it wasn't very well written. The book, the *Times'* critic said, raised the inevitable scenario: "Supposing a vicious demagogue used such techniques. Suppose scholarly types in love with their own cleverness arranged false crises and paid people to act them out, each episode designed to make their hero seem more heroic? The possibilities are endless."[52] Other reviewers struggled with these same questions. "Could it happen? Who knows? Maybe it happened in 1960," wrote a reviewer for the *Chicago Tribune,* but either way, he concluded, Burdick had "made simulation as real as a toothache."[53]

Pool carefully clipped reviews of *The 480* from newspapers, and even from the *American Behavioral Scientist,* a scholarly journal edited by his old Stanford friend Al de Grazia. Burdick, said de Grazia's reviewer (likely Pool himself), had founded a new genre: behavioral science fiction. The review concluded, "To get angry over having Professor Burdick's new novel foisted on us is futile."[54] That might have been balm to Pool's wounds, but he was less pleased by two stories he marked up with pencil and tucked together into a folder, both from the *Times:* the laudatory review of *The 480* and a review by investigative reporter Ben Bagdikian of a new book called *The Invisible Government,* about the CIA, a book whose claims included the assertion that "the Center for International Studies at the Massachusetts Institute of Technology is, in fact, a C.I.A. operation." Noted Bagdikian, "This could be true."[55]

But Pool also added to that folder one last essay, a full-bore attack on Eugene Burdick, also from the *Times,* and written by Sidney Hyman, a University of Chicago professor who had been a speechwriter for Kennedy. Hyman classed *The 480* with *The Ugly American* and *Fail-Safe* and claimed that all three novels were unpatriotic: *The Ugly American* had unfairly indicted the American diplomatic corps; *Fail-Safe* had all but revealed

315

military secrets; and the claims that Burdick made in *The 480* about Simulmatics' work for the Kennedy campaign in 1960 were lies. Hyman said he'd talked to Kennedy's campaign director, Lawrence O'Brien, and that "if Mr. O'Brien ever heard of the Simulmatics Corporation, it has since passed out of his mind."[56] (This denial is interesting, but not to be credited. Correspondence between Simulmatics and O'Brien, along with each of the 1960 reports, survive in O'Brien's papers.)

The 480's sensationalism made Simulmatics look effective but sinister; the Kennedy campaign's denials made Simulmatics look ineffective and duplicitous. Irving Kristol, then a forty-four-year-old editor at Basic Books, reviewed Burdick's book in an essay called "Computer Politics." "The actual Simulmatics Corporation," Kristol wrote, "is a struggling little company which, despite the fact that it worked on a few problems for the Kennedy organization in 1960, has since had a difficult time making ends meet."[57] What was all the fuss about?

The unruly 1964 Republican convention opened in San Francisco's Cow Palace on July 13. The American party system of the 1950s, in the days of the early voting studies, when few Americans knew the meaning of the words "liberal" and "conservative" and

316

fewer still could tell which party was aligned with which ideology, died in the summer of 1964. Americans began to move to the poles.

College students from all over the country spent that summer, Freedom Summer, trying to register black voters in the Jim Crow South. They also protested the rightward movement of the Republican Party. At the Cow Palace, students from the University of California, Berkeley, protested outside the convention hall. They protested in vain. The conservative movement had first raised its banner in 1955, when William F. Buckley Jr. launched *National Review* and pledged to try to stop history. In 1964, for the first time, conservatives took over a Republican National Convention. They damned moderation, booing Nelson Rockefeller when he called for it from the rostrum. They damned calls for decorum. They mocked the press, attacking it for liberal bias. "You know, these nighttime news shows sound to me like they're being broadcast from Moscow," said one delegate, riding in an elevator with Chet Huntley and David Brinkley, the anchors of *NBC Nightly News*. They applauded calls for halting the civil rights revolution: more than seven in ten of the GOP delegates voted against a platform plank that would have affirmed the constitutionality of the 1964 Civil Rights Act. The party of Lincoln, the Great Emancipator, had become a different party altogether.

Richard Nixon, defeated in the California governor's race two years before, delivered the speech nominating Barry Goldwater. Nixon, a political survivor, gritted his teeth and endorsed the Arizona senator. "Mr. Conservative," Nixon said, "is the man who . . . after the greatest campaign in history, will be Mr. President." On July 15, on the first ballot, Goldwater handily defeated Rockefeller, as the Republican Party took a hard turn to the right. The people Rockefeller had run against as the Right's "lunatic fringe" had won. On the day Goldwater was to accept the nomination, he was cornered by a reporter who asked him whether Democrats would use his vote against the Civil Rights Act against Republicans in the upcoming campaign. "After Lyndon Johnson — the biggest faker in the United States?" Goldwater asked. "He opposed civil rights until this year. Let them make an issue of it. He's the phoniest individual who ever came around."[58] In his acceptance speech, Goldwater denounced moderation. "Extremism in the defense of liberty is no vice," he said, and "moderation in the pursuit of justice is no virtue."

Goldwater had staked much of his campaign on Johnson's unwillingness to commit more fully to the war in Vietnam. Johnson had remained uncertain about what to do and wanted to wait until after the election. The North Vietnamese had been making steady gains. The government in Saigon had been riven by infighting and corruption and worse. The defeat of South Vietnam appeared imminent. "Have we got anybody that's got a military mind that can give us some military plans for winning that war?" he asked McNamara. "We need somebody over there that can get us better plans than what we've got because what we've got is what we've had since fifty-four."[59] His old friends, men outside the White House, advised him to pull out of Vietnam. "We should get out," Richard Russell of Georgia, chairman of the Senate Armed Services Committee had told him in December 1963. In May 1964, after listening to McNamara testify before his committee, Russell warned Johnson, "I'm not too sure he understands the history and background of those people out there as fully as he should." Get out, Russell said. Bombing North Vietnam won't work: "You ain't gonna stop these people."[60] Johnson ignored him.

McNamara laid plans for escalating American involvement after the election, but in the meanwhile, Johnson wanted to find an excuse to flex American muscle in Vietnam. That chance came in early August. On August 2, in a brief summer lull between the Republican convention and the Democratic one, North Vietnamese torpedo boats attacked the

USS *Maddox,* a navy destroyer in the Gulf of Tonkin off the coast of Vietnam. The United States elected not to respond, until a second attack on August 4. It is almost certain that this second attack never took place.

"I personally would recommend to you, after a second attempt on our ships, that we do retaliate," McNamara told the president.

George Ball, undersecretary of state, remained one of the only dissenters, objecting, passionately, again and again, to American engagement in Vietnam. "Once on the tiger's back," he said, "we cannot be sure of picking the place to dismount."[61] Johnson ignored him.

Instead, Johnson took McNamara's advice and asked Congress to pass a resolution authorizing the use of force. He stopped short of asking for a declaration of war. But the United States engaged in one, all the same. For the first time, Johnson sent American forces to bomb North Vietnam. The first American fighter pilot captured by the North Vietnamese, twenty-six-year-old Everett Alvarez Jr., was shot down on August 5. He would spend more than eight years as a prisoner of war. Two days later, Congress passed the Tonkin Gulf Resolution.

By August 24, when the gavel fell, convening the Democratic National Convention, the United States had entered a wider war, with or without a declaration. *The 480* had climbed to No. 7 on the *New York Times* bestseller list. The Democrats met in Atlantic City; if the power of the Republican Party had moved to the west, and especially to California, Democrats hoped to hold on to voters in the South. But the party has also failed to meet the challenge of racial equality among its own delegations. Most delegations from southern states, states where blacks were effectively unable to vote, were all white. In protest, black Mississippians, led by civil rights activist Fannie Lou Hamer, had formed the Mississippi Freedom Democratic Party, to challenge that state's all-white delegation. Hamer, born to sharecroppers, had been an organizer for the Student Nonviolent Coordinating Committee since 1962, fighting for the right to vote. She had a wide face and the voice of a preacher and she wore a floral dress. Her body bore the scars of segregation, and the war wounds of the fight for civil rights. She'd grown up picking cotton. She'd suffered from polio. She'd undergone forced sterilization. Fannie Lou Hamer's call before the Credentials Committee for the sixty-eight delegates of the Mississippi Freedom Democratic Party to be seated in place of Mississippi's all-white delegation proved the most stirring and important speech of the convention. Lyndon Johnson, hoping to keep her speech off television, delivered a press conference while she spoke, but the networks broadcast her speech later that day.

Fannie Lou Hamer told the nation that day what she'd endured, trying to register to vote. She'd lost her job; she'd lost her home. She'd been arrested. In jail, she'd heard the sounds of other women and men being beaten, tortured. "We're going to make you wish you was dead," a Mississippi state patrolman told her. She'd been forced onto a bunk bed and beaten by two men, for hours, till both were exhausted from exertion.[62] "All of this is on account if we want to register, to become first-class citizens," said Hamer, her voice rising. "And if the Freedom Democratic Party is not seated now," she said, "I question America. Is this America, the land of the free and the home of the brave, where we have to sleep with our telephones off of the hooks because our lives be threatened daily, because we want to live as decent human beings, in America?"[63]

The Mississippi Freedom Democratic Party was not seated, and more and more Americans began to question America. But the party pledged never again to seat a delegation elected by a whites-only vote. On August 27, the last day of the convention, Robert F. Kennedy took the stage to honor the memory of his brother and received a twenty-two-minute ovation. Near tears, he quoted from *Romeo and Juliet:* "When he shall die, take him and cut him out into the stars, and he shall make the face of heaven so fine that all the world will be in love with night and pay no worship to the garish sun." John F. Kennedy had been in life a political realist; in death he became an idealist.

In the streaked shadow of a shattered Camelot, Lyndon Johnson accepted his party's nomination, with Minnesotan Hubert Humphrey as his running mate, having defeated a behind-the-scenes campaign to put Robert Kennedy on the bottom half of the ticket. With a giant banner of the dead president behind him, Johnson claimed the succession. "Let us here tonight, each of us, all of us, rededicate ourselves to keeping burning the golden torch of promise which John Fitzgerald Kennedy set aflame," he said, in his tight, clipped voice.[64] And he would hold that torch. But it would set the nation aflame.

Days later, students returned to the University of Berkeley for the start of the fall semester. Many of them had been engaged in political struggle over the summer, registering voters in the Jim Crow South, protesting outside the Republican National Convention in San Francisco. On arriving back on campus, they were told that they could no longer distribute political pamphlets on a twenty-five-foot brick walkway where students had always done their politicking, beneath the campus's Sather Gate, near the corner of

323

Bancroft Way and Telegraph Avenue. The university's trustees had ordered the ban, under pressure from California conservatives who'd been outraged by the student protests outside the GOP convention.[65]

Unsurprisingly, the attempt to suppress the students backfired. Students read the ban as an attack on the civil rights movement. They began protesting, in the name of the civil rights movement, in the name of free speech, and in opposition to what they believed higher education in the United States had become: a factory that treated students like bits of data, to be fed into a computer and spat out. On September 21, seven hundred students staged a sit-in at Berkeley's Sproul Hall. Graduate students joined the movement, too, drawing nearly their greatest numbers from Eugene Burdick's own department, Political Science.[66] "Dear Mom," wrote one freshman, in a letter home, from the Sproul Hall sit-in, "what started as a demand for free speech and advocacy has changed to include the whole meaning of education. There is anger at being only an IBM card, anger at the bureaucracy, at the money going to technology."[67] They were campaigning against a lot of things, and those things included the automation of education. They carried signs that read, "WE STUDENTS AREN'T I.B.M. CARDS — DON'T PROGRAM OUR MINDS."[68]

324

University president Clark Kerr hadn't ordered the ban on politicking at Sather Gate, but he hadn't been willing to override it, either. He quickly became the focus of the students' grievances. Kerr, an economist much admired for his fortitude in the 1950s in the face of McCarthyism, had recently given a series of notable public addresses in which he'd described the modern research university as a corporation, part of the "knowledge industry," a place where not education but "the production, distribution, and consumption of knowledge" happened, for the sake of economic growth.[69] From the students' vantage, Kerr had aligned the University of California with the military-industrial complex warned of by Eisenhower. He had elevated engineering and computers above art and poetry. He had abandoned long-held intellectual principles, as well as principles of university and civic life: academic freedom, the freedom of expression, the vitality of dissent, the importance of individuals and of the humanities. As a sign of their rejection of everything Kerr stood for, the protesters wore around their necks IBM punch cards — the cards with which they were required to register for classes — punched out to read "FREE SPEECH." They carried signs that read, "I AM A UC STUDENT. PLEASE DO NOT BEND, FOLD, SPINDLE, OR MUTILATE ME." And when Mario Savio, a senior philosophy major who'd spent the summer registering voters in Mississippi, climbed on top of a police car in Sproul Plaza, where thousands had gathered, he cried out to a crowd aching to hear an indictment of the age of the machine, his voice rasping with exhaustion: "There is a time when the operation of the machine becomes so odious, makes you so sick at heart that you can't take part; you can't even tacitly take part, and you've got to put your bodies upon the gears and upon the wheels, upon the levers, upon all the apparatus and you've got to make it stop."[70] Make it stop.

Kerr, a decent, moral man, struggled over how to respond. In November, police arrested eight hundred students in a single day. That month, Lyndon Johnson defeated Barry Goldwater in a rout: he won forty-four states and more than 60 percent of the popular vote. The nation's conservative insurgency had been, seemingly, crushed. This defeat did not quiet the unrest at Berkeley. "At Cal, you're little more than an IBM card," Savio told a reporter.[71] In December, Kerr tried to quiet the protests by holding a meeting in the campus's Greek Theatre, with little success. When Kerr finished his address, Savio climbed onto the stage, but as he approached the microphone, he was dragged away by three campus policemen.[72] Meanwhile, a sea of students bearing witness to the stifling of

Mario Savio waved signs that mocked IBM's motto: "THINK."[73]

Eugene Burdick wasn't much on campus that fall; he was out promoting *The 480*. He wasn't especially sympathetic with the Free Speech Movement and what went along with it, mayhem he called "the FSM ruckus."[74] He sided with Kerr. Like many midcentury white liberals, he agreed with the students' demands. He also shared their rage at the machine. But he was distressed by the nature of their protests.[75] Privately, he and his closest colleagues regretted "the tactics of the ultra left."[76]

Publicly, Burdick didn't say much about the protests. His book tour must have been exhausting. He likely went to both the Republican and the Democratic conventions, even though he was unwell, suffering from gout, diabetes, and a worsening heart condition. He'd long since all but stopped teaching. In January 1965, he requested a medical leave of absence.[77] He was back in Berkeley for the spring semester, but not offering any classes. Mainly, he was trying to rest.

By early 1965, Johnson, who'd never sought U.S. involvement in Vietnam, had decided to commit, fiercely, furiously. Rejecting warnings from Democrats in the Senate, and from Humphrey, he proceeded with a strategy of escalation.[78] "I am not going to be the President who saw Southeast Asia go the way

China went," he'd resolved earlier. In March 1965, the United States began bombing North Vietnam, in Operation Rolling Thunder. The first U.S. ground troops landed, in Da Nang. In April, Johnson raised the number of U.S. troops in Vietnam to sixty thousand and, for the first time, committed to ground forces. He hid the escalation from the American people. He lied. The bombs rained down from planes. The boots of marines hit the ground. Their bodies dropped in the fields. They came home in steel coffins. And for what? In a top secret memo drafted by McNamara's assistant secretary of defense on March 24, 1965, the causes for the United States fighting a war in Vietnam were broken down, by the numbers: 10 percent to help the people of South Vietnam "enjoy a better, freer way of life"; 20 percent to keep South Vietnam from the Chinese; and 70 percent "to avoid a humiliating U.S. defeat."[79] Johnson wanted an end to the war, but the war had only begun. The United States would be mired in Vietnam for ten more years.

And already, already, knowing so little about what was actually happening, much of the American public had had enough. In particular, many faculty and students at American universities, conscripted for so long in this cause, had had enough. Two weeks after Marines landed in Da Nang, at the University of Michigan, where undergraduates from the left had formed the Students for a Democratic Society, faculty members held what they called a "teach-in," the first of the war, meeting in lecture halls with more than three hundred students, to explain and reckon with and talk through what was going on, so far away, in their name. The civil rights movement was driven by sit-ins, at lunch counters and diners and on buses. The antiwar movement would be driven by teach-ins. They spread across the country. Students for a Democratic Society called for a march on Washington. Twenty-three thousand people flooded the National Mall, standing between the Washington Monument and the Lincoln Memorial, where, less than two years before, Martin Luther King Jr. had made that land sacred, speaking about a dream.

Beginning with the lunch-counter sit-ins in Greensboro in 1960, students had become, for the first time in American history, the leaders of national political movements, movements that reached from campus to campus, from the Student Nonviolent Coordinating Committee to the Free Speech Movement to the SDS. But the biggest antiwar gathering of that year was a teach-in held in May at the University of California, Berkeley. Thirty thousand people gathered for thirty-six hours. "If we wish to take a strange country away from strangers," Norman Mailer told the crowd at Berkeley, "let us at least be strong enough and brave enough to defeat them on the ground." Mario Savio spoke at the Berkeley teach-in, indicting the war, and so did baby doctor Benjamin Spock. Folksinger Phil Ochs sang "I heard many men lying / I saw many more dying" from his song "I Ain't Marchin' Anymore." Nevertheless, the American military was marching: that summer, Johnson sent fifty thousand more men to Vietnam.

Eugene Burdick had at first agreed to speak at the Berkeley teach-in, one of only two people willing to defend the Johnson administration. At the last minute, he withdrew.[80]

Burdick was falling apart. And so was the country. The Ballantine Ale man, with a thirst for that manlier brew, the James Bond of the American academy, had grown wan and weak. His Cold War was nearly over, his brand of liberalism dying.

But for Simulmatics, things were beginning, at last, to look up. The Department of Defense's Advanced Research Projects Agency had written to the company, requesting a proposal on "using dynamic modeling as a tool in dealing with the problems of counterinsurgency."[81] Ithiel de Sola Pool set to work. By the summer of 1965, Simulmatics would have an office in Saigon.

Burdick never knew about that operation. "All goes miserably in Berkeley," he wrote wearily in June. "I talked with a professor from Stanford recently, and asked him what was doing down there. He said, 'Well, we just get up in the morning and read what is happening in Berkeley.' "[82]

On July 26, 1965, during an ill-advised tennis match, Eugene Burdick died of a heart attack. He was only forty-six. He left the American scene unaware of what the United States was actually doing in Vietnam, where the Simulmatics Corporation would make more money than it had ever made before, fighting a war the United States would never win.

■ ■ ■ ■

PART THREE:
HEARTS AND MINDS

■ ■ ■ ■

Robots did the dropping. They had no
conscience, and no circuits which would allow them to imagine what was happening to
the people on the ground.
— Kurt Vonnegut, *Slaughterhouse-Five,*
1969

CHAPTER 10
ARMIES OF THE NIGHT

Vietnam is the greatest social-science laboratory we have ever had!
— Ithiel de Sola Pool, 1966

Simulmatics in Saigon, 1966. In the front row, Maureen Shea is fourth from the left; Cuc Thu Duong is second from the left. In the back row, Sam Popkin is third from the right.

"Your girl in Saigon made it," Maureen Shea wrote to her family on August 16, 1966, when she arrived in Vietnam, a land of indigo bays and yellow-green jungles and rice paddies and clay-hardened roads lined with mop-headed banana trees. She was twenty-three and tiny, pint-sized, adorable, in sundresses and scuffed sandals, with a pageboy haircut and a sly smile. After college, she'd moved to New York; she heard about the job from a guy in the apartment next door. She was the only person who interviewed; she got the job. She'd spent a year in Paris during college; she spoke French. Simulmatics did not suggest that she learn Vietnamese. She packed peanut butter.[1]

She flew to Vietnam by way of Clark Air Base, in the Philippines, on a military plane headed for Tan Son Nhut, an air base first built in the 1930s by the French, as a landing strip. Even from the sky, Vietnam bore the traces of the Chinese, and the French, and the Japanese, and the French again, and then the Americans, each another failed rule. During the Second World War, the occupying Japanese army had turned the French landing strip into a transport base. After the war, with Vietnam once again under France's imperial authority, the French had enlarged it into a military base. In the 1960s, as the United States escalated its fateful involvement in Vietnam, Tan Son Nhut would become one of the busiest airports in the world, the jet engines raging like monsoons. But it was the noise of helicopter rotors, *th-WHACK, th-WHACK, th-WHACK,* that would become the enduring sound, the haunting roar, of that long war.

Maureen Shea's plane landed and skidded to a stop. "It was a little hard to put our pretty jet down among all the Air Force bombers and helicopters but we did it," she wrote in her first letter home. From the airport she was taken by car down an avenue the French called Rue Catinat but the Vietnamese had renamed Tu Do: Freedom Street. She passed bicycles and motorized rickshaws ("God knows what they burn in the motors but it spouts straight black smoke"), Solex motorbikes, Renault taxis, Peugeots and Citroëns, and U.S. Army trucks, all of them stampeding like so many antelopes, gazelles, water buffaloes, wildebeests, zebras, and rhinoceroses across a savannah. "The French were kind enough to leave their fine sense of traffic," Shea wrote home. "Every street is like the Arc de Triomphe."[2]

Shea spent her first night at the Saigon Palace, on Tu Do Street, "in the good bar section of town," sharing quarters with a lizard and a rat.[3] The next day, she toured the city. *Washington Post* war correspondent Ward Just described Saigon that summer as a collage of "Vienna in the 1930's, London in the 1940's, and Algiers in the 1950's."[4] Shea was young and earnest, but she wasn't wide-eyed. There were three prices, she noticed, for everything: "Vietnamese, French, and American, ascending in that order." She ordered in French.[5]

She could hear shelling in the distance, thudding like summer thunder. Tanks rumbled through the streets, stirring up clouds of dust. Sandbags barricaded the military buildings of both the US and the GVN, the Government of Vietnam. Everything had an acronym. ARPA (Ar-pah) occupied offices at 4A Ben Bach Dang, on the pier, by the river. The building that housed the Joint United States Public Affairs Office, JUSPAO (Jus-pow) sat at the corner of Le Loi and Nguyen Hue Streets.[6] RAND rented a two-story French villa at 176 Pasteur, defended by a concrete wall ten feet tall and topped with barbed wire.[7] IBM's headquarters occupied a guarded villa at 115 Minh Mang Alley, across from the Milan Hotel and just off Truong Tan Buu Street; the IBM staff set up a croquet course in a courtyard that flooded during monsoon season, at which point the IBMers called the courtyard Lake Watson. MACV (Mac-vee), the headquarters of Military Assistance Command, Vietnam, also known as Pentagon East, started out in a villa on Cong Ly Street but moved to a bigger structure near the airport, maybe to accommodate its suite of IBM computers, run by MACV's Data Management Agency, which printed more than fourteen hundred pounds of reports per day, carted around like baskets of rice.[8]

The Americans' computers shed their packaging, snakes sloughing off their skin. When the air force replaced its IBM 1410, a successor to the machine the *New York Times* used on Election Night in 1962, the new system arrived in eighteen enormous wooden crates. Eight days later, the empty crates had disappeared: Vietnamese families, the poorest among the war's refugees, had carried them off and made them into houses — shanties — their doorways stenciled "IBM" in the company's trademark blue.[9]

By the time Maureen Shea reached Vietnam, much of the Simulmatics team had already arrived. Ithiel de Sola Pool, forty-eight, his hair thinning, had landed in Saigon on June 26.[10] "He's a charmer," Shea told her parents after she met him.[11] Pool had been asked to head the Department of Defense's Behavioral and Social Science Program. But he'd said no, wanting, instead, to work for the Pentagon in Saigon.[12]

He'd staked the future of Simulmatics on Vietnam. This seemed, at the start, a good bet. Simulmatics would make more money, its first year in Vietnam, than it had made in any other year in its history.[13] It planned to penetrate Vietnamese minds. It planned to win Vietnamese hearts.

The U.S. war in Vietnam was the first war waged by computer. Computers had been used in the Second World War to target missiles and break codes, but never before had they been used to plan and conduct war.[14] IF/THEN. If x number of troops and y number of tanks, then z number of casualties. If x number of radio broadcasts and y number of pamphlets, then z number of defectors. Run the simulation, again and again and again, an endless loop for an endless war.

This war, this computer war, was Robert McNamara's war. McNamara would say, about proposed operations, that this one had a 65 percent chance of success and that one a 30 percent chance, as if predictions could be made with this sort of precision. George Ball once teased him, suggesting that maybe the first really had a 64 percent chance, and the second a 29. McNamara was not amused.[15]

McNamara looked like a nineteenth-century banker, with slicked-back hair and spectacles and rounded shoulders, and not for nothing did Barry Goldwater call him "an IBM machine with legs."[16] He aimed to reduce military affairs to a computational science. McNamara, who'd trained at Harvard

Business School, had been president of the Ford Motor Company, which he'd run by way of computer-based systems analysis. He'd brought this approach to the Department of Defense.[17] "He wields that computer and those figures like King Arthur wielded Excalibur," one of Johnson's assistants said.[18] As if it were magic.

McNamara married his commitment to systems analysis to a theory of what had come to be called "counterinsurgency." Cold Warriors like McNamara and Pool believed the central problem of the Cold War to be political revolutions in developing nations, which they considered to be insurgencies incited by the Soviets or the Chinese. A campaign of counterinsurgency could be waged militarily, by winning battles and capturing prisoners and defeating guerrillas, and it could be waged by way of behavioral science, or what McNamara liked to call the winning of "hearts and minds."[19] Pool believed that the best way to deal with insurgency wasn't to counter it but, instead, to predict and avert it. In 1961, he'd written the Kennedy administration to propose a project on the prediction of insurgency in developing countries, with an eye to the administration's priority at the time: halting Communist revolution in Latin America. Simulmatics had then undertaken a series of behavioral science experiments, beginning with a contract for a massive counterinsurgency project in Latin America. In 1963, Simulmatics sent three people to Caracas, where they worked at the Centro de Estudios del Desarrollo, an affiliate of the Universidad Central de Venezuela and a longtime collaborator with MIT's Center for International Studies. They aimed to do nothing less than to simulate the entire economy of Venezuela.[20] The plan was to feed all available economic data about Venezuela into a computer in Cambridge, run the simulation program, and, on its evidence, offer the government of Venezuela a program for economic development that would serve to inoculate Venezuela from the influence of Communist revolutionaries.[21] Politically and diplomatically, the program threatened to be explosive: it rather resembled espionage. "Everybody here is very nervous," a miserable and homesick Alex Bernstein wrote from Caracas to his wife, back in Brooklyn with their new baby.[22] This was not the sort of work he'd had in mind when he'd programmed an IBM 704 to play chess.

In 1964, following a report produced by a committee led by Pool, the U.S. Army announced the launch of Project Camelot, named in homage to the slain president, the largest behavioral science project in American history, Simulmatics' Venezuela study writ large. Several people who'd worked for Simulmatics, or soon would, including James

Coleman and Al de Grazia, helped design the project, but its proposal cited the work of only a single scholar: Ithiel de Sola Pool.[23] The army explained, "Project Camelot is a study whose objective is to determine the feasibility of developing a general social systems model which would make it possible to predict and influence politically significant aspects of social change in the developing nations of the world."[24]

Project Camelot was to political insurgencies what SAGE was to nuclear attacks: an early warning system. It enlisted behavioral scientists in the work of gathering data and devising computer models that would allow them "to achieve early detection and prevention of the predisposing conditions" for insurgency.[25] Budgeted at $6 million, and targeted initially at Latin America, Project Camelot was expected to grow to $50 million (something like $400 million in 2020 money), enough for the United States to win what the project's director called not the arms race but "The Minds Race."[26]

Alex Bernstein, who didn't stay in Venezuela long, had been right to be nervous. Modeling large-scale social systems by way of computer simulation had all sorts of objectives and ambitions, but Project Camelot appeared intended to direct the economic and political development of other countries with the aim of suppressing not only revolution but political expression itself. When word got out about the existence of the project, it set off a political firestorm. Early in 1965, after a Chilean anthropologist told colleagues in Santiago about the project, the Chilean Senate denounced it. So did Americans, most notably the Vietnam War critic Senator J. William Fulbright, who pointed out that "implicit in Camelot as in the concept of 'counterinsurgency' is an assumption that revolutionary movements are dangerous to the interests of the United States and that the United States must be prepared to assist, if not actually to participate in, measures to repress them."[27]

American academics denounced the project too, condemning, at once, imperialism, computer modeling, and behavioral science undertaken on behalf of the military. Speaking before the American Anthropological Association, anthropologist Marshall Sahlins delivered a speech called "The Established Order: Do Not Fold, Spindle, or Mutilate" (a nod to the anti–punch card motto of the Berkeley Free Speech protest), in which he called the premise of Project Camelot "a snare and a delusion" and warned that "the cold war researcher is potentially a servant of power, placed in a sycophantic relation to the state unbefitting science or citizenship."[28] Sahlins, in short, took aim at Pool. In July 1965, the Defense Department, unable or unwilling to defend Project Camelot, terminated it.

That didn't end the uproar. Later that year, Congress held hearings on "Behavioral Science and National Security," which made clear that while Simulmatics' counterinsurgency work remained little known outside the defense and intelligence communities, it was well known inside the Pentagon. During the hearings, Seymour Deitchman, special assistant in counterinsurgency at the Department of Defense, fielded questions from Representative Dante Fascell, a Florida Democrat and chair of the committee. Fascell pressed Deitchman on the use of computers for "analysis in the field of behavioral sciences for military purposes."

DEITCHMAN: The use of computers in behavioral science is something that is very much under study now.

FASCELL: Where?

DEITCHMAN: In the Defense Department.

FASCELL: That is what I was asking about.

DEITCHMAN: Specifically, the use we see right now, which is a fairly obvious one, is the one that you have addressed yourself to.

FASCELL: Which is?

DEITCHMAN: Storage and retrieval of

information which already exists in many places, books, papers, and so on. When it comes to analysis, it is not at all clear that you can use computers. We are giving a lot of thought to this.

FASCELL: There is an industrial organization called Simulmatics which might give you some sort of argument.

DEITCHMAN: We are very familiar with that.[29]

Project Camelot convinced many behavioral scientists to stop taking money from the Department of Defense. Not Pool. The anxious young man who'd been denied a security clearance because of suspicion that he was a Communist, years before, had long since settled into his convictions: he wore his anti-Communist credentials like so many ribbons and medals. In 1966, a turning point in U.S. involvement in Vietnam, Pool, for all his gentleness of manner, fiercely committed himself both to the cause of counterinsurgency and to the role of behavioral scientists in pursuing that cause. In an angry and overheated essay, "The Necessity for Social Scientists Doing Research for Governments," Pool insisted that nothing was more noble or wise or more true to both science and citizenship than projects like Camelot.

Describing the social sciences as "the new humanities of the Twentieth Century," Pool argued that while statesmen in times past had consulted philosophy, literature, and history, statesmen of the Cold War era were instead obligated to consult the behavioral sciences. The "McNamara revolution," he argued, had "remade American defense policy in accordance with a series of ideas that germinated in the late 1950's in the RAND corporation." Given a choice "between policy based on moralisms and policy based on social science," he was glad to report that the secretary of defense had rejected the humanities and morality in favor of behavioral science and rationality.[30]

The more unpopular the war in Vietnam, the deeper Pool's commitment, and the closer he grew not only to the Pentagon but to the White House. He'd sometimes bring his young son with him when he went to 1600 Pennsylvania Avenue; he'd park him in the Oval Office. Adam would sit in Lyndon Johnson's chair and pretend to be the president. When Pool broke his leg during a Political Science Department baseball game and couldn't travel to Washington for meetings, National Security Adviser McGeorge Bundy and Secretary of Defense Robert McNamara, on a visit to Cambridge, instead held the meeting in the Pools' backyard on Irving Street, the street where, two doors over, Adlai Stevenson and John F. Kennedy had visited Arthur Schlesinger's house, only a few short years before, in another America.[31]

Armies win wars by gaining territory and measure progress by the movement of battle lines. In a counterinsurgency operation, the battle isn't really for territory, which makes progress difficult to measure. In Vietnam, McNamara's Department of Defense decided to measure military progress by the number of insurgents killed, the counting of the dead, a scheme of quantification that encouraged indiscriminate slaughter, including the massacre of civilians. But McNamara's Department of Defense struggled to come up with a way to count the proportion of the population inoculated against communism, the number of hearts and minds won. For that, it called on behavioral scientists at places like RAND — and Simulmatics.

Behavioral science devoted to counterinsurgency operations fell under the administration of the Department of Defense's Advanced Research Projects Agency, specifically under a program known as Project Agile, launched in 1961. With funding from Project Agile, RAND had in 1964 sent a team of researchers to Saigon to interview North Vietnamese and Vietcong prisoners and defectors for a "Motivation and Morale" study that attempted to provide an answer to McNamara's question "Who are the Viet Cong and what makes them tick?" Early results of the study, known as the M&M, suggested that the Vietcong army was a far more formidable force than anyone in the American military seemed, thus far, to have suspected.[32]

The M&M study might have persuaded Johnson to abandon Vietnam. It did not. Instead, Johnson decided to ignore it. George Ball, Johnson's undersecretary of state, read the M&M study and concluded that the war in Vietnam could not be won. In 1965, Ball wrote a memo for Johnson pointing out that RAND had shown that the government in Saigon was a travesty and that the Vietcong were formidable. "A deep commitment of United States forces in a land war in South Viet-Nam would be a catastrophic error," Ball wrote. "If ever there was an occasion for a tactical withdrawal, this is it." He was the very last person in the administration to cite the M&M study in an official document.[33]

Still, the M&M study gave McNamara pause. When General William Westmoreland asked for one hundred thousand more troops, McNamara advised Johnson that he faced a choice between negotiating a peace and sending in more troops, and warned him that the chances of victory were no better than one in three. Johnson decided to send in more troops.[34]

But the management of more troops required faster computers that needed better simulations. Late in 1964, ARPA had informed Simulmatics of a new initiative, one that would explore the possibility of taking computer simulation to the next level by using dynamic modeling for the purposes of counterinsurgency. Dynamic modeling is more complicated than the kind of work Simulmatics had been doing, computer-program simulation: Simulmatics created a simulated world, an imaginary population, and then ran queries, testing the effects of various changes, one at a time. Dynamic modeling, by contrast, created a simulated world and set it in motion. Johnson had bombed North Vietnam. He'd sent in the marines. He'd sent in army combat troops. And still the United States was losing the war. ARPA wanted computationally minded behavioral scientists to devise a machine that could think through a war that had left the Johnson administration stumped. ARPA explained, "The aim is to let the experimenter see the modeled processes unfold before his eyes, watch the effects of changing the values of variables and parameters, try out alternative model structures in quick succession and, in short, explore the modeled processes freely and think about them creatively with the aid of advanced computing equipment and techniques."[35] The aim was to devise a real-time war game.

RAND, from its villa in Saigon, launched a new study, a "Factor and Content Analysis" of the situation in Vietnam. Joel Edelman, a young scientist from Columbia who'd been hired into RAND's Aero-Astronautics Department, was among those people charged with drafting what he later described as "a computer program that would hopefully be able to tell us what the trends were and make some predictions about how we could predict how things were going — how could we assess progress." RAND researchers gathered all sorts of data, on combat incidents, kill ratios, body counts, voting trends, weather, the phases of the moon, and newspaper reports (these they "found a way to code"). "You name it, we had the data for it," Edelman said. They threw in even "the Vietnamese version of the kitchen sink." Then they wrote a program bigger and more complicated than anything RAND had ever attempted, a prediction machine, a Vietnamese People Machine.

Within the brittle logic of FORTRAN, systems analysis proved unable to contemplate the illogic of counting motives, the immeasurability of suffering, and the immorality of war. RAND's program ran on an IBM 7090; it kept crashing. And the results it spat out, Edelman said, looking back, were precise but not necessarily accurate: "We could tell to the second decimal place from the results what was going on, but we had no idea whether it was true." He found the whole experience so distressing that, for more than ten years afterward, he refused to use a computer.[36]

They counted and they counted. "We're killing these people at a rate of ten to one," General Westmoreland told Senator Fritz Hollings, from South Carolina, when Hollings visited Saigon. Westmoreland was making those numbers up, but to Hollings, even if those numbers were right, they were still wrong. "The American people don't care about the ten," answered Hollings. "They care about the one."[37] Westmoreland could not hear this, could not fathom this.

"Every quantitative measurement we have shows we are winning the war," McNamara had declared as early as 1962. American troop withdrawal will begin in 1966, he predicted in 1965. "All the measurements pointed that way." But McNamara's numbers meant nothing, reported the *Washington Post*'s Ward Just, who'd been injured in a grenade blast in Vietnam in 1966 and left the country the next year. Just wrote about McNamara witheringly: "What he wanted was a new math, more likely a new calculus, to measure change and increments of change, to measure progress or the lack of it in a war without front lines." But there simply was no way to measure what McNamara most wanted to measure, wrote Just, asking, "What is the quantitative measurement of passive resistance?"[38] How long will the North Vietnamese last? When will we win? These questions had no answers.

Vietnam would be the test of McNamara's policy, and of RAND and Simulmatics' behavioral science: decision by numbers, knowledge without humanity, the future in figures. It would fail. It would also endure. In the twenty-first century, it would organize daily life, politics, war, commerce. Everything.

Early in 1966, Ithiel de Sola Pool, dashing through the airport in Washington after attending a series of meetings, bumped into the distinguished Columbia University historian Fritz Stern, a Polish-born scholar of the rise of Nazism, and stopped to say hello. When Stern explained that he'd been in the city to protest the war, Pool confessed himself shocked, objecting, "But Vietnam is the greatest social-science laboratory we have ever had!"[39]

At Simulmatics, that laboratory had been years in the planning. Seymour Deitchman, who, since 1964, had headed counterinsurgency operations at ARPA, realized that between the controversy over Project Camelot and rising opposition to U.S. involvement in Vietnam on college campuses, few university scholars were any longer willing to accept work from the Department of Defense. That made a company like Simulmatics especially attractive: it offered ARPA the services of some of the nation's best behavioral scientists without requiring contracting directly with a university.[40]

Early on, Bernstein had sent Deitchman a proposal for research on urban insurgency. "In recent years there has been a great deal of research done on counter insurgency in jungle areas," Bernstein wrote. "Most of the guerrilla actions with which we have recently been faced have, indeed, been in the tropical countryside. We cannot assume, however, that this will always be the case. . . . It is probably only a matter of time before we find ourselves having to fight or support fighting in a place like Saigon or Calcutta or Manila." He proposed a computational analysis of surveillance photographs taken from the air. ARPA did not fund this project.[41] Instead, in grants awarded over four years and involving millions of dollars, ARPA chiefly contracted with Simulmatics to study the Department of Defense's psychological warfare operations. In Vietnam, as in the United States, Simulmatics never had its own computer equipment. Instead, the company undertook research and prepared reports designed to help ARPA make better use both of its own computers and of the dynamic modeling programs being written at RAND.

The bulk of Simulmatics' research consisted

of opinion surveys. Simulmatics scientists were supposed to go into the field to conduct interviews and administer questionnaires, recording the interviews on reel-to-reel tape and bringing the tapes back to Saigon, where administrative aides like Maureen Shea would type up transcripts and the research staff would tally the findings. ARPA expected that Simulmatics' leading scientists would write the reports.

Pool directed the research for the entire operation and intended to spend the summers, when MIT was out of session, in Vietnam. His work began in earnest in the summer of 1966. He left Cambridge on June 16, with his wife, Jean, and eight-year-old Adam; they stopped in Hawaii, for a holiday, to see the dolphins and swim, and then visited Kyoto, to see palaces and temples. Pool then flew to Saigon, while Jean and Adam went on to Hong Kong; Pool visited them there on weekends, or else met them in the Philippines. He chronicled his time in Vietnam in the frequent, loving letters he sent to his family and, more formally, in reports he sent to Ed Greenfield in New York and to Seymour Deitchman in Washington.

Pool, a well-traveled man, did not like Saigon: "It's hot, it's harried, it's drenched, it's run down and depleted by war." But he also found the city "moving and exhilarating" because of what he viewed as the great success of the American war effort. "The idealism of the Americans here is impressive indeed," he wrote. "They work 12 hours a day seven days a week and with remarkable good will to the Vietnamese. Most stunning of all are the many evidences of how little genuine VC support there is."[42] We are winning, Pool reported. We are winning, he told himself.

Within days of Pool's arrival in Vietnam, a Simulmatics field team landed in Saigon and began heading out to the countryside to conduct interviews.[43] Or, rather, to oversee the Vietnamese interpreters, mainly college students, who conducted the interviews. "Polling and survey research in Vietnam turn out to be much more feasible than originally thought or than most people think," the field team reported, that first summer. "As in any country, there are areas of reserve, but people have little hesitation about voicing their criticisms of government or their own desires."[44]

More astute American observers disagreed with this assessment, none more keenly than the young redheaded American reporter Frances FitzGerald. FitzGerald, whose mother had been Adlai Stevenson's longtime mistress, reported from Vietnam beginning in 1966, when she was twenty-six. (Stevenson had died of a heart attack the year before, while walking down the streets of London with FitzGerald's mother.)[45] "Young men from RAND and Simulmatics bounded about the countryside in Land Rovers studying 'upward mobility among village elites' or 'the interrelationship of land reform with peasant political motivation,'" she wrote in *Fire in the Lake,* her prizewinning book about the war. FitzGerald found that she was able to get Vietnamese peasants to talk to her because they figured she couldn't possibly have been a secret agent, partly because she was a woman and partly because she didn't arrive in a military jeep. The same wasn't true for the young men from RAND and Simulmatics. "The peasants must have thought, 'Here's another American, who's probably CIA.' If you went around with a military escort, it was just hopeless. People just shut down."[46] FitzGerald found the young men from RAND and Simulmatics naïve, self-important, and bumbling, and their research useless. She mocked their ridiculous predictions: " 'Of course,' they would say with a slight swagger as they emptied the clips from the Swedish K submachine guns, 'if the GVN realized the RF-PF potential, the lower-level Viet Cong village hierarchies would disappear in a matter of weeks.' "[47]

Not all the young men from RAND and Simulmatics were so cocky. Sam Popkin, one of Pool's most promising students, went to Saigon in 1966. He looked to a lot of people like a big, nervous kid, something he had in common with most of the soldiers and sailors and marines who went to Vietnam. He knew that Pool's military work was controversial. But he admired Pool's sense of fairness, his intellectual curiosity, and his sense of humor. Once, when a leftist member of the faculty came up for promotion and some of Pool's colleagues complained that he was too radical, Pool said, "Look, he just wrote the *Communist Manifesto.* We're only voting to make him an assistant professor. If he writes *Das Kapital,* we'll give him tenure."[48]

Popkin also thought, before he got to Vietnam, that all the Americans he'd meet there would be dedicated to helping the South Vietnamese in their struggle to remain independent from the North. His brother, a student at the University of Wisconsin, ended up in prison over his opposition to the war. Popkin didn't believe in the war, either. For him, researching in Vietnam was a way to avoid fighting in Vietnam. He'd finished college; he hadn't started graduate school, which meant that he could be drafted. By going to Vietnam as an employee of Simulmatics, he got a deferment from the Pentagon. He thought that Simulmatics' research would help persuade the Johnson administration to change course and stop dropping bombs.[49]

"No one who has been here more than one month and has been at all involved with the

Vietnamese has an easy answer to any question," Popkin wrote home from Saigon not long after he arrived. In the middle of the city one day, he spent an afternoon watching *Batman,* the campy mid-sixties television show, dubbed into Vietnamese, on a twenty-one-inch television set inside a miniature pagoda, with a delighted crowd. He wrote home, "I set a new personal one day record for 'most hearts and minds captured by unarmed american not dispensing money.' "[50]

In Saigon, Popkin heard a lot about Daniel Ellsberg, a longtime RAND analyst and former marine who was working for Edward Lansdale. (Lansdale, a CIA officer who had run a psychological warfare operation from 1954 to 1957, and who wrongly believed himself to be the inspiration for the American in Graham Greene's 1955 novel, *The Quiet American,* was undeniably the inspiration for Edwin Hillandale, the hero of Eugene Burdick's 1958 novel, *The Ugly American.* "Every person and every nation has a key which will open their hearts," says Burdick's Hillandale.)[51] Ellsberg had briefly visited Vietnam in 1961, finished a PhD in economics at Harvard the following year, and in 1964 joined the Defense Department as a Vietnam analyst, a dedicated Cold Warrior.[52] In 1966, thirty-four, Ellsberg was known to head out to the countryside on patrol, carrying a gun and grenades, searching for the thrill of combat, sometimes on LSD. "Stop playing soldier," Lansdale warned him. "If you weren't careful around the heart of darkness," Popkin later said, it could change you. "It was easy to turn into an Ellsberg."[53]

Popkin didn't turn into an Ellsberg. Popkin was just plain scared: sweet, bewildered, anxious, and shaken. He and his five Vietnamese interpreters were once in a village when it was shelled by the South Vietnamese. The shells killed the district chief's wife and wounded his two children; Popkin managed to get them out by medevac.[54] He had very quickly come to see: something had gone very wrong in Vietnam.

Cuc Thu Duong began working as an interpreter for Simulmatics in 1966. "Miss Cuc," her American colleagues called her. Pool described her as "our best interviewer."[55] She was twenty-six and slight and wore her hair in a chin-length wave. She was a Buddhist. She was engaged to a Harvard-trained former adviser of Prime Minister Diem named Sieu The Luong. They had to wait to get married because she was still in mourning; her father had died not long before she began working for Simulmatics. Her father had been a revolutionary, opposed to the French. But he'd also opposed the Communists. Cuc had followed his lead. "I thought I too was working for peace," she said, "and peace not with

Communists." People she knew in Saigon hated her, despised her, for working for the Americans. Cuc hated the Communists. She thought the Americans could help defeat them. But she came to conclude: Simulmatics wasn't helping.[56]

Pool had ordered a Basic Vietnamese Language Program, made available by the army — sixty audiocassettes — but there is little evidence that any of the Americans who worked for Simulmatics spoke more than a few words of Vietnamese.[57] (Pool's misrepresentations on this score would later lead to an investigation by ARPA. Pool had claimed, for instance, that one American researcher had been an exchange student at the University of Saigon for three years when, in fact, she had studied there for less than a year and, in a restaurant, proved unable to order in Vietnamese.)[58] Instead, Simulmatics hired Vietnamese translators and interpreters, like Cuc Thu Duong.

Simulmatics staffers, both Vietnamese and American, wore a uniform, the blue of a Cub Scout uniform, with a badge on the shoulder, a sunflower-yellow oval embroidered with a depiction of the reels of a tape recorder, a Vietnamese rice hat, and the words SIMULMATICS SAIGON.[59] With their uniforms, the likeness between the American and Vietnamese members of the Simulmatics staff ended.

Neither Cuc Thu Duong nor any of the Vietnamese interpreters employed by Simulmatics were consulted about the research, or even given any explanation of it. Cuc found this maddening. When interview teams went out into the field — to villages and hamlets — the scientists, in their quasi-military blue shirts, arrived by military helicopter or U.S. Army truck and usually stayed in officers' quarters. Ward Just said that the Simulmatics scientists gave everyone the impression that they'd woken up in their villa in the morning, put on their blue safari suits, flown to this place or that, interviewed a few people, and intended to get back to Saigon for dinner.[60] The Vietnamese translators stayed in barracks and lived like soldiers.[61]

Cuc found the questions she was supposed to ask the villagers inane. "All of us want certain things out of life," she was supposed to say to her interview subjects, reading from the Simulmatics questionnaire. "When you think about what really matters in your own life, what are your wishes and hopes for the future? In other words, if you imagine your future in the best possible light, what would your life look like then, if you are to be happy?"[62]

Most people answered, "We wish for peace."

"But what does that mean?" Cuc was supposed to ask. "What is peace? Is peace being with us, or is peace being with the Viet Cong?"

They would not say. The only thing they would say was what they thought the interviewers wanted them to say. What else could they say?

Cuc Thu Duong believed it to be useless. Simulmatics researchers landed, they interviewed, they climbed onto a helicopter or scrambled back into an army truck. They were strangers. No one would speak frankly to a stranger. They learned, she thought, absolutely nothing.[63]

Little that Simulmatics undertook in Vietnam went smoothly. ARPA's records on its contracts with Simulmatics contain complaint after complaint, most involving charges of ineptitude laid against the behavioral scientists heading the projects, who were not the behavioral scientists ARPA had been led to believe Simulmatics would provide.

Aside from Pool, none of the scientists who'd founded Simulmatics agreed to go to Vietnam. Instead, Pool and Greenfield rounded up second-rate scientists and nonscientists: oddballs, has-beens, and outcasts. One of Simulmatics' first Saigon-based projects, a study of the South Vietnamese Regional and Popular Forces, was led by a psychologist from the University of Texas named Philip Worchel. The Regional and

Popular Forces, or RF/PF, known as the Ruff Puffs, were bands of militiamen who had joined neither army but instead defended villages and hamlets. Worchel and his team interviewed thirteen hundred Ruff Puffs, their wives, and villagers, over six months, administering psychological tests and conducting interviews, working with translators who were usually Vietnamese college students. Popkin heard rumors that Worchel was a hack who'd been pushed out of his last job.[64] Notwithstanding complaints about the Ruff Puff study, ARPA, even more desperate than Simulmatics on the matter of staffing, named Worchel its director of behavioral science research in Saigon.[65]

Another of Simulmatics' early ARPA studies was conducted by a New York psychoanalyst named Walter Slote, a friend of Ed Greenfield's. (Naomi Spatz, who met Slote at a party she went to with Greenfield, told me he was mostly famous for sleeping with his patients.)[66] With ARPA funding, Simulmatics paid Slote to fly to Saigon and in July 1966 he began to psychoanalyze four Vietnamese dissidents. "Let me emphasize again, this is not for the military; this is for social research," Slote told his subjects. Once, when someone else from Simulmatics interrupted an interview, the subject broke it off, telling Slote he was pretty sure that the other guy was from the CIA.

Slote wasn't a secret agent. He was a Freudian.

Nearly every day in July and early August 1966, Slote conducted sessions with four subjects, whose names were redacted in his reports: a twenty-six-year-old Vietnamese student leader being held against his will in a pagoda outside Saigon ("He seemed to be searching rather actively for a father, both in me and in others"); a captured Vietcong leader imprisoned at the Police Interrogation Center, who told Slote he was being tortured (Slote asked him about his dreams and about sex); a senior Buddhist monk (who was celibate); and a fifty-six-year-old intellectual (a French-educated mathematician turned writer), who had been imprisoned, for years, by Diem.

He questioned them. He gave them Rorschach tests. He administered a modified Thematic Apperception Test. Slote recorded all of his conversations with these subjects. The man identified in Slote's reports only as a fifty-six-year-old intellectual is known to have been Ho Huu Tuong. Ho Huu Tuong had studied at the University of Marseilles and had first been arrested, by the French, in 1932. Like Ithiel de Sola Pool, Ho had been a Trotskyite in the 1930s. But by the 1950s, he'd abandoned both Trotskyism and Marxism. In prison again in 1957, starving and skeletal, he'd nearly lost his mind. Slote extracted from Ho his entire life story, though he was chiefly interested in Ho's early childhood and in his dreams. Ho told Slote about a dream he had involving a meeting with Krishna, Buddha, Moses, Confucius, Lao Tze, Jesus, Muhammad, and Marx. "The first to speak was Marx," Ho said, "in order to recite his accusastions against me" (for abandoning Marxism). Ho then described how, later in the dream, he became a monkey. Slote asked him about being a monkey. "This is a symbol," Slote said. "What comes to your mind about this monkey?"

Ho Huu Tuong and Slote's other subjects — men who had been imprisoned and tortured — were, unsurprisingly, evasive. Slote, by contrast, was expansive. He told his subjects he opposed the war.

> SLOTE: I don't think that one country can just import a form of government from another country. It has to be indigenous, true for that country, and perhaps based on the nature of the people. I certainly don't think we in the US are qualified to decide what is the best form of government for the Vietnamese people. I think that is a decision for the Vietnamese people to make.
> DISSIDENT: That is right, and that is the reason why I fight.[67]

■ ■ ■

This, however, is not the reason Slote cited in his analysis of the motivations of Vietnamese dissidents. In his final report, "Observations on Psychodynamic Structures in Vietnamese Personality," Slote mainly discussed how all four of his subjects "had set up the United States as the all-powerful, all-giving father image — the institutionalized substitute for all they had missed in their childhood."[68]

Vietnamese men, women, and children were dying, starving, being shot, bombed, burned, and napalmed. American soldiers were being shipped home in boxes, coffins, and bags. And the U.S. government was paying an Upper West Side Freudian analyst to explain that the Vietnamese, as a people, had Oedipal issues.

Pool insisted that Slote had done fine work. ARPA proved less enthusiastic. "Not well received," read one evaluation's wry understatement.[69] The more formal assessment deemed the Slote report "methodologically deficient in so serious a way" as to be useless, because "selecting four deviant personalities from within a culture and studying them hardly permits constructive generalizations to be made as to what is normal in that culture."[70] Also, Greenfield had apparently agreed to pay Slote something close to his New York salary of $50,000 a year; working for Simulmatics, Slote took only a 20 percent cut in his professional salary, which meant he made more than almost everyone at ARPA.[71] (Greenfield had likely also been a patient of Slote's. He saw a series of analysts, including a particularly nutty therapist his children called Dr. Coconuts.)[72] And, as Garry Quinn, Simulmatics' program manager at ARPA, later wrote Pool, if he'd known the salary Simulmatics had promised Slote, he'd never have agreed to send him to Saigon.[73] "Even I wouldn't charge that much to work for you," Quinn told Pool.[74]

Slote finished his study just before Maureen Shea, the new typist, arrived, so he carried his research materials back to the States, to be typed up at Simulmatics' New York offices. The company had moved from a suite of upstairs rooms at 501 Madison Avenue to a warren-like brownstone at 16 East Forty-First Street, a stone's throw from the New York Public Library. Ann Penner, twenty-four, had been hired there as a typist. (In offices, clever, well-educated, and observant young women like Penner and Shea, unable to pursue other careers, nearly always ended up taking jobs as typists.) Like the rest of the staff at the New York office, Penner was opposed to the war.

Penner took a look at the cards Slote used

to administer the Thematic Apperception Test on his four subjects. They were special, Vietnamese versions of the test cards, with the European features on the faces painted over with Asian features.[75] Penner didn't have the security clearance required to see Slote's reports, much less type them. Nor did anyone in that office. Meanwhile, at Simulmatics in Saigon, someone from the office leaked Slote's report to the press — specifically, to the *Washington Post*'s Ward Just.[76] The one person who could not have leaked the reports was Maureen Shea, who'd only just arrived. Also, Just said, Maureen Shea was a "tomb of secrets."[77] Walter Slote, apparently, was not.

August 1966, the season of Maureen Shea's arrival in Saigon, was a hinge time, the end of one kind of war, the beginning of another. Only in the middle of 1966 did the Vietnam War — Americans' Vietnam War — become what Ward Just called "a serious war."[78]

Before 1966, notwithstanding the teach-ins on scattered college campuses, Americans had expressed little sense of urgency about U.S. involvement in Vietnam. No high-level American official had given up his career to join the military, as Just pointed out, and the war seemed well more than arm's length away: "Saigon was itself a microcosm of this, with cocktail parties and miniskirts on the one hand and the bombing of billets on the other."[79] When 1966 began, there were 184,300 U.S. troops in Vietnam and 2,344 had died. Before 1966, 10,000 very young men had been called up by the draft every month. Most of the men who got sent were poor; middle-class kids and college students received educational deferments or got out some other way. All that was about to change. The number of young men called up to the draft, every month, was raised from 10,000 to 30,000. An educational deferment could not spare you after the age of twenty-four. That meant that, for the first time, middle-class white men — boys — were sent to Vietnam. By the end of 1967, there would be 485,600 U.S. troops in Vietnam, and 20,057 would have died.[80] Also, American reporting on the war was getting tougher.

Legendary war correspondent Martha Gellhorn reached Vietnam about the same time as Maureen Shea. The hard-hitting Gellhorn, fifty-seven, had published her first article in the *New Republic* in 1929, when she was twenty. In Europe, she'd reported on the rise of Adolf Hitler; she'd been among the first American reporters to write about Nazi concentration camps. In the 1940s, she'd been married to Ernest Hemingway. In 1966, she reported on Vietnam for the *Ladies' Home Journal.* But there was nothing ladylike about it.

370

"The Red Cross Amputee Center in Saigon is a corrugated tin shed, crowded to capacity and as comfortable in that heavy, airless heat as an oven," Gellhorn reported, writing with an uncanny intimacy. There was no nobility in war, for Gellhorn, and no purpose to it, either. It was hideous and brutal and needless and unforgivable. At the Red Cross Amputee Center in Saigon, most of the maimed civilians were waiting for prosthetics because the Red Cross had run out of arms and legs, she explained. Gellhorn asked the men, women, and children how they'd been hurt. She made a list. "Six had been wounded by Viet Cong mines. One had been caught in machine-gun cross fire between Viet Cong and American soldiers, while working in the fields. One, a sad reminder of the endless misery and futility of war, had lost a leg from Japanese bombing in World War II. One, the most completely ruined of them all, with both legs cut off just below the hip, an arm gone, and two fingers lopped from the remaining hand, had been struck down by a hit-and-run U.S. military car."[81] This war, Gellhorn wanted the readers of *Ladies' Home Journal* to know, was awful, pointless, criminal, and unredeemable.

Two weeks after Maureen Shea reached Saigon, Simulmatics equipped her with a gun for her first foray into the countryside with the field team. "I didn't have the heart to tell them that the only thing I'd ever shoot would be my own hand," she wrote home. She boarded an eight-seat plane to fly to the southern coastal city of Phan Thiet, where the 101st Airborne was stationed, while the Simulmatics staff conducted interviews with Vietcong defectors nearby. On Shea's second night in Phan Thiet, an eighteen-year-old soldier from the 101st got drunk and, driving to town, swerved off the road and ran over eight Vietnamese people, killing three children and an old woman and injuring three more people, who, Shea said, weren't expected to live. If they had survived, they might have ended up at the Red Cross Amputee Center in Saigon — assuming they were lucky, assuming they were very lucky. Shea had, for the first time, glimpsed the war.[82] Minds change in glimpses.

That fall, *New York Times* correspondent Neil Sheehan confessed that his own view of the war had changed, after long and painful observation. He'd been in Vietnam from 1962 to 1964, when there were 17,000 American servicemen in Vietnam, and he'd returned for one more year in 1965, when that number had risen to more than 300,000. In October 1966, when he looked "at the bombed-out peasant hamlets, the orphans begging and stealing on the streets of Saigon and the women and children with napalm burns lying on the hospital cots," he wondered "whether the United States or any nation has the right to inflict this suffering and degradation on another people for its own ends."[83]

Maureen Shea wondered, too. She'd grown close to Cuc Thu Duong. "Miss Cuc is great because aside from being a nice, fun person, she is a great yakker and we get all information on the Vietnamese way of life from her," Shea wrote home. Cuc taught Shea how to sew an *ao dai,* the traditional Vietnamese woman's dress; and how to cook Vietnamese food; and what the order of the procession means at a Buddhist funeral.[84] Cuc kept her views about the war to herself.

Shea, unlike Cuc, had been told the nature of at least some of Simulmatics' research. "Our purpose is to give recommendations for improving the Chieu Hoi program and for increasing the number of defectors," she wrote her family. "Chieu Hoi means open arms."[85] The Chieu Hoi program was a psychological warfare operation whose purpose was to convince Vietcong soldiers to defect and to join the South Vietnamese by promising them that they would be welcomed with open arms. It involved old-fashioned propaganda methods, like the air-dropping of leaflets. It also involved campaigns of force and coercion, like bombing, defoliation, and crop spraying, designed to reduce enemy soldiers to a state of starvation and desperation. These same campaigns reduced civilians to a state of starvation and desperation, which only incited and strengthened the North Vietnamese people's hatred for Americans. Simulmatics interviewed Vietcong defectors living at Chieu Hoi centers as part of a series of projects whose ultimate aim was the construction of a model of the Vietnamese mind.[86]

To that end, Simulmatics provided advice to the Department of Defense on the Strategic Hamlet Program.[87] Simulmatics also urged the development of "a Hamlet Data Bank," a project that the Department of Defense adopted not long afterward, as the Hamlet Evaluation System, a data bank of everything known about every hamlet, updated constantly, updated instantly, a dynamic model.[88]

The Strategic Hamlet Program served as only the latest in a series of decades-old "pacification" schemes under which military forces urged and eventually forced South Vietnamese people off their land and into fortified hamlets, to be defended by the joint forces of the United States and the Army of the Republic of Vietnam, or ARVN (Ar-vin). Frances FitzGerald described Strategic Hamlets as having been built on top of the ruins of older hamlets that had been built on top of the ruins of still older hamlets, their mud-walled forts enclosed by fences built by the fathers and grandfathers of the farmers whose families were crowded inside.[89]

Simulmatics, meanwhile, had grown so much that it sought more spacious quarters. After Pool left Saigon, he went to Washington to court the favor of the National Security Council, which included Robert Komer, a former Harvard Business School student and longtime CIA analyst, and Walt Rostow, Pool's longtime friend and MIT colleague, who had replaced McGeorge Bundy as national security adviser. "Dr. Pool went down to Washington to see Komer and Walt Rostow and all at the White House were VERY impressed with the progress report," Shea wrote home in October. With the work seemingly going well, and likely to secure more funding, Simulmatics engaged Miss Cuc and her mother to look for a villa, a lavish one, to become the new Saigon headquarters.[90] They found a place not far from the American-style Caravelle Hotel. Everyone moved in. Shea took to calling it the La Residence Grise.[91]

La Residence Grise became something of a scandal. Simulmatics' Saigon office was headed by Charles Ramond, a Louisiana-born Madison Avenue advertising man and friend of Ed Greenfield's who'd provided advice to Simulmatics, early on, when it was selling its services to advertising agencies. Ramond had brought along his wife, Mary, who, like him, had not the least qualification to conduct behavioral science research in wartime for the Department of Defense. The Ramonds were extravagant, and so was the villa, whose acquisition had not exactly been approved by ARPA. "If you could see the hate letters flying across the ocean between Dr. Ramond and Simulmatics you would be set up with one of the finest humor books ever to come out of a typewriter," Shea wrote home.[92] Also, Ramond wasn't supposed to have his wife there. Simulmatics, operating under a military contract, was expected to abide by military budgets and military rules, which included a prohibition on family members. That didn't stop Ramond, or Philip Worchel or yet another Simulmatics scientist, Lee Wiggins, from bringing their wives. (Ramond and Wiggins also put their wives on the Simulmatics payroll.)[93]

On November 1, 1966, the South Vietnamese celebrated National Day with a parade marking the third anniversary of the overthrow of Ngo Dinh Diem, who'd been assassinated only three weeks before JFK's murder. The staff of Simulmatics decided to celebrate National Day as if it were the Fourth of July, by sunbathing and grilling hot dogs. The North Vietnamese celebrated by bombing Saigon. Worchel's research staff, a new band of young men, had just arrived for a second round of the Ruff Puff study. They "were somewhat aghast to come upon a place where they were washing the blood off the sidewalks," Shea wrote home.[94] Inside the villa, the Ramonds were serving cocktails. Outside, people were dying in the streets, blasted, dismembered, splattered.

Ed Greenfield landed at Tan Son Nhut in November 1966. He was forty-nine, heavier, his hair longer, his lapels wider and his ties, too. The Simulmatics staff greeted him at the airport with a sign that read: "E. GREENFIELD IS A CAPITALIST EXPLOITER OF THE INTELLECTUAL PROLETARIAT." Maybe they wanted better pay; maybe this was a joke. No one any longer remembers.

Shea called Greenfield "our great white leader." The flimflam man, the busy, busy American businessman, the swagger. He appeared to want nothing so much as to get away as quickly as possible.

"Have I had any mail or cables?" he barked, talking fast and even faster. "How long does it take to get to Bangkok? How long does it take to get to Hong Kong? How far is it to Cambodia? Can you get to Singapore?"[95]

Most of Greenfield's friends opposed the war. His wife opposed the war. Greenfield didn't like the war, either. This wasn't what he'd founded his company to do. He'd wanted to help liberal Democrats win elections and Ralston Purina sell dog food. He'd wanted to convince smokers to switch cigarette brands. Counterinsurgency? This wasn't what he wanted to be doing. Capitalist exploiter of the intellectual proletariat? He wanted to get out of there.

Also, he'd begun drinking more, too much, while his lawyers negotiated his divorce. He wanted partial custody of their four children, and he wanted Patty to agree to sell their house on Twenty-Second Street (the house on the beach at Wading River was hers) but she didn't want to sell, so he'd tormented her by flaunting his life at her. "Party at Ed Greenfield's house at 22nd Street near 8th Avenue," Al de Grazia wrote in his diary in 1965. "Naomi Spatz, Ed's mistress, was there (the Madame is at the Beach)."[96] It was cruel.

In Saigon, Greenfield drank so much that Maureen Shea had to hide the liquor bottles. He blurted out nonsense. He staggered. He slurred his speech. He had plans, he always had plans, this time for a Southeast Asian Simulmatics subsidiary, based in Thailand. He talked and he talked and he talked. He would solve everything! Shea wrote home that he held forth on "the past and future of Simulmatics, and the fact that now that he had come these 10,000 miles he understood all the problems — ho, ho, ho!!!"[97] She was not amused.

Pool flew to Saigon that November, too, for a Simulmatics summit. He didn't stay at the villa; he stayed with Daniel Ellsberg; they'd been friends then. "He was a very charming guy," Ellsberg later said about Pool. Still, "I thought of him as the most corrupt social scientist I had ever met, without question."[98] Greenfield and Pool fought at the summit, fiercely. Shea told her parents only that it was "quite testy," sparing them all the "gory details."[99]

When Greenfield got back to New York, he kept on trying to bully his wife out of the house on Twenty-Second Street.[100] "Ed came back from Saigon more pliant on the children and more aggressive on selling the house," Patty wrote to her brother that fall.[101] She found it difficult to fight him. He was exhausting, had always been exhausting.

From Pool and Greenfield's meeting in Saigon, it became clear that they no longer saw eye to eye. The Simulmatics operation in Saigon was Pool's baby — his laboratory — not Greenfield's. But for a great many of its screwups — for the Ramonds and the extravagance, for the surplus wives, for Walter Slote — Pool must have blamed Greenfield. Whatever dream Greenfield had about what Simulmatics could be, how it could help create a better America, it had become as mired in Vietnam as the rest of American liberalism.

Maureen Shea's tour of duty came to an end not long after it began. She spent Thanksgiving Day 1966 at Daniel Ellsberg's, with Pool and a half dozen other guests, celebrat-

379

ing one more American holiday very far from home.[102] Whatever the battle between Pool and Greenfield had been, Pool had won. "We are in the process of negotiating a new contract with ARPA which would set up the Simulmatics Field Unit in somewhat the same way that the Rand unit is now structured," Pool wrote the week after Thanksgiving, full of optimism.[103]

Shea was ready to leave. She began sending presents home for Christmas. She hosted a party at the villa for all the Vietnamese people who worked for Simulmatics, the translators, the cooks, the maids, and the drivers, a dinner with American food, having "planned the party for a time when the Ramonds would be away because though we had repeatedly asked them to give such a dinner, they had not done so." Shea made spaghetti and served cherry pie with vanilla ice cream.[104]

On January 14, 1967, five months after she had first arrived in Saigon, Shea began her voyage home, to an angrier America.[105] How long will this war last? When will the United States win? Simulmatics had not been able to answer those questions.

There's a story told about what happened, that year, at the Pentagon.[106] "There is no computer that can tell the hour and the day of peace," Johnson had said in 1966.[107] But apparently there was, an enormous computer housed in an enormous room in the basement of the Pentagon — the story appears to have conflated computers in Saigon with computers in Washington — a computer into which McNamara's men fed punch cards containing every possible fact known about Vietnam. The number of troops, ships, planes, helicopters. The size of the population. A census of water buffaloes, the price of rice, the flammability of thatch. The body count. The kill ratio. The density of the peasant mind. The weight of a human heart. On a Friday in 1967, McNamara's men fed the last punch cards into the maw of this giant computer. They asked it a question: "When will we win in Vietnam?"

The machine hummed. The machine whirred. Its lights blinked. It hummed and thrummed and blinked and whirred, all weekend. On Monday, McNamara's men came back. The output tray contained a single punch card. It read, "You won in 1965."

CHAPTER 11
THE THINGS THEY CARRIED

> Liberal academics had no root of a real war with technology land itself, no, in all likelihood, they were the natural managers of that future air-conditioned vault where the last of human life would still exist.
> — Norman Mailer, *The Armies of the Night,* 1968

It was raining in New York on the morning of April 15, 1967, a Saturday in springtime, when the first antiwar protesters began gathering at the Sheep Meadow in Central Park. They were students, they were children, they were businessmen and housewives and old people and new people, even newborns, and strangers and friends and veterans and draftees, and amputees. Struggling to light matches in the damp, they burned draft cards, fires hissing in the mist. They came by subway, they spilled out of buses, they arrived on foot. They carried Purple Hearts. They rolled on wheelchairs, they hobbled on crutches, missing limbs. They carried cam-

Thomas Morgan and his daughter at an antiwar march in New York, 1967.

eras. Some carried newspapers, tucked under arms. They carried flags and banners and signs; they carried purses and snacks and transistor radios and umbrellas and ponchos and flowers, tied in bundles. In their pockets, they carried change. They carried children on their shoulders and babies in their arms and, in their wallets, photographs of the dead, soldiers they'd loved, and lost, sons and fathers and brothers and husbands. They pushed strollers. They did not carry computers.

In 1967, Simulmatics grossed $1 million, about $8 million in 2020 money. Seven out of every ten of those dollars came from the Department of Defense for work in Vietnam.[1]

In 1967, the number of combined American, South Vietnamese, Korean, Thai, Australian, and Filipino forces in South Vietnam climbed to 1,300,000, one soldier for every fifteen civilians.[2] The Simulmatics scientists, housed in a villa in Saigon, amounted to a very small band, a handful of men, the littlest platoon. They carried tape recorders. They carried questionnaires. Some carried guns. They held assumptions. They had dodged the draft. They carried pens and notepads marked SIMULMATICS CORP. They wanted to help. They were not sure they were helping. They carried doubts.

In 1967, the American antiwar movement became a national campaign of untiring, anguished dissent. The decade had begun with college students staging a sit-in a Greensboro lunch counter, a demonstration that had stirred the conscience of the nation, followed by Freedom Rides, freedom marches, a Freedom Summer, the Free Speech Movement, the March on Washington, Fannie Lou Hamer at the Democratic National Convention. Then, in 1965, the midpoint of the decade, the year of the march to Selma, the teach-ins had begun, scattered discussions of the Vietnam War on college campuses, in lecture halls, in seminar rooms, on football fields. But in 1967, Martin Luther King Jr. broke his silence about the war and talked about what tied the civil rights movement and the peace movement together. "We have been repeatedly faced with the cruel irony of watching Negro and white boys on TV screens as they kill and die together for a nation that has been unable to seat them together in the same schools," he said.[3] Vietnam had made his dream a nightmare.

On April 15, 1967, while the rain fell, and grew heavier, the hundreds of thousands of people who had gathered in a meadow in Central Park left the park for the streets and began to march down Fifth Avenue and across to the United Nations. A sea of protesters flooded twenty blocks; the first marchers reached the U.N. while the last were still in the park. Thomas Morgan, Simulmatics' onetime PR man, marched that day with his daughter, Kate. He'd long since left the company, but his friendship with Ed Greenfield had fallen apart over Simulmatics' work in Vietnam, for Johnson's war, for McNamara's war, for an immoral war.

Veterans carried a sign that day that read, VIETNAM VETERANS AGAINST THE WAR; after April 15, they'd form their own antiwar organization. And Martin Luther King Jr. marched that day, in a long wool coat, buttoned up against the wind. Outside the U.N., he climbed up onto a small stage and began to speak, his words rolling out like a sermon. The cause of civil rights and the cause of peace, he said, have "the same moral roots":

the struggle for the equality of races is a struggle, too, for the equality of nations. STOP THE BOMBING, he cried, and the crowd cried back, STOP THE BOMBING. "The thunder of our voices will be the only sound stronger than the blast of bombs," he said.[4] They shouted louder, pelted by rain.

Civil rights leaders disavowed King's involvement in the antiwar movement, arguing that it diluted and diminished the urgency of their cause. King did not relent. "It is my belief that justice is indivisible," he said. "My concern is for all mankind."[5] Hundreds of thousands of people marched in San Francisco on April 15, too. And after that day, antiwar organizations all over the country formed a coalition under the banner of the National Mobilization Committee to End the War in Vietnam, led by fifty-one-year-old David Dellinger, a pacifist who'd been a conscientious objector in the Second World War. They would march every fall and every spring, until the war finally came to an end. One day soon, some of them would direct their protest at a new target: the Simulmatics Corporation.

Except for Ithiel de Sola Pool, the behavioral scientists who'd founded Simulmatics in 1959 refused to go to Vietnam, even though Pool kept asking. "With my Simulmatics hat on, I have been looking hard for people, but people who are willing to go to Vietnam," he wrote Jim Coleman in 1967. "Right now I need, above all, a man, probably a psychologist, who is good on experimental design and tests and measurements, and who is willing to spend most of his time in Vietnam over a period of ten months or a year."[6] Coleman was not interested. Nor, it seems, was he willing to give Pool names.

"I am still trying to staff the Saigon TV study," Pool wrote to Bill McPhee, in Colorado. "Are you sure you don't want to participate in it?"[7] McPhee was sure.

Pool kept promising ARPA that Simulmatics would send the best and the brightest, and then it kept sending something rather less. Searching for someone "to direct the establishment of a psychological warfare center in Vietnam," he asked the political scientist Al de Grazia if he'd be willing to go to Saigon. Pool and de Grazia had been friends since they were undergraduates and then graduate students together at the University of Chicago. They'd been at Stanford together, too. But after de Grazia was denied tenure at Stanford, he'd become something of an academic exile. He was still teaching, at NYU and Princeton, but he was best known as the most prominent American defender of the psychiatrist Immanuel Velikovsky, who believed that ancient Egypt and other ancient civilizations had been influenced by collisions

387

between the earth and other planets.

Pool, by that point, had grown desperate. A lot of people at ARPA had had it with Simulmatics, for all kinds of administrative mismanagement, including leaks to the press, a near "mutiny" among those employed in Saigon, a failure to submit required reports, and a general lack of appreciation for chain of command. Then there were the security lapses, which, notwithstanding ARPA's wrist slapping, continued, unabated.[8] Wrote one official at ARPA in December 1966, understating the case, "I am somewhat distressed to find that we are seriously considering additional funding for Simulmatics at this time."[9]

Greenfield approached de Grazia in January 1967, asking him, as de Grazia phrased it, "to set up and direct a paramilitary action research center" in Saigon. Greenfield offered a whole list of perks. De Grazia would be paid handsomely, $30,000 the first year, with a bump-up if he stayed on, plus Simulmatics stock options and bonuses.[10] And still de Grazia hesitated, mainly because every time he met with Greenfield to talk terms, Greenfield got drunk.

"Ed drank constantly for eight hours," de Grazia wrote in his diary about the night before the two men were to meet with Garry Quinn, Simulmatics' project manager from ARPA. For all his delusions about Velikovsky, de Grazia could see through Greenfield: "He loves himself in a gargantuan way and forces his environment into his personality. I enjoy him, even like him. He had better watch out, though. His old friends and associates in Simulmatics Corporation are now wary and sour. A couple of errors more and he may find himself with a company but without key clients and associates. This at a time when the Company could begin to run a good profit on a good volume of work."[11]

After hesitating, de Grazia at last decided to accept. "I move toward the day when I shall be winging away to Vietnam," he wrote in his diary.[12] Some of his seven children all but disowned him over his decision. De Grazia was undaunted.[13] A lot of American families fell apart over Vietnam. "Your sons and your daughters are beyond your command," Bob Dylan sang.[14] Robert McNamara's son, Craig, away at prep school, hung a Vietcong flag on the wall of his room and, when he got to college, joined the antiwar movement.[15] Pool's son Jeremy, in the Harvard class of 1967, and passionately opposed to the war, never really forgave his father for Simulmatics' involvement in Vietnam. They got through it by never talking about it. When the subject of the war came up, Pool would tell his son, "You don't know what I know. I'm privy to all kinds of information that you can't possibly know. So there's no point in arguing because your opinion is uninformed; we're not coming from positions of equal knowledge."[16] Other Simulmatics fathers and sons fought those battles differently. Ed Greenfield's son, Michael, was fifteen in 1967; they yelled.[17]

None of the Simulmatics scientists had sons fighting in Vietnam. Pool's oldest son, Jonathan, volunteered for the Peace Corps in Turkey. He wrote home to tell his father about the ruins he'd been visiting and that he was considering applying to graduate school.[18] Other people's sons, many other people's sons, wrote different letters home: they carried news of terror. "They carried all they could bear, and then some, including a silent awe for the terrible power of the things they carried," Tim O'Brien would write about American soldiers in Vietnam, in *The Things They Carried.*[19]

"My squad got into a helluva fray, and lost (momentarily) one member. ME!" Second Lieutenant Marion Lee Kempner, a marine platoon leader, wrote home in September 1966. "I am all right. I am all right. I am all right. . . . Please, now, I am all right. . . . P.S. I am all right!!" Kempner recovered from his wounds and went back out to fight. Two months later he was killed in a mine explosion near Tien Phu.[20] His family carried his remains.

■ ■ ■ ■

In April 1967, as hundreds of thousands of New Yorkers went to the streets to protest the war and Al de Grazia packed his bags for Saigon, a series of reports from Vietnam began to appear in the *New York Review of Books.* They were written by Mary McCarthy.

McCarthy, fifty-four and four times married, was a prominent essayist well known for her best-selling 1963 novel, *The Group,* the story of the plight of eight Vassar classmates, often read as a fictional companion to Betty Friedan's *Feminine Mystique,* which appeared the same year and came out of Friedan's research on her Smith classmates.

Reporting from Vietnam, McCarthy blamed the war on behavioral scientists. She didn't name Simulmatics, but she described its work precisely. Nor did she mention Kim Le, a twenty-four-year-old Vietnamese woman she'd met in Saigon. McCarthy walked with difficulty, using a cane. Kim Le escorted her around Vietnam. Kim Le had worked for Simulmatics.

Born outside Hanoi in 1942, Le was the daughter of a provincial judge. In 1946, her family had moved to the jungle to fight alongside the forces of Ho Chi Minh, against the French. But at the partition in 1954, opposing communism, they fled to the south.

Le, selected for an American education program, had gone to the University of Tennessee to study social work, returning to Vietnam after she graduated in 1966. She wanted to help save South Vietnam from the ravages of communism. She worked for Simulmatics for six months. She was not impressed. "You can't conduct psychological warfare when you don't know the culture," she later said. "It was so naïve." She had meanwhile become a crucial guide for American reporters, including Ward Just and Frances FitzGerald. McCarthy particularly liked her because Le told her she'd read and loved *The Group.* They spent a great deal of time together.[21]

In the *New York Review of Books,* McCarthy argued that while World War II had carried physical scientists, with their atom bomb, to prominence, the Vietnam War had elevated the less sound and still less scrupulous science of the human mind. Behavioral scientists were more dangerous than physicists, she thought, because "conceivably you can outlaw the Bomb, but what about the Brain?" In the 1950s, she explained, describing projects like Pool's ComCom, "the behavior of the enemy was studied under university microscopes, with the aid of samples furnished by defectors to the Free World. Practical experiment, however, was not really feasible until the war in Vietnam provided a laboratory for testing the new weapon, an academic B-52." With this observation, she made use of an explosive metaphor that Simulmatics itself employed, in calling the People Machine "the A-bomb of the social sciences."[22]

Al de Grazia arrived in Saigon on April 19, 1967, as Simulmatics' new director of Southeast Operations. He'd agreed to "most of the next year and a half in the field."[23] In truth, he spent mere weeks there. He tried to hire new men. Simulmatics had sent him a list of names. He crossed them out, one by one. "Get cholera shot," he wrote on a to do list. "Call Bill McPhee." McPhee was still not interested.[24] Greenfield would call McPhee sometimes, and they would scream at each other.

In Saigon in April, de Grazia began by making arrangements for Simulmatics to move to a new and still more lavish villa, a former hotel with twenty-nine rooms.[25] He worked on proposals for new contracts. (ARPA contracts from the 1966 funding cycle ended on June 30, 1967, and new contracts were to begin July 1, 1967.) And then, at a time when a rising number of Americans had grown outraged by the application of behavioral science to a war many Americans did not support, he proposed building a new facility for Simulmatics, a state-of-the-art paramilitary and psyops center.

ARPA, on reviewing the new proposals, which ran to the tune of half a million dollars, 85 percent of which was overhead, concluded that they looked distressingly similar to Simulmatics' 1966 proposals, research for which ARPA had already paid but Simulmatics had never completed. "I would imagine most government procurement specialists take a very dim view of buying data twice," complained Colonel John Patterson, head of ARPA in Saigon. Patterson described ARPA as being placed in the unenviable position of "trying to control, direct and contain the mass of protoplasm called Simulmatics." Nothing about the organization ever seemed to work, including its staffing: "To all appearances Simulmatics is now composed of nothing but Sandy Koufaxes, all pitchers."[26]

On April 28, 1967, while de Grazia was in Saigon, the *Washington Post*'s Ward Just stopped by Patterson's office and left a message: "I'd like to talk with you tomorrow afternoon — story about ARPA & Simulmatics."[27] De Grazia left Saigon on April 30.[28] That day, the *Washington Post* ran a splashy story about ARPA by Just, full of leaked information about Simulmatics, including the disclosure of a classified, Simulmatics-recommended plan to drop radios in the Vietnamese countryside "as part of an expanded psychological warfare effort." He derided the "swarms of analysts" the Pentagon had sent to Saigon, who, he implied, had discovered virtually nothing.[29]

Patterson was furious. He blamed Simulmatics' "avarice for publicity" and said he was pretty sure about the source of the leak: "personable, affable, engaging Dr. Al de Grazia, in his St. Bernard dog like congeniality."[30] This did not improve Simulmatics' standing at ARPA. Pool, defensive, insisted that Saigon was "a fish bowl" and that Simulmatics could not possibly be held responsible for Ward Just's articles.[31] Patterson was unmoved.

By 1967, much of the press had turned from reporting what U.S. government officials said about the war to reporting that U.S. government officials appeared to be lying about the war.[32] Privately, reporters expressed still greater doubts about the American mission. "All of it is madness," Just wrote to Frances FitzGerald. "We have gone from not knowing anything about the workings of this country, to plunging into every scrap of detail the computers can assemble."[33]

Just had learned of a MACV program, the Hamlet Evaluation System, a system first recommended by Simulmatics the year before. McNamara wanted to know just how successful the Pentagon's pacification program was: What was the state of the hearts and minds of the peasants confined to "strategic hamlets" (which, by 1967, were referred to by American antiwar protesters as "concentration camps")? In October 1966, following Simulmatics' initial field reports, McNamara had charged the CIA with coming up with a plan to measure the situation in some 12,500 strategic hamlets in forty-four provinces. The CIA devised a "matrix" of eighteen indicators under six factors to be written onto a Hamlet Evaluation Worksheet and then entered into punch cards. The program was the brainchild of Robert Komer, who'd been put in charge of the pacification program. "I guess I'm really a manager," Komer said. "I'm trying to manage a great big subsidiary corporation."[34]

Launched in January 1967, the Hamlet Evaluation System involved some 250 advisers and Vietnamese researchers who gathered data in their districts. Then some 200 analysts at MACV headquarters applied an algorithm to these figures to come up with a score for each hamlet.[35] ARPA contracted with Simulmatics to evaluate the program.

American officials in Vietnam were losing their minds, Just wrote FitzGerald in April 1967, describing four who'd gone mad in just the last month. One military official, asked to report on the latest developments, said, "I was out on the road the other day . . . I was out on the road the other day . . . and . . . I was out on the road . . . and . . . I was out on the road . . . ." Wrote Just, "He went on like that for a few minutes, and then man in charge gently cut him short and he sat down quietly and went back to thinking whatever he had been thinking about."[36]

Madness everywhere. Back in New York on May 1, 1967, de Grazia debriefed with Greenfield. "It turned out poorly," de Grazia wrote in his diary, "for he was somewhat drunk and surly and I lost patience with him, letting myself feel for the first time an open dislike of his boorishness and ways of doing business." The next night, Pool came down from Cambridge to meet with both men. Greenfield, de Grazia reported, "raised a quite unnecessary row, attempting all the while to enlist me against Pool and resenting my attempts at pacification." The meeting with Pool turned out poorly, too, Pool and Greenfield fighting over whether New York or Cambridge ought to run Simulmatics' affairs, as they related to Vietnam. De Grazia concluded, "I doubt that Ed can be trusted in the clutch."[37]

The problem with running Simulmatics' Vietnam operations out of Cambridge had to do with the students. MIT, long a fairly quiet campus, had lately become a site of dissent. An MIT chapter of Students for a Democratic Society had begun organizing. "You will find quite a heated situation when you get back here," Pool warned Sam Popkin, who was with his girlfriend in China, where she was doing her doctoral research. "The country is getting more and more polarized about the war in Viet Nam. The SDS and other groups have been attacking this department."[38]

Pool had rented space for Simulmatics off-campus — fairly far from the campus — in a little two-story house at 930 Massachusetts Avenue, halfway between Central and Harvard Squares. Hoping to avert further leaks, security lapses, and other mishaps, Pool asked Seymour Deitchman and Garry Quinn to come up to Cambridge from Washington to meet with everyone on Pool's team who intended to travel to Vietnam that summer, to make sure they understood all of the protocols.[39] Meanwhile, in early May, Pool sought support at the Pentagon.[40]

At the Pentagon, though, McNamara's private doubts about the war had only grown. On May 19, he wrote a memo to Johnson: "There may be a limit beyond which many Americans and much of the world will not permit the United States to go. The picture of the world's greatest superpower killing or seriously injuring 1,000 non-combatants a week, while trying to pound a tiny, backward nation into submission on an issue whose merits are hotly disputed, is not a pretty one. It could conceivably produce a costly distortion in the American national consciousness and in the world image of the United

States."[41] In June, McNamara would commission a secret report, a documentary history and analysis of the U.S. involvement in Vietnam going back to the administrations of Truman and Eisenhower, parts of which would be written by Daniel Ellsberg. It would come to be known as the Pentagon Papers.

Flush with new contracts and new recruits, Simulmatics' Saigon office began the summer of 1967 in full force. "The Simulmatics Field Team has grown to alarming proportions," Pool reported.[42] "Our 'villa' here is now full," he wrote home in July. "It's a funny feeling looking over this large 29 room building with some 50 people of all sorts working in it and realizing that I'm somehow in charge. . . . We have 3 cars & drivers, 5 maids & cooks, translators, interviewers."[43]

Nearly the whole field team was new. This kind of turnover was not uncommon. "There was a timeless quality to the American effort," FitzGerald wrote. "Each year the new young men, so full of vague notions of 'development,' so certain of their own capacity to solve 'problems,' so anxious to 'communicate' with the Vietnamese, eagerly took their places in this old, old war." These were the young men of Simulmatics. "Only the faces of the young men and the numbers of the hamlets changed year after year," FitzGerald wrote.[44]

The young men carried theories of human behavior. They carried solutions to human problems. The older men carried briefcases. Colonel William Arnold, who replaced Patterson as head of the military's research efforts in Vietnam, called the Simulmatics' scientists "Academic Holiday Briefcase Directors." They'd turn up in Saigon during summer break, or winter break, and maybe even for spring break. They'd spend a few days, then go home, leaving their graduate students to do the work. And "a pool of research assistants without scientists is similar to a pool of typists without writers," Arnold pointed out.[45]

They struck reporters as terribly callow.[46] "The old-timers would scoff at their naïveté," FitzGerald wrote.[47] They carried illusions. "Ithiel appeared to me like one of those Americans who came thinking that they could get a grasp on Vietnam the way no one else could," FitzGerald later said.[48]

"Just back from Saigon," Pool would write at the end of his summer in Vietnam in 1967. "Amazing progress in 2 1/2 months."[49]

De Grazia had come back, too, full of loud righteousness and mad schemes.[50] When ARPA notified Simulmatics of yet another security violation, de Grazia told Deitchman to "watch the higher priorities of the struggle and avoid the chicken shit."[51] His own higher priority had to do with building a new population center in South Vietnam through the capture and forced resettlement of VC (Vietcong) and NVM (North Vietnamese) forces. He wrote a memo to Arnold proposing a new and staggering initiative he called Project Renaissance: "Project 'Renaissance' calls for a multi-force, multi-agency operation to evict 3,000 VC-NVM forces from a productive area 'A'; to 'rescue' 20,000 persons in a presently VC-NVM area 'B,' depriving enemy forces of resource-bases; to resettle the 'refugees' in a prepared new urban-rural model community with agriculture and light industry; and to Chieu Hoi the abandoned VC to come over to their kinfolk in the model community."[52] Pool endorsed this proposal, noting, "Al de Grazia and I remain deeply intrigued by the possibilities of a preliminary look at where and how whole areas could be secured by the settling of 20,000 to 50,000 people in a strategic location."[53] De Grazia wrote a press release announcing this plan. "The Americans in Vietnam are about to execute a unique and sweeping maneuver in the near future, inviting the wholesale transfer of people out of the support bases of the Viet Cong," he imagined himself announcing. "It will be the biggest air carry feat since the Berlin Air Lift."[54]

Pool spent much of his time training Vietnamese students to conduct interviews: "We pretend to be Vietnamese peasants and they ask us questions. We try to give uncooperative answers to see what they will do."[55] In the work of interviewing, Simulmatics was engaged in a rivalry with RAND, which in 1967 produced *Two Analytical Aids for Use with the Rand Interviews*. (The interview program had been criticized from within. RAND analyst Anthony Russo, who would go on, with Daniel Ellsberg, to leak the Pentagon Papers, said that "material on torture of prisoners or brutal treatment of civilians" was removed when the interviews were transcribed from tape to type.)[56] Simulmatics had been less systematic than RAND in its work in 1966, work that had been criticized for its shortcomings, and Pool was attempting to introduce more systemizing to Simulmatics' data gathering and data analysis.

ARPA had complained, yet again, not only that Simulmatics' lead scientists were not the top scientists they had been promised, but also that these second-rate scientists did not actually conduct any research. Pool set about to right this situation. In the summer of 1967, he went to the Vietnamese countryside, to lead his field team. From the coastal city of My Tho, on the Mekong Delta, he traveled to the rural district Cho Gao, in a convoy of two army trucks. The first truck carried Pool and his six-person team. The second truck carried armed soldiers. The trucks, he wrote

402

home, "drove rapidly honking everyone else aside for the seven miles to Cho Gao." They drove through banana groves and rice fields, past lookout towers and barbed-wire fences. They took a ferry across a canal to the district headquarters, a fortified compound. He was quite close to combat, as close as he had ever been.[57]

Simulmatics conducted a number of studies in the summer of 1967 involving interviews with the same population: 829 villagers in 82 hamlets in 11 provinces. Pool himself did not conduct the interviews; they were conducted by American graduate students and Vietnamese interpreters. "I'll be sitting in this headquarters for 5 days while the team goes out to 4 hamlets around here to interview," Pool explained from Cho Gao.[58] "I'm writing this letter sitting in an American prefab trailer kitchen," he wrote home on another trip to the countryside that summer. "The trailer is the guest house at the headquarters of the 4th Division up here."[59]

The Vietnamese interpreters complained that the Americans, led by Pool, spent their time with officers and soldiers, "telling war stories," while the Vietnamese had to do all the work, with virtually no knowledge of what they were supposed to be doing, and very little faith in the work.[60] Later, de Grazia would summarily fire all of the increasingly discontented Vietnamese interviewers, claiming that they'd been neglecting their duties, though a later investigation did not support his charges. Aggrieved, they held a "Hate the U.S." banquet — and then they took their story to the press.[61]

ARPA found the conclusions Simulmatics drew from its field research dubious and its methods questionable. The problems began with the questionnaire itself. Written in English, it made little sense in Vietnamese, and the difficulty of conducting the survey was made worse by Simulmatics' decision not to explain the nature of the study to the Vietnamese interpreters. Also, with more than a hundred questions, it took more than two hours to administer. Sometimes, the respondents just walked away in the middle of the interview.[62]

Then there was the matter of research design. One set of questions involved radio-listening habits; it found that 80 percent of the villagers listened to the radio. Only 18 percent of the villagers' families owned a television set. One part of the interview involved presenting the villagers with a list of thirteen things, including both appliances and livestock, and asking them which of those things they owned and which they wanted. One conclusion: people who owned radios wanted more modern things than people who didn't own radios.[63] It is also true that in the United States there are more swimming pools in the suburbs than in the city, but whether anyone ought to fund a quantitative research project to come to that conclusion is harder to say.

Pool cared most about the TV study. The idea was to evaluate the effectiveness of television as a tool for counterinsurgency by studying "TV Hamlets," Vietnamese hamlets where televisions had been installed.[64] ARPA's Lieutenant J. Stephen Morris, who had a PhD in psychology, worked with Pool on the project and entirely lost faith in him as a scientist.[65] Pool did not appear to know what a null hypothesis is, Morris complained. "He didn't know why he couldn't conclude that TV was having no effect if he couldn't measure it."[66] Other investigations of Simulmatics' design and conduct of this research proved equally damning.[67] A military official in Saigon who met in the Simulmatics villa with the Vietnamese interviewers who worked on the study concluded that it "sounded as though someone had taken a book of rules about scientific methodology, then systematically violated each one." It was as if the study had been "performed so ineptly that it almost had to have been done purposely." The proposal said the purpose of the study was to "assess what changes in information, attitudes, and behavior result from placing TV sets in Vietnamese villages" but, as this assessment pointed out, this is not a researchable question: "In essence, it would involve studying all things about all Vietnamese."[68] So far from being a way to study all things about all Vietnamese, Simulmatics' research appeared to some people at ARPA as a way to study nothing at all.

Simulmatics had one other, pressing assignment: to evaluate McNamara's Hamlet Evaluation System, the massive bank of computerized data, first suggested by Simulmatics in 1966, that had been compiled, updated, and analyzed, day by day, on 12,500 strategic hamlets in forty-four provinces, and overseen by Robert Komer. The absurdity of this effort had already attracted the attention of the press, beginning with Ward Just's *Post* article in April 1967. In the *New Republic,* William Lederer, the polyglot former intelligence officer who'd written *The Ugly American* with Eugene Burdick, reported that he'd interviewed many of the personnel working on the program — it took more than 250 people simply to provide the data, and another 100 people to analyze it — and found that only one in a hundred officers who filled out the forms knew even a word of Vietnamese, and that the interpreters understood only a few hundred words of English. The interviews themselves, Lederer thought, were useless, since the subjects told the interviewers only what they wanted to hear. Lederer had watched film footage of Hamlet Evaluation System interviews with a South Vietnamese friend who told him that the Vietnamese interpreters changed the questions the Americans instructed them to ask; the peasants didn't give answers to the questions the Vietnamese interpreters asked; and the Vietnamese interpreters changed the peasants' answers when they translated them for the Americans, anyway. Lederer wrote, "It is through similar daisy chains of misinformation — only far worse — that Komer's computer is operating." McNamara had seduced the American public into an ill-gotten optimism about the war, playing a "numbers game." And he'd done the same thing, Lederer charged, with this "computer ruse."[69]

None of the concerns raised by Lederer found a place in Simulmatics' very well funded, intensively man-powered study of Komer's computer-run Hamlet Evaluation System. Pool, who drafted Simulmatics' report, concluded that the Hamlet Evaluation System was working well. ARPA was not pleased, since it had wanted suggestions for improvement.[70] Philip Worchel, at MACV, read the draft and, finding it "poorly designed, improperly analyzed, and superficially written," recommended that "under no circumstances should this study be distributed."[71]

By August 1967, ARPA had nearly had

enough. Simulmatics' reluctance to tell ARPA anything it didn't think the Department of Defense wanted to hear piled on top of further complaints about Simulmatics. ARPA considered a proposal to continue the TV study to be padded, its budget stuffed with fluff. It deemed another proposal to be "not the work of responsible researchers."[72] Al de Grazia had to go.[73] Then there was the matter of the scientists' salaries. Quinn wrote Pool, "Ithiel, may I say frankly that I regard your own salary as a key peg point?"[74] Someone, Pool complained to Komer, was out to "crucify Simulmatics."[75]

Simulmatics undertook one last, disastrous study in Vietnam. In the summer of 1967, Ithiel de Sola Pool and Al de Grazia headed out to an eighty-room monastery on ten lush acres just outside Saigon, to talk psyops.[76] Ten very old priests cloistered in the monastery raised pigs, rabbits, and guinea pigs and grew bananas, papayas, and rice. They were refugees from North Vietnam. Most of them had been there since 1954, when the Geneva Conference divided the country at the seventeenth parallel, and Catholics had fled to the south.[77]

Pool and de Grazia had come to see Nguyen Van Thu, forty-six, a visitor himself, a young priest with slicked-black hair and a grave stare. He worked for Simulmatics.

Nguyen Van Thu was born in a fishing village in Thanh Hoa Province, on the coast of the South China Sea, the son of a fisherman and one of ten children. He'd gone to seminary in North Vietnam during an era of terrible famine. He'd changed his name to Joseph Hoc and been ordained in Rome in 1949, where he'd gone as part of a delegation seeking ways to halt the spread of communism.[78] After his ordination at the Vatican, he'd traveled to the United States, where he studied at Worcester's College of the Holy Cross, Catholic University, and Stanford, earning a doctorate in sociology in 1953. Most people he met in the United States had never met someone from Vietnam before; when he said Mass, many of his American parishioners — finding him to be so small — mistook him for an altar boy.[79]

Hoc's relatives had been persecuted by the Viet Minh, his entire family imprisoned for possessing a papal blessing Hoc had sent them from Rome. His sister had been starved to death.[80] He expected to return to Vietnam as a missionary — in "search of human souls," as he put it — despite the growing violence against Catholics; Communists had burned his old seminary. But he remained steadfast in his faith; as he wrote to Dorothy Wayman, his American sponsor (a woman he referred to as "mother"), "If one day, you hear the Communists kill me, you can be proud of your son, mother, shedding his blood for the cause of Christ." He'd been devastated when he was stopped in Rome, on his way home, and told it was not safe to return. "I lost all hope of getting back," Hoc wrote Wayman.[81]

In December 1954, Communist forces began an assault on Hoc's home village. "Nothing is more dreadful than to sit here, doing nothing, and be aware of the suffering and death of my dear ones," he wrote.[82] But by 1965, when he began teaching at Boston College, Hoc was again hopeful that he might at last return to his native country.[83] He finally found passage back to Vietnam in 1967, as a contractor for Simulmatics. The company had hired him to conduct a study called "Testing New Psychological Warfare Weapons in Viet Nam." On his way from Boston, he'd stopped in Rome, where he'd met for twenty minutes, in a private audience, with the pope, for his blessing.[84]

Hoc argued, with cause, that earlier research in Vietnam had failed because of the way Americans conducted their interviews. "Certain standard American styles of public opinion interviewing create the impression among a Vietnamese peasant population of a police interrogation and result in questions that are hard to translate into ordinary colloquial Vietnamese," he explained. As a Vietnamese scholar, he promised to explore methods of warfare more suited to his people. He proposed to build a "living laboratory," a network far better than the men who submitted reports to Robert Komer's computer-run Hamlet Evaluation System, Vietnamese villagers who would test his new psychological warfare weapons.

Acting, essentially, as a spy, Hoc tapped into Vietnamese social networks in six hamlets, two controlled by the government, two by the Vietcong, and two contested. In each of these villages, Hoc found informants ("participant-observers," he called them), whom he hired to test his psychological warfare weapons. "A psychological warfare instrument such as a letter or a rumor or a prophecy was disseminated in these hamlets," he reported. "The participant-observers wrote reports on the conversations that they heard concerning the tested instrument."

Hoc began collecting the first of nineteen thousand field reports in October 1967.[85] He planned six projects. Most of them failed.[86]

Hoc proposed something called the Sorcerers Project, he explained, because "Vietnamese villagers believe in prophecy and the power of holy men to foretell the future." (Not unlike secretaries of defense who vest the same power in IBM machines.) For this project he intended to circulate false magic, the way another sort of psychological warrior might circulate, for instance, fake news. Alas, "the sorcerers did not say what they were supposed to say."

Another operation, Hoc's Chain Letter Project, involved giving villagers envelopes containing four copies of a letter and asking them to help get the letters to the Vietcong. Somehow, this never came off. Hoc's Folksinger Project involved composing new folk songs intended to "inject dislike and hatred of the VC." But North Vietnamese attacks immobilized all of Hoc's folksingers. Hoc's Prophesy Project was actually executed: five thousand copies of a booklet containing "a prophesy that the Viet Cong will be defeated in 1969" were distributed to the villages. Alas, villagers found this prophecy difficult to credit.[87]

ARPA, unhappy with Hoc's work, rejected his report, citing deficiencies in research design and in reporting. Its conclusion: "The project was lacking in methodological rigor."[88] Also: "The objectivity of certain statements unsupported by data is questioned." ARPA classified Hoc's report SECRET, filed it away, and ordered him to stop distributing it.[89] Pool sympathized, writing to Hoc, "It may be a long time before your work is appreciated."[90]

On October 21, 1967, American antiwar protesters marched not on the White House or the Capitol but on the Pentagon itself, taking the battle to the generals. Tens of thousands of people had come to Washington the night before. They hitchhiked. They hopped on trains. They came in caravans of buses, through the night. They slept in churches, they slept in hostels, they slept on strangers' couches, they slept on the hard stone steps of the Lincoln Memorial. They carried blankets. In the early morning, they huddled together, against the autumn cold. Helicopters patrolled overhead, as if they were patrolling at Da Nang or over the Mekong Delta. At the Pentagon, U.S. marshals and military police officers spent the night erecting barricades and fences, stacking sandbags, as if the nation were under siege. They wore helmets. They carried rifles. They carried bayonets. They carried tear gas.[91]

The march had been planned in May, by the National Mobilization Committee to End the War in Vietnam, the coalition that had organized as a result of the march in New York in April. David Dellinger had asked twenty-nine-year-old Jerry Rubin to plan a march in Washington. Rubin, who'd dropped out of the University of California, Berkeley, in 1964, had organized the teach-in at Berkeley in May 1965, the meeting at which Eugene Burdick had been too unwell to speak. Rubin had also helped found the Youth International Party, or Yippies, with Abbie Hoffman. He believed in the political power

413

of pranks. Rubin had come up with the idea to lead a march to the Pentagon.[92]

By midmorning, more than a hundred thousand people had gathered at the Lincoln Memorial. They carried daisies. They carried roses. Organizers had shipped in some two hundred pounds of flowers.[93] The rally began at eleven o'clock in the morning. Peter, Paul and Mary sang wailing, mournful ballads; Phil Ochs sang antiwar anthems. Then came the speeches — from celebrities, mainly. Baby doctor Benjamin Spock and bad boy novelist Norman Mailer, who wrote a book about the protest, called *The Armies of the Night.* "We consider the war Lyndon Johnson is waging as disastrous in every way," Spock said. The crowd roared, "Hell no! We won't go!" Ella Collins, Malcolm X's sister, rose to the stage and said, "This is the first time I have witnessed white people and black people rocking in the same boat." At one-thirty, a trumpet sounded, the call for the march to begin to move. About fifty thousand protesters peeled off and began to walk toward the Arlington Memorial Bridge. They crossed the Potomac. By about four o'clock, they'd reached the North Parking Lot, some hundred feet away from the Pentagon.

Rubin's hope had been to encircle the Pentagon and attempt to "levitate" it, a stunt designed to bring attention to what protesters called "the war machine." The Beat poet

Allen Ginsberg climbed onto a flatbed truck, using it as a makeshift soundstage, and chanted, "Ommmmmm." But that's about as close as anyone came to levitating anything. More than two thousand marshals and MPs defended the building, in military formation. Protesters stepped up to armed soldiers and pressed the stems of flowers down the barrels of their guns. An iconic photograph captured one of the demonstrators, a young blond man in a bulky turtleneck sweater, poking a carnation into the muzzle of a rifle. Mailer and Dellinger and dozens more tried to break through the line and were arrested. There were plenty of students among the protesters, but Mailer, in his book about the protest, captured the complicity of the university in the war effort. "Liberal academics had no root of a real war with technology land itself," he wrote. Instead, he went on, "in all likelihood, they were the natural managers of that future air-conditioned vault where the last of human life would still exist."[94] The same judgment became the signature position of Noam Chomsky, a frizzle-haired thirty-eight-year-old MIT professor of linguistics, also arrested in D.C.: he'd lately emerged as an intellectual leader of the antiwar movement and would soon take aim at Simulmatics itself. When police began making arrests, more protesters began tearing through the barricades, some even reaching the steps to the Pentagon. By seven o'clock, the crowd, still some twenty thousand strong, had advanced to within thirty yards of the building, where they sat down. More than seven hundred were arrested.[95]

Daniel Ellsberg, whose views about the war had begun to change since his return from Vietnam, joined the march from the Lincoln Memorial. He hadn't wholly committed. From the rally in the parking lot, he later said, he peeled away and went inside the Pentagon, and into McNamara's office, "to get a better view." He and McNamara watched the protest out McNamara's window, not saying a word.[96]

Five days later, on October 26, thirty-year-old navy pilot John McCain, flying on a bombing mission, was shot down over Hanoi. He would spend five and a half years as a prisoner of war.

In early November, McNamara submitted a memo to Johnson, urging him to stop the bombing in the North, freeze troop levels, withdraw from Vietnam, and get out of the war. Johnson refused. By the end of the month, McNamara had submitted his resignation.

The Johnson administration undertook a massive publicity campaign, trying to fortify public support for the war. "Do not be disturbed at news reports about USA," Father Hoc's American sponsor wrote to him in Saigon. "It is only perhaps 10% of our people who are so cowardly or misled by foreign propaganda to speak against our men trying to help Viet Nam."[97] On November 3, Ed Greenfield, reporting to Simulmatics' investors, said, "We are now considered one of the 'best contractors' in Viet Nam."[98] Ithiel de Sola Pool continued to profess optimism about the war, a view not widely shared among his colleagues at MIT. Despite the classified nature of the research Simulmatics was undertaking in Vietnam, Pool wrote up detailed notes after his visits to Vietnam and distributed them to friends and colleagues, dutifully advancing Johnson's public relations campaign: the war was going very, very well.[99]

At the end of November, Greenfield and Pool went to the Pentagon to meet with ARPA's Seymour Deitchman, who had just returned from a trip to Saigon. There'd been no end of complaints about Simulmatics, allegations of negligence, malfeasance, and even fraud. One internal report chronicled, project by project, "monumental mismanagement and misrepresentation."[100] Al de Grazia's firing of the Vietnamese interpreters, which had threatened to blow up into a public scandal, had been the last straw.[101] "Let the record speak for itself," a final assessment concluded. "Simulmatics reflects discredit not only upon itself as an organization — it appears more a sham — but upon behavioral research in general."[102]

Deitchman, having taken stock of each of Simulmatics' ARPA projects, decided to terminate all of them, except Father Hoc's psychological warfare weapons experiments, which would be allowed to wind down.[103] Greenfield retreated to Simulmatics' office in New York. Pool went back to Simulmatics' office in Cambridge. Simulmatics' office in Saigon was shut down, its reports shredded, its equipment auctioned, everything else trashed.

In Saigon, the laboratory the scientists of Simulmatics had made of a whole country was disassembled, the experiments they had conducted on the minds of an entire people abandoned. But their work didn't end. Instead, they carried it home.

CHAPTER 12
THE FIRE NEXT TIME

<u>To do what</u>? To do things like estimate the number of riot police necessary to stop a ghetto rebellion in city X that might be triggered by event Y because of communications pattern K given Q number of political agitators of type Z.
— MIT student protest pamphlet, 1969

Ann Penner and Marcellus Winston at Wading River, 1971.

Ann Penner, who wore her long dark hair straight and parted down the middle, was twenty-four when Ed Greenfield hired her to work at Simulmatics' offices in New York. She'd finished college, gone to England, and then spent a while in Berkeley demonstrating on behalf of free speech, raging against the machine. She'd found out about the job at Simulmatics at the New York State Employment Agency when she was looking for temporary work. At first, she filled in for Ed Greenfield's secretary, who'd gone off on a long holiday in Ireland. She did a little typing and answered the phone. She paid attention to everything. Simulmatics, she noticed, had its finger in every pie.[1] Sam Popkin, who sometimes worked in the New York office, brought her back some silk from Saigon, which she had sewn into an *ao dai.*[2] Penner opposed the war, hated the war, marched against the war, protested the war. But, like most people who worked for Simulmatics in New York, she had very little idea about what Simulmatics was actually doing in Vietnam. One day, a team of government agents in black suits and black shoes barged into the office, asked for all of Simulmatics' reports from Saigon, got out a stamp and a rubber stamp pad, stamped every report SECRET, piled them all back into a file cabinet, put a lock on the cabinet, and marched out the door.[3]

In 1967, Simulmatics' New York offices, on East Forty-First Street, occupied three stories of a narrow, four-story building whose floors were covered in crummy carpeting. The whole place seemed pasted together, like a youth hostel or a hippie commune or a vegan co-op. Greenfield, brooding, used to walk around the halls with the lights off, looking mysterious and elegant. After his regular secretary got back from Ireland, he called Penner into his office. "We really like you," he said. "What can you do?" "I can write," she said, "and do design." She'd majored in English and had studied art. He hired her. She wrote and designed copy. She sat at a Formica desk and watched the door.

All sorts of people crossed that threshold, a who's who of 1960s celebrities. James Baldwin, Richard Pryor, Jane Fonda, Tom Hayden, Martha Gellhorn, Louis Farrakhan.[4] Greenfield knew scores of people, especially writers and activists on the left. He made a particular point of getting to know black intellectuals. He met Claude Brown in 1965, right about when Brown published his best-selling autobiographical novel, *Manchild in the Promised Land,* about growing up in poverty in Harlem. Greenfield reeled him in. He had him over for Passover. Every Seder, Greenfield would go around the table and ask everyone to answer the same question: "What does freedom mean to you?"[5] Brown loved him.

"Now the President of Simulmatics is one of my dearest friends, a Jewish fellow," Brown told Robert F. Kennedy in 1966, during hearings in the Senate on federal government's role in the problems of urban America. "He is one of the few white people that I could become close to, you know, in a week's time," Brown said. "I have really become a friend where I could say, 'Okay, you white Nigger,' jesting, of course, joking."[6] That was Greenfield, all over.

Robert F. Kennedy, having expediently changed his address from Massachusetts to New York, had been elected a senator from New York, and he had been asking Claude Brown questions, and the Senate had been holding hearings, because riots had been breaking out in American cities, erupting ever since 1964. That year, after an off-duty policeman shot a fifteen-year-old black boy from Harlem, a crowd gathered on the night of the boy's funeral and went to the police precinct at West 123rd Street, where they clashed with police officers in violence that spread to Brooklyn, lasted six nights, and involved some four thousand New Yorkers. More than a hundred people were injured, one killed, and nearly five hundred arrested.

Protests broke out in Rochester two days after the violence ended in Harlem. Rochester was then a one-company town, but its growing African American population suffered from a 14 percent unemployment rate at a time when the city's one company, the Eastman Kodak Company, had six thousand job openings. The city's schools and housing were segregated, and Rochester offered less public housing than any other city in the state. The riots that summer started after police officers arrested a black man during a block party. The city declared a state of emergency and called in the National Guard. The Rochester police brought in dogs and arrested nearly nine hundred people, a massive number, given the city's size. Five people were killed.[7]

Violence broke out in South Central Los Angeles one year later, in a neighborhood called Watts, after police pulled over a black man driving a car. The LAPD was notorious for its brutality. The violence lasted for six nights and involved an astonishing thirty-five thousand people. A thousand people were injured. Thirty-four were killed. "If I've got to die, I ain't dying in Vietnam," said one protester. "I'm going to die here."[8]

Lyndon Johnson shepherded the passage of the 1964 Civil Rights Act and the 1965 Voting Rights Act, declared a War on Poverty, and planned to build a Great Society, with federal aid for economic development. Conservatives read the riots as an indictment of everything he had done and hoped to do.

Greenfield had a plan of his own, Claude Brown told the Senate. Greenfield had the idea that there were mathematical geniuses all over Harlem, in every ghetto in the country, kids who were running numbers games, hustlers with a natural aptitude for business. "What he wanted to do was get all the communities, from all the so-called depressed areas throughout the nation, people with whom he could communicate," Brown said. "And the guy is great, you know, at communicating across ethnic groups." He'd connect these kids, hustlers, with local business owners, in entrepreneurial training programs. But he couldn't get the Johnson administration to pay for his scheme. "I am certain Greenfield would have been willing to do this throughout the country," Brown said. "But the Poverty people turned it down flat."[9]

But the poverty people hadn't turned down all of Ed Greenfield's proposals. Simulmatics found other uses for the research it had done in Saigon, and the reverse was true, too: the research Simulmatics had done as part of the war in Vietnam came out of the work it had done as part of the war on poverty.[10]

"Saigon is like a stewing Los Angeles, shading into Hollywood, Venice Beach, and Watts," Mary McCarthy wrote in the *New York Review of Books* in 1967.[11] Simulmatics' own road to Saigon went through Watts, and Harlem, and Rochester. Simulmatics had entered the work of counterinsurgency not only by way of Ithiel de Sola Pool's ties to the Pentagon but also by way of urban studies. Simulmatics aimed to solve the "Negro problem" in American cities by building a new simulation, not a People Machine but a Riot Prediction Machine.

Simulmatics ran Great Society projects out of two divisions in its New York offices: the Educational Division, headed by Jim Coleman, and the Urban Studies Division, headed by a sociologist named Sol Chaneles. For the Educational Division, Coleman undertook projects like a kids' game called Ghetto, a simulation of ghetto life; the goal of the game was to devise a means of escaping poverty.[12] Much of the work in the Educational Division was done by Marcellus Winston. Winston, a young black man from a big family in Washington, D.C., was a rising star. In 1951, he'd been the first African American to attend Hotchkiss, the fancy prep school. Then he'd gone to Harvard and, after that, to the Sorbonne, on a Fulbright. He described himself as a "writer and community organizer" who'd "published essays on the problem of the Negro in America and on urbanization."[13] Ann Penner and Gus Winston met at Simulmatics and moved in together in 1966, not long before the Supreme Court decided, in *Loving v. Virginia,* that laws that prohibited interracial marriage were unconstitutional.

In 1966, Coleman had become a controversial figure, as the author of *Equality of Educational Opportunity,* an extraordinarily ambitious study known as the Coleman Report. Preparation of the report had been mandated under the terms of the 1964 Civil Rights Act, which required, within two years' time, the publication of a study "concerning the lack of availability of equal educational opportunities for individuals by reason of race, color, religion, or national origin in public educational institutions." A statistical study of 650,000 students and teachers in more than 3,000 schools, Coleman's 737-page report found that the more significant indicator of educational achievement was the family's educational background. He also found that integrating kids from different backgrounds improved educational outcomes for the children of less well-educated parents. Coleman understood the report as an argument for increasing efforts to desegregate American public schools — the report would undergird arguments for busing — but critics on the left dismissed the report as racist, especially after Coleman publicly argued against busing, predicting, correctly, that it would lead to white flight.[14]

In 1967, around the same time Greenfield convinced Al de Grazia to run Simulmatics' operations in Saigon, he hired Sol Chaneles to run Simulmatics' Urban Studies Division, which sought contracts from federal, state, and municipal governments to conduct simulations, making models of things like crime, unemployment, and traffic. Chaneles had a PhD in sociology from NYU, with a specialty in criminal justice, and had worked for the New York Department of Corrections.[15]

But the real impetus for Simulmatics' Urban Studies Division came from Daniel Patrick Moynihan. In 1964, Pool had participated in a conference on "computer methods in the analysis of large-scale social systems" held at the Harvard-MIT Joint Center for Urban Studies, where Moynihan would shortly become the director.[16] Shortly afterward, Pool and Alex Bernstein submitted a Simulmatics proposal involving the simulation of labor markets to Moynihan, who was then assistant secretary of labor. "I hope it titillates," Greenfield wrote Moynihan.[17]

Greenfield may well have known Moynihan since the 1950s, when they both worked for Averell Harriman. Moynihan, long-limbed, pale, and imperious, having finished a PhD in history in 1961, had spent the early 1960s working in the Kennedy and Johnson administrations. In 1965 he wrote *The Negro Family: The Case for National Action,* known as the Moynihan Report, a controversial account that was seen by critics as describing the black family as pathological. In 1966, he'd been appointed director of the Harvard–MIT Joint Center for Urban Studies. He moved into a house in Cambridge a few blocks from Pool.[18]

In the spring of 1967, Moynihan secured a contract for Simulmatics with the Eastman Kodak Company, in Rochester, New York.[19] That summer, Simulmatics hired Moynihan as a consultant, for a term of three years, and appointed him to the research board. He was in the New York office often.[20] Moynihan, then forty, was to be paid a minimum of $10,000 a year. He was also given a sizable amount of stock.[21]

Moynihan and Coleman teamed up on a Simulmatics project on urban design (which was another way of talking about "urban renewal").[22] But in joining Simulmatics, Moynihan had one particular research interest, above all: he wanted to know whether the computer simulation of urban problems could be used to predict race riots.

This project began in Rochester. Since the riots there in 1964, community, civil rights, and labor organizers had pressed for change, including the institution of job training programs and fair employment practices at the Eastman Kodak Company.[23] The company had made few changes. And as the summers of 1965 and 1966 had seen still more rioting in American cities, executives at Kodak had grown worried.[24]

How Eastman Kodak came to hire Simulmatics is a little murky, but Coleman, who'd initially trained as a chemical engineer, had worked for Eastman Kodak in the 1950s, before going to graduate school.[25] And Moynihan had known the head of Kodak since college.[26]

Eastman Kodak hired Simulmatics to study whether "violence, incendiarism and looting" could "be predicted sufficiently in advance of its occurrence in order to limit their scope and intensity." In July 1967, Simulmatics sent a six-person team to Rochester — "a sociologist, a novelist, a journalist, a social case-work researcher, a former Federal investigator and a typist," two black men, one black woman, two white men, and one white woman. The team was headed by Peter Shulman, a Greenwich Village artist who'd been hired as deputy director of the Urban Studies Division. They spent one week in Rochester, during which they interviewed eighty people. Shulman wore army fatigues when he went out to interview. He talked to guys who were making Molotov cocktails. "We are not afraid to die, we are dying every day," they told him. Shulman's team did not use computers to undertake any of their calculations, and they did not write a simulation program. Still, they prepared a report with a very specific prediction: "By noon, Friday July 21, the data produced by the research team supported an unqualified prediction of violence to occur about 11 P.M. on Sunday night July 23."

Shulman reported to Sol Chaneles, who reported to Daniel Patrick Moynihan, who on that Friday afternoon alerted the governor of New York about the predicted day and hour of the riot. According to Simulmatics, the riot occurred, exactly as predicted, but police were there to keep the mayhem to a minimum. Over the weekend, someone evidently gave the entire report to the New York State Police, because on Monday, Rochester police detectives came to Simulmatics' New York office and asked Peter Shulman for the names of the guys who were making Molotov cocktails; Shulman refused to tell them.[27]

Simulmatics took the Rochester study, and its effectiveness, as a proof of concept. "The first question raised by this small-scale research effort: can a riot be predicted so as to permit the limitation of its scope and intensity? must be answered in the affirmative."[28] Its Urban Studies Division decided to tackle a new problem, on a national scale: counterinsurgency at home.

During the long, hot summer, the summer of 1967, race riots broke out in Newark and Detroit and all over the country, 159 in all. In July, Lyndon Johnson called for a national commission to conduct special investigation into "civil disorders." Chaired by Illinois governor Otto Kerner, the Kerner Commission was charged with answering three questions about the spate of race riots in the United States: "What happened? Why did it happen? What can be done to prevent it from happening again and again?"[29] The commission's Media Analysis Task Force was supposed to answer one more question: "What effect do the mass media have on the riots?"[30] Pool, just back from his summer in Vietnam, studying the effect of radio and television on Vietnamese peasants, bid on the media analysis project in September.[31] Simulmatics' Urban Studies Division submitted its proposal in October.[32] It won the contract, with a budget of $221,000 (roughly $1.7 million in 2020 money).[33]

Riot prediction had been an object of Pool's scholarship since 1942, when he'd first conducted content analysis with Harold Lasswell, to spot signs of impending revolutions. Reporters writing about his work on Project ComCom had described its objective as the construction of a riot prediction machine at least since the summer of 1965, the summer of Watts. "IN THE WORKS: A COMPUTER TO PREDICT RIOTS, REVOLUTIONS" was the headline of a story about Pool published in the *Boston Globe* that year.[34] Heading into the summer of 1966, riot prediction was everywhere, a desperate prophecy. "Large-scale rioting by frustrated Negroes was forecast here today," said reporters from Washington, D.C.[35] As if political unrest were like a hurricane.

City by city, city planners had begun building their own riot prediction machines. Detroit's Social Data Bank, "first of its kind in the nation," according to the city's mayor, was launched in July 1966. It stored data "on 25 different social factors such as crime, juvenile delinquency, welfare load, health problems, requests for legal aid, truancy and dropouts." Monthly reports were meant to tell city planners whether aid programs were working and where more aid was needed. "Could the Social Data Bank predict riots?" one reporter wanted to know. Maybe it can, answered its designer.[36]

Civil rights activists had little use for such schemes. "I will not predict riots," James Farmer, head of the Congress of Racial Equality said on CBS's *Face the Nation* in 1965. "No one has enough knowledge to know that there will be riots." The real problem, he pointed out, was that no one was really doing what needed to be done to prevent riots, by addressing the problems that led to them.[37]

"I am not going to predict rioting here," Martin Luther King Jr. told the press in Cleveland in June 1967. "This will be determined by the progress and responsiveness by those in positions of leadership."[38] But the fantasy of computer-aided riot prediction endured, as widely and passionately held as the twenty-first century's fantasy that all urban problems can be solved by "smart cities," and "predictive policing," and that ongoing civil unrest and racial inequality and police brutality can be addressed by more cameras, more data, and more computers, and, above all, by predictive, what-if algorithms.

The study Simulmatics conducted for the Kerner Commission consisted of two parts. Both parts borrowed from methods Simulmatics had used in Vietnam. One part, headed by Chaneles, involved sending three-man teams to interview "ghetto residents" in seven cities: Tampa, Atlanta, Newark, Detroit, New Haven, Cincinnati, and Milwaukee. The other part, headed by Pool, involved collecting all the newspaper, radio, and television coverage of the riots in fifteen cities and submitting it to a quantitative content analysis. Both parts involved the collection, preparation, and coding of data for computer analysis.[39]

To do the work, Chaneles and Pool hired teams of young people, mostly in their early twenties. Some were PhD students. Clarence Major, a talented young black poet, started out working on the coding of the newspaper, radio, and television coverage of the riots and then moved over to conducting interviews. The black people he interviewed disputed the very idea of a riot. "One of the main complaints I heard in Milwaukee was that police were making little or no distinction between people demonstrating peacefully and those breaking the law," Major later said. "People I talked to said they were cracking heads everywhere."[40] Police brutality didn't quite fit into the Simulmatics codebook.

Chaneles, following the method he'd devised in Rochester, put together interview teams consisting of two black men and one white man. In total, across the seven selected cities, Simulmatics researchers interviewed 567 blacks and 191 whites. Their instructions were to go to black neighborhoods and interview people they were to find in specific locations: "2 pool halls, 4 bars, 2 barber shops, 2 bowling alleys, 4 stores (2 grocery and 2 small clothing stores), 2 school 'hangouts,' street corners, 2 community action groups (e.g. Poverty Programs, etc.), 2 civil rights groups (e.g. NAACP, CORE, etc.) and other similar locations where people congregate." During the very same months that Joseph Hoc, from a monastery outside Saigon, gathered three reports a day from his informants in six Vietnamese hamlets, Chaneles, in New York, gathered reports from his three-man teams in seven American cities. "What do you watch on television?" Pool's interviewers asked Vietnamese peasants in eleven provinces in South Vietnam. "What television programs do you watch?" Simulmatics' researchers asked African Americans in Tampa, Atlanta, Newark, Detroit, New Haven, Cincinnati, and Milwaukee. "Did you or anyone you know go to a riot after you saw television pictures?"[41]

Simulmatics' work on riot prediction in American cities bore an eerie similarity to McNamara's giant computer program, the Hamlet Evaluation Study, which, after all, aimed to predict insurgency. In the United States, Simulmatics attempted to determine the "Mood and Atmosphere Within Riot Community" and to assign it a number ("1-Calm, restrained; 2-Tense; 3-Angry; 4-Fearful, nervous; 5-Apathetic; 6-Euphoric (Carnival); 7-Friendly, good-willed; 8-Chaotic; 9-Orderly; 0-Other").[42] Joseph Hoc's codebook for his Vietnam psyops participant-observers is not dissimilar.

There were more likenesses, too. The Kerner Commission, not unlike the Advanced Research Projects Agency, was unimpressed with Simulmatics' research, and for similar reasons: poor design, lack of expertise, and misrepresentation. Chaneles reported in November that Kurt Lang, a noted sociologist, "has been working on the study for 3 days per week" as "Senior Social Scientist."[43] But according to Lang, he had turned down an offer to work on the project, had had nothing to do with it, and, having looked over the interim report, found it reprehensibly sloppy.[44] Nor was he alone.[45] A former reporter who was supposed to have oversight over the project but had been largely left out of it complained about the work: "I just felt they didn't know what the hell they were doing."[46] Moynihan, too, was embarrassed. "Get the hell out of there," he told Peter Shulman. Moynihan got him a job at RAND, and Shulman went to Vietnam.[47]

In spite of the many allegations made about the shortcomings of Simulmatics' research, at least some of its findings informed the Kerner Commission's deliberations. During a meeting of the commission on November 6, 1967, a sociologist from Boston University told the commission, "Until we can predict riots with at least as much reliability as the weather forecast predicts rain or snow, it will be very difficult rationally to devise policies of control."[48] Nor was the Simulmatics report the only material submitted to the commission aimed at the prediction of riots. A behavioral scientist from the Law Enforcement Study Center at Washington University in St. Louis submitted to the commission a white paper in which he'd attempted to combine all of the data in his hands —

including income, crime, and population figures from 184 cities with a population of fifty thousand or more — to come up with a predictive scheme. He purported to divide cities into those that had already had riots and those that were in "pre-riot tension conditions." Many of his predicted riots did not occur. Riots occurred in cities where he had not predicted them. The author was undaunted ("Failure to predict actual riots in the cases above shows considerable localization; intensive analysis may well determine that the composite variables are not yet sufficiently detailed to account for older non-white migration patterns").[49]

The data gathering conducted by the Kerner Commission was part of a larger Johnson administration project on the management of race relations. In February 1968, the *Washington Post* reported that "a computer in the Justice Department's basement" would in the summer become one of "the nerve centers of Federal riot-watching activity." Building on data gathered by the FBI and from Criminal Division and Civil Rights Division lawyers, the idea was "to move the Government from riot reaction to riot prediction." The *Post* reported that no data had yet been given to the Justice Department from the Kerner Commission but hinted that this new riot prediction machine would synthesize all the available information: "Highest hopes are for the machine's 'cross-reference' powers, helping to pull together data on cities, men and situations." Ithiel de Sola Pool clipped the story from the newspaper and tucked it into his files.[50]

Very few predictions made in 1967 about Vietnam or the United States came true in 1968. All through the second half of 1967, the Johnson administration ran an "Optimism Campaign," a public relations blitz that aimed to convince the American public that the war in Vietnam was nearly at an end — and that the United States would win. Everyone involved in the war in a position of authority, Pentagon-wise, was asked to pitch in. "I have made a prediction that we have seen our last major battle," declared one general at a White House press conference in August. In September, officers assured soldiers in the field that "we would surely win the war within the next six months." Pool did his part, too. "I don't know what people are talking about when they refer to the stalemate in Vietnam," he wrote in September, in notes he circulated to friends, family members, and colleagues. "There has been striking progress over the last year."[51] The enemy's strength had plummeted, reported General William Westmoreland that fall. MACV printed a pamphlet for the press, called *1967 Wrap-Up: A Year of Progress*. Near the end of the year, a senior White House official told a reporter for the *New York Times,* "Forget the war. The war is over."[52]

Who, though, could really say? Joseph Hoc believed *he* could say. "I want to be able to predict and even control the enemy's actions," Hoc had told ARPA. By January 1968, he had more than nineteen thousand field reports, too many to tally by hand.[53] He asked ARPA for time on the IBM 360 at MACV, to conduct a proper analysis.[54]

Meanwhile, North Vietnam announced that it would hold a ceasefire from January 27 to February 3, during the Vietnamese New Year, a holiday known as Tet. South Vietnam forces were to observe a seven-day truce. On January 25, Pool was at the Pentagon for meetings with ARPA, seeking new funding for Father Joseph Hoc, when, in a faraway hamlet in Vietnam, one of Hoc's informants had a conversation with a villager Hoc referred to as Mr. R.B. "Guerrillas and local regular forces will not celebrate Tet but are instead preparing for an attack in Dinh Tuong during the Tet holidays," Mr. R.B. said. "There will not be a truce at all."[55]

This was one prediction that came true. Five days later, the Vietcong and the North Vietnamese army began the assault known as the Tet Offensive. In the largest campaign of the war, Communist forces attacked more than a hundred cities and towns, including

Saigon, caught altogether unaware: two hundred American MACV officers had been attending a pool party.

The Tet Offensive, by demonstrating the enemy's strength, made clear that the Johnson administration had been lying about the nearness of victory. By the end of February, American support for the war, never high, had sunk. On CBS News, Walter Cronkite looked sullen. "For it seems now more certain than ever that the bloody experience of Vietnam is to end in a stalemate," Cronkite said. "To say that we are closer to victory today is to believe in the face of the evidence the optimists who have been wrong in the past."[56]

In the aftermath of Tet, the press questioned whether all of the work of prediction — the RAND reports, the Simulmatics studies, Komer's Hamlet Evaluation System — had been of any use at all, since no one had seen the Tet Offensive coming. In a letter to the *New York Times,* Pool refused to back down, arguing that Tet ought to be understood, instead, as an urgent call for *more* resources for computer-driven, data-based prediction programs.[57] But ARPA would never again fund a Simulmatics project. Simulmatics' Urban Studies Division was a shambles. So, too, it seemed, was the nation itself, frayed, bruised, and bloodied.

On February 29, a weary Robert McNamara left office, three months after handing

Johnson his resignation, and the Kerner Commission issued its report on civil disorders. The report blamed American race riots on structural racial inequality in housing, employment, and education. (The commission ignored Pool's content analysis of press coverage of the riots in favor of Simulmatics' interviews with people who lived in black neighborhoods.)[58] "Our nation is moving toward two societies, one black, one white — separate and unequal," the *New York Times*'s Tom Wicker wrote in the report's introduction.[59]

Johnson ignored the report. He had the weight of his legacy on his mind. He was facing reelection. As the 1968 Democratic primaries began, he was met by challenges from within his own party, both from the left, by antiwar candidate Eugene McCarthy and Bobby Kennedy, who would campaign against "Johnson's War," and from the right, by Alabama segregationist George Wallace. On March 31, Johnson appeared on live television, sitting before an American flag, to deliver a speech about the war in Vietnam. At the very end of the speech, out of nowhere, he made an announcement that staggered the nation: "I shall not seek and I will not accept the nomination of my party for another term as your president." A lot of people thought they had misheard him, it was so unexpected, so unbelievable. He'd told almost no one of his decision before he made his announcement, not even his cabinet. But it was true. Lyndon Johnson's political career had begun in 1937, in the age of radio. It ended in 1968, in the age of color television. He said he wanted to spend his remaining time in office negotiating for peace in Vietnam.[60]

He might have forged that peace. But four days later, on April 4, 1968, in Memphis, Martin Luther King Jr. was assassinated, exactly one year after he'd first spoken out against the war in Vietnam. He'd been slated to a deliver a sermon, the following Sunday, at his own church, the Ebenezer Baptist Church in Atlanta. The sermon was to be called "Why America May Go to Hell."[61]

And it did seem to be going to hell. "Martin Luther King dedicated his life to love and to justice between fellow human beings, and he died because of that effort," Robert F. Kennedy said, shaken, when he heard the news.[62] Two months later, on June 5, after winning the California primary, Kennedy was assassinated in Los Angeles.

Norman Mailer watched the news on television. " 'No,' he cried. 'No! No! No!' " (Mailer liked to write about himself in the third person.) "He felt as if he were being despoiled of a vital part of himself, and in the middle of this horror noted that he screamed like a pig, not a lion, nor a bear."[63] It was almost impossible to believe. People sank to their knees in the street. John F. Kennedy, Martin Luther King, Robert F. Kennedy. What more, asked the nation, like Job. Arthur Schlesinger wrote in his diary, "We have now murdered the three men who more than any other incarnated the idealism of America in our time."[64] Who, or what machine, could have predicted this political massacre?

In early June, Ed Greenfield laid off everyone left at Simulmatics' New York office — the Urban Studies Division, the Educational Division, everyone. He could no longer pay their wages. On June 17, Simulmatics' board met in New York. Greenfield offered to resign but, in the end, did not resign. He was supposed to visit Patty the next night. She'd asked to meet him. He didn't show up.[65]

The separation had been hard on her — Ed hadn't made it easy — but Patty Greenfield had been trying to rebuild her life. "I am slowly, but very slowly, getting to the point of thinking alone," she'd written at the end of 1967.[66] She was writing for her local newspaper, the *Chelsea-Clinton News,* and working there as an editor, too. She profiled neighborhood schools, like P.S. 51 in Hell's Kitchen, a kid-sized United Nations, a collection of kids from all over the world.[67] She'd begun to write a memoir. She was president of the PTA for her local public

443

school.[68] She was raising money for a nursery school. She'd been going to civil rights demonstrations and antiwar marches, taking along her three girls, all of them holding hands, like paper dolls. Maybe she'd been planning to work for Bobby Kennedy's campaign. By June, she was getting ready for a summer on the beach, at the old beach house, the big, rickety Victorian, with the fireplace of stones pulled from Long Island Sound.

The men of Simulmatics had mostly parted ways. But not the women. After those seaside Simulmatics summers together at Wading River, their own Peyton Place, Patty had stayed close to the Simulmatics wives, even as so many of their marriages fell apart. The Greenfields, the Morgans, the Colemans, the McPhees, the de Grazias: all their marriages collapsed in the 1960s. Lu Coleman, who'd dropped out of college to marry Jim Coleman, left him not long after she went back to school; she finished her degree in 1966, went to graduate school, and then to seminary.[69] Minnow McPhee, who'd started her own preschool, had left Bill, who, while still teaching at the University of Colorado, continued to struggle with manic depression. When Minnow's daughter Wendy got pregnant in her last year in college, Minnow sent her to New York, to live with Patty at the townhouse at 349 West Twenty-Second Street, in Chelsea. Wendy, who was twenty-one, planned to give the baby up for adoption and needed to be far away from Colorado, to hide her pregnancy: five years before the Supreme Court's decision in *Roe v. Wade,* this is how these things were done. Michael, sixteen, was away at school. Wendy spent most of her time with the three Greenfield girls, part big sister, part babysitter. Ann was fourteen; Susan, thirteen; and Jennifer, not yet four. They'd flop down together on a big king-sized bed. Wendy invented a fortune-telling game, using a deck of playing cards. They'd hold spooky seances, by candlelight, and ask the cards questions, as if they were a Magic 8-Ball. One of the girls asked, "Will my mother live beyond forty?" The cards said, "No."[70]

By March, Patty had gotten concerned about Wendy, who'd started spotting.[71] Eventually, she had a miscarriage. "I can only say, 'thank God,' " Patty wrote to Michael, away at school, sure that losing the baby would be easier on Wendy than giving it up.[72] Wendy went back to Colorado in April, just after Martin Luther King was shot, and the nation fell into mourning, and the world began spinning. After Robert Kennedy was shot, on June 5, Patty nearly collapsed. She could not bear it, the hate, the murderousness, the loss, the leaderlessness. At night, in the dark, she'd play music from the record player in the drawing room at the back of the

445

house and dance, her Martha Graham modern dance, a dance of political despair, as if she were a ghost. It felt as if the house were haunted.[73]

The night Ed was supposed to meet her but didn't turn up, she drank alone and then went to bed, in her bedroom on the second floor. She'd been known to sleepwalk. Sometime during the night, she got up and walked out to bedroom's balcony and, somehow, fell off the balcony and landed in the garden. She hit her head, began to bleed, and became unconscious. In the morning, when the girls were getting ready for school, they assumed their mother was upstairs, in her bed, asleep. The dog, a shepherd named Chutney, was barking like crazy, but dogs sometimes do that. She wasn't found until that afternoon. She died in the hospital the next day.[74] She was forty years old. An autopsy declared that she'd died of a subdural hematoma and contusion of the brain; her death was ruled accidental.[75] A lot of people, including Ed, blamed Ed.

At the memorial service, Michael read a poem by Dylan Thomas, "And death shall have no dominion." It called back those long summers at Wading River: "No more may gulls cry at their ears / Or waves break loud on the seashores." The principal of the children's elementary school delivered the eulogy; she talked about Patty's intelligence, her warmth, and her generosity. "Let's use my house," Patty would say when the school-children needed a place to rehearse a play or the parents needed a place to meet. Her brother spoke at her graveside. "She hated no one," he said.[76]

That summer, Simulmatics' founders and chief stockholders — Greenfield, Pool, McPhee, Bernstein, Coleman, and Abelson — met in New York and decided to sell the company.[77] Not long afterward, when Green-field, who was in no condition to do anything, failed to effect a sale, Coleman resigned in disgust, forfeiting all his stock, twenty-five thousand shares. "Frankly, Ed, you had an opportunity, and muffed it," he wrote Green-field.[78] Creditors came calling. "I just cannot understand why you cannot pay me the money you owe me," Daniel Patrick Moyni-han wrote Greenfield, furious. "I think it is disgraceful that I should have to beg you for my quarterly. It is now 24 days past due."[79] Everyone who still had money invested in Simulmatics lost it.

Ed Greenfield sank like a ship. The children moved into Naomi Spatz's apartment, but, after a while, Spatz left. Friends took charge of the children. Greenfield could hardly take care of himself. He lived for a long time in an unlit room, stocked with cartons of liquor and cigarettes.[80] Ann Penner married Gus Winston at the Greenfields' house in Wading

River, but not long after that, Ed Greenfield was sleeping on the floor of the living room of the Winstons' house.[81] He'd always pick himself up and start again. But still he'd say, day after day, "I ruined everything and everyone."[82]

Something close to an obituary for Simulmatics appeared not long after Patty Greenfield's death, in a book written by *New Republic* reporter James Ridgeway. "Simulmatics looks like nothing more than a dummy corporation through which Pool runs his outside Defense work," Ridgeway reported. (Among Ridgeway's sources for this takedown was Moynihan, who talked to him off the record.)[83] Simulmatics' influence, Ridgeway said, had come to an end. "Simulation companies are not so popular as they once were; their proprietors are often regarded as cultists, and the generals who were persuaded to hire them by liberals in the Kennedy and early Johnson administrations are sour on the whole business."[84]

Richard Nixon, though, was still interested in the computational political work that Simulmatics had pioneered and that some people believed had contributed to his defeat in 1960. Nixon's hour had finally come. After his defeat in the gubernatorial race in California in 1962, he'd moved to the right, nominating Barry Goldwater in 1964 and learning from Ronald Reagan's success at winning the

California governor's office in 1966 by running as a conservative. In 1968, Nixon expected to win the Republican presidential nomination, and with Johnson out of the race and the Democratic Party in disarray, he expected to become the next president of the United States.

In May 1968, a mysterious Simulmatics competitor called the Simultron Project pitched its services to Nixon in a typewritten memo sent to Nixon aide H. R. Haldeman. Somehow, the Nixon campaign had gotten copies of the confidential reports Simulmatics had prepared for the DNC and Kennedy in 1960. Maybe that information had come by way of Haldeman, a Nixon campaign manager who'd been a Madison Avenue ad man for more than twenty years and had first offered his services to Nixon in 1952 (Haldeman's father had contributed to the slush fund for which Nixon had had to apologize in his "Checkers" speech). Haldeman had worked on Nixon's campaign in 1956; he'd been trained by the longtime Republican political consultants Clem Whitaker and Leone Baxter.[85]

But it's also possible — and more likely — that the Simulmatics reports came from Pool himself, and it's even possible that, even while Pool was providing advice to the Humphrey campaign, he was trying to sell simulation services to the Nixon campaign. Ever since

Nixon had helped him get a security clearance, Pool had considered himself to be in Nixon's debt, and in 1972, Pool would publicly announce his defection from the Democrats and campaign for Nixon's reelection. The Simultron Project memo is unsigned, but it sounds very much like Pool, and the services it offers are very much the services of Simulmatics. Also, Pool was in Washington the week Nixon and his aides received and discussed Simultron's proposal: he might have delivered it in person, along with the 1960 Simulmatics report. Then, too, a forensics comparison of the Simultron Project memo with a document authored by Pool and known to have been produced on Pool's typewriter (a memo to the Humphrey campaign) proves that the two documents were typed on the same model of machine; the two documents also exhibit similar misalignment and spacing characteristics, suggesting that they may have been produced by the same typewriter.[86]

Whoever was behind the Simultron Project, Nixon's staff seemed dubious. "Before the simulmatics idea is finally canned, if it is to be," Nixon aide Patrick Buchanan wrote in a memo to Nixon, would the president mind reading the proposal? The Simultron Project offered its services to the Nixon '68 campaign at a budget of $300,000 (more than $2 million in 2020 money). "Simulation is not magic," Simultron admitted, but by breaking the electorate into voter types and policies into issue clusters, it could provide Nixon with "the most precise data possible upon which to make the crucial decisions necessary in the months ahead." Nixon scrawled his thoughts at the bottom of Buchanan's memo: "This is on balance just an organized effort to get info that any decent campaign organization should have at its fingertips." He decided to spend $40,000 (about $300,000 in 2020 money) on the Simultron Project, but no more.[87]

In 1960, the Democrats had hired Simulmatics to figure out what the Democratic nominee should say about civil rights to win the "Negro vote." Nixon believed that he already knew what to say about African American voters, and to whom. His campaign had plenty of data. He'd run the numbers. He'd already decided to court the white vote — he'd call this new coalition the American "silent majority" — and let the Democrats have black voters.[88] To that end, he predicted riots. "In Watts and Harlem and Detroit and Newark, we have had a foretaste of what the organizations of insurrection are planning for the summer ahead," he said in a radio address, looking ahead to the long, hot summer of 1968. He promised to end the mayhem and usher in a regime of law and order.[89] He predicted chaos, from which only he could save the nation.

Nixon didn't need the Simultron Project. Between the rise of the Simulmatics Corporation and its fall, American political campaigns had come to be led by computer-run data analysis, with or without the company that pioneered it. They knew all about Simulmatics' 480 voter types. After reading Kevin Phillips's data-crunching 1969 book, *The Emerging Republican Majority,* Nixon told Haldeman, "Go for Poles, Italians, Irish, must learn to understand Silent Majority . . . don't go for Jews & Blacks." Months later, Nixon read the opposition's data-crunching book, *The Real Majority,* by Democratic strategists Richard M. Scammon and Ben J. Wattenberg, whose insights Democrats ignored but Nixon took to heart. The average voter, Scammon and Wattenberg's numbers told them, is a forty-seven-year-old Catholic housewife from Dayton, Ohio, married to a machinist; "she has a mixed view about blacks and civil rights because before moving to the suburbs she lived in a neighborhood that became all black." Nixon told Haldeman: "We should aim our strategy primarily at disaffected Democrats, at blue-collar workers, and at working-class white ethnics" and "set out to capture the vote of the forty-seven-year-old Dayton housewife."[90] And that's exactly what they did.

By 1965, a century after the end of the Civil

War, a century of lynchings and bombings and beatings, a century of sit-ins and marches and mourning and preaching, civil rights had been won, voting rights had been guaranteed. But the closer African Americans had come to being able to vote, the more furiously political consultants had labored to divide and segment the electorate by ideology and by race. Simulmatics had designed the People Machine in 1959 to predict the "Negro vote." After 1965, when African Americans finally, fully, entered the electorate, the parties, newly sorted, began to move to the poles of the political spectrum. A computer network whose first messages were sent in 1969 would widen and deepen those divisions; a half century later, those divisions would become almost impossible to bridge. Voter by voter, issue by issue, the American divide had been simulated, and then it had been automated. *Simulmatics.*

CHAPTER 13
AN OCTOPUTER

> Do you mind if I ask you a personal question?
> — HAL 9000, in *2001: A Space Odyssey,*
> 1968

Nineteen sixty-eight, a year of anxiety, terror, agony, and uncertainty, was a year obsessed with the future and awash in predictions. The day before Martin Luther King Jr. was killed, Stanley Kubrick's film *2001: A Space Odyssey,* with its sinister IBM-like HAL computer, debuted in theaters. HAL did much more than store data and run programs: HAL talked. That same month, April 1968, *Science & Technology* magazine dedicated a special issue to "The Communications Revolution," the revolution that would come when computers would be able to communicate with one another across a single, vast network. MIT professor J.C.R. Licklider, who had earlier headed the behavioral science division at ARPA, predicted the emergence of "online interactive communities." These

An MIT SDS poster, indicting Pool and Simulmatics, c. 1971.

would all be to the good. "Life will be happier for the on-line individual because the people with whom one interacts most strongly will be selected more by commonality of interests and goals than by accidents of proximity."[1] As if people would be happier if they kept to their own kind, like tribes.

Ithiel de Sola Pool contributed an essay to this special issue of *Science & Technology*, too. Pool predicted that the communications revolution, this brave new "on-line" world, would usher in a new age of hyperindividualism, in which, for instance, customized news feeds would mean that people would see only the news they wanted to see. Pool also predicted, with uncanny prescience, that this change would transform politics. "In the coming atomized society, the information the citizen gets will arise from his own specific concerns," he wrote, anticipating that "when everyone can select his own fund of information, the political problem of gathering an effective body of support behind a particular rational issue or candidate will become very different and very much more difficult."[2] Party politics was already yielding to interest group politics, but Pool saw that interest group politics would soon yield to a politics of self, every citizen a party of one.

Pool's predictions about many things were far wide of the mark, especially his predictions about the future of Vietnam, which rested on his misperceptions of its past and its present. In 1968, when Pool, Daniel Ellsberg, Arthur Schlesinger, and Frances FitzGerald presented papers at a conference at the Adlai Stevenson Institute of International Affairs, Pool began his remarks by objecting to the title of the conference, "Vietnam: Misconceptions in Action." It was, he said, "like being asked whether one has stopped beating one's wife." He believed there had been no misconceptions.[3] And yet Pool's predictions about emerging technologies were bold and true and wise. No one, in the 1960s — or, for that matter, in the 1970s and 1980s — would turn out to have been more right.

The week Robert F. Kennedy was shot, a book of predictions appeared in bookstores. "Amazing predictions of what life will be like 50 years from today," ads announced.[4] The book was called *Toward the Year 2018,* and it included forecasts about the state of all sorts of fields in fifty years' time. The director of automation research at IBM predicted the development of "very small, portable" computers and was particularly confident that, no matter what the pace of technological change, between 1968 and 2018, "the political and social institutions of the United States will remain flexible enough to ingest the fruits of science and technology without basic damage to its value systems." An electrical engineer from Bell Labs argued that face-to-face communication across long distances would be available everywhere by 2018 and that "the transmission of pictures and texts and the distant manipulation of computers and other machines will be added to the transmission of the human voice on a scale that will eventually approach the universality of telephony." Its consequences, he believed, were harder to see. "What all this will do to the world I cannot guess," he admitted, with a becoming modesty. "It seems bound to affect us all."

Pool's essay in *Toward the Year 2018* offered by far the sharpest assessment, spot-on. "By 2018 it will be cheaper to store information in a computer bank than on paper," he predicted, foreseeing that information that in 1968 was saved in paper form — "tax returns, social security records, census forms, military records, perhaps a criminal record, hospital records, security clearance files, school transcripts . . . bank statements, credit ratings, job records" — would in 2018 be stored on computers, computers that could communicate with one another over a vast international network. People living in 2018 would be able find out anything about anyone, without ever leaving their desks. "By 2018 the researcher sitting at his console will be able to compile a cross-tabulation of consumer purchases (from store records) by people of low IQ (from school records) who have an unemployed member of the family (from social security records). That is, he will have the technological capability to do this. Will he have the legal right?" Pool had no answer to that question, instead demurring: "This is not the place to speculate how society will achieve a balance between its desire for knowledge and its desire for privacy."[5]

Pool predicted so accurately because he knew so much. Few people, anywhere, had been as closely involved with each aspect of the new data and communications infrastructure as he had been, or for as long.

The data-driven twenty-first century that Pool saw coming began with what Simulmatics called "massive data," the aggregation of many smaller sets of data that would come to be called "big data."[6] In the early 1960s, behavioral scientists keen to make better predictions by detecting patterns impossible to see with smaller sets of data began seeking formal means to aggregate their data. In 1963, "A Proposal to Establish a National Archives for Social Science Survey Data," published in Al de Grazia's journal *American Behavioral Scientist,* cited the research conducted by Simulmatics on the 1960 election as the best example of the kind of work that could be done with big data.[7] Two years later, the report of a Social Science Research Council Committee on the Preservation and Use of Economic Data, chaired by Yale economist Richard Ruggles, warned that "without appropriate data, the economist with a computer would be in the same position as a biologist with a powerful microscope but no biological specimens." The Ruggles Report led to a formal proposal that the federal government establish a "National

Data Center."[8]

Establishing a National Data Center made a great deal of sense. The implementation and assessment of programs created by Johnson's Great Society required that the federal government collect all sorts of data — including economic, social, health, employment, voting, housing, and population data — in order not only to provide services but also to ensure and demonstrate that those services had been provided fairly and equitably, under the oversight provisions of a whole swath of legislation, including the Civil Rights Act, the Voting Rights Act, and the Fair Housing Act. A peer institution to the Library of Congress (which holds books) and the National Archives (which holds manuscripts), the National Data Center would serve as a central depository of computerized information from federal agencies, encompassing everything from the Social Security Administration to the Census Bureau, from federal prisons to public schools.[9]

But by the time Johnson established a task force to figure out how to build a National Data Center, members of Congress had begun to raise concerns that this new institution would compromise citizens' privacy. In the summer of 1966, the House Special Subcommittee on Invasion of Privacy, chaired by New Jersey Democrat Cornelius E. Gallagher, held three days of hearings about the proposed National Data Center. As New York Republican Frank Horton pointed out, the privacy of citizens whose information was held by federal government agencies had chiefly been secured by the very fact that no such center existed. "Good computermen know that one of the most practical of our present safeguards of privacy is the fragmented nature of present information," Horton said. "It is scattered in little bits and pieces across the geography and years of our life. Retrieval is impractical and often impossible. A central data bank removes completely this safeguard."[10]

Horton had a point. Computers could store more information than ever before. Methods for retrieving information were getting better, and faster, every day. A Senate hearing on the National Data Center reported that the federal government held, in separate agencies, "more than 3 billion records on individuals, including 27.2 billion names, 2.3 billion addresses, 264 million criminal histories, 280 million mental health records, 916 million profiles on alcoholism and drug addiction, and 1.2 billion financial records."[11] What would happen if all these records were glommed together? "The citizen concerned about the erosion of his privacy has until now had some consolation in knowing that all these records about his life have been widely dispersed and often difficult to get at," wrote

461

the best-selling social critic Vance Packard, who appeared before Gallagher's committee, and who later raised an alarm in an essay in the *New York Times Magazine*. "But today," he warned, "this consolation is vanishing."[12]

The proposal for a National Data Center had failed to address any of these concerns.[13] That left the way open for critics to call for the abandonment of the plan. Still, much of the animus against the National Data Center was plainly partisan: conservatives hated Johnson's Great Society and the rising authority of the state, and a National Data Center served as a powerful symbol of government overreach. Gallagher was a Democrat, but nearly everyone who applauded him for fighting the proposal was a conservative. Ordinary Americans from all over the country who wrote to thank him called the proposed National Data Center a "Gestapo scheme," "creeping socialism," and worse. "This looks like Communist infiltration!" declared one Californian. "Guess we might just as well move to Moscow," wrote a man from Pennsylvania.[14]

The *New York Times* said the data center risked an "Orwellian nightmare." The *Wall Street Journal* called it an "incipient octopus." "BIG BROTHER NEVER RESTS," ran the headline in the *Indianapolis Star*. "A Giant Peeping Tom" was the name given to the data center by one paper. Priests and ministers and rabbis spoke out about it from churches and synagogues. Even the Daughters of the American Revolution passed a resolution of opposition.[15]

The debate over the National Data Center marked the first time in American history that the public had engaged in a sustained conversation about the consequences of the aggregation of personal data on computers. University of Michigan law professor Arthur Miller sounded an alarm in the pages of the *Atlantic,* warning of a grave danger to privacy: "a central computer might become the heart of a government surveillance system that would lay bare our finances, our associations, or our mental and physical health to government inquisitors or even to casual observers."[16] In an essay in *Playboy,* another legal scholar warned of the rise of "data surveillance."[17] But it was an anonymous law review article that offered the most poignant and prescient warning: "The fabric of human social relationships, dependent upon each person's having only limited knowledge about other people, could be rent by leaks from the data center."[18]

Each of these commentators missed something important, something that had come up at the hearings before Cornelius Gallagher's committee only to be dismissed, not least because members of Congress tended not to have the most sophisticated knowledge of the latest developments in computer technology, or even of the not-so-latest developments. This remained the case for decades. In 2018, at hearings about Facebook's handling of users' personal data, eighty-four-year-old Senator Orrin Hatch asked baby-faced, thirty-three-year-old Facebook CEO Mark Zuckerberg, "How do you sustain a business model in which users don't pay for your service?" Zuckerberg blinked three times, apparently shocked that Hatch could know so little about so basic a matter. "Senator," he said, "we run ads."[19]

During the hearings Gallagher held in the summer of 1966, Paul Baran, a very sharp computer scientist from RAND, said that it didn't matter what Congress decided about whether or not to build a National Data Center, because data would be aggregated, would become big data, with or without the federal government. Computers, Baran tried to explain to the committee, would soon all be connected to one another, in a vast network of networks. As for information held by the federal government, "whether the information is centralized in one central data bank or whether it is spread around the country doesn't make a darn bit of difference," Baran said. "The result is the same." The main thing, Baran advised, was to set up ethical guidelines, safeguards, and rules. What is data? To whom does data belong? What obligation does the collector, or holder, or analyst of data hold over the subject of the data? Can data be shared? Can it be sold?[20]

Congress didn't answer these questions, or even debate them. Gallagher's committee doesn't seem to have understood what Baran was saying about the coming network of networks, any more than Orrin Hatch in 2018 understood what Mark Zuckerberg said about how Facebook works. Instead of shifting the debate to the nature and ownership and privacy of data itself, Congress simply tabled the proposal for a National Data Center. By 1968, the idea had been shelved, another casualty of the disintegration of Johnson's Great Society. No building was ever built, no agency ever established, no federal regulations on data devised. And yet, exactly as Paul Baran had predicted, a de facto national data center was established soon enough, by way of the linking of computers across federal agencies and, eventually, across corporations, without any regulatory regime whatsoever.[21] They had kicked the question down the road.

"This is not the place to speculate how society will achieve a balance between its desire for knowledge and its desire for privacy," Pool wrote in 1968. If not then, when?

In 1968, after the defeat of the National Data Center, Pool and his MIT colleague J.C.R.

Licklider drafted a proposal for a giant "data library." They called it Project Cambridge and pitched it to ARPA, submitting a proposal with a giant budget, $7.6 million (about $56 million in 2020 money). Pool hoped to build Project Cambridge on the ashes of the Simulmatics Corporation. Like the National Data Center, it became the subject of fierce public debate. Big, splashy photographs of Pool and Licklider appeared together on the cover of the glossy magazine *Scientific Research,* the "newsmagazine of science," with a giant red question mark between them: "Will Project Cambridge Go?"[22]

Licklider, fifty-four, had the slightly bedraggled look of a beloved high school math teacher. He was famously visionary, and just as famously kindhearted. He is widely credited with coming up with the idea for the Internet. Trained in both psychology and engineering, he'd taught at MIT in the 1950s, then worked at the federally funded Lincoln Laboratory, on the SAGE early warning system. After that, he'd been employed at the Cambridge research firm Bolt Beranek and Newman, where he'd contributed to the very sort of research that would make possible MIT's ambitious, ARPA-funded computer research program, Project MAC, the Project on Mathematics and Computation. He imagined the Internet in a series of extraordinary essays, as well as in a

1962 memo in which he proposed the development of what he jokingly called an "intergalactic computer network," a system that would link all the world's computers into a single system. With Licklider's paper, the human conception of computers took a crucial turn, from understanding computers as places that could store information and machines that could analyze it to understanding computers as communications systems.

While Licklider was at Bolt Beranek and Newman, he wrote a report published as *Libraries of the Future,* a beautiful and hopeful set of speculations about the nature, scope, and acquisition of knowledge in the digital age. Licklider imagined what libraries would look like in the year 2000. To help his reader picture what he pictured, he conjured a scene set in 2000 in which a person sits at a computer console and attempts to get to the bottom of a research question merely by undertaking a series of searches. Nearly all of what Licklider described in *Libraries of the Future* later came to pass: the digitization of printed material, the networking of library catalogs and their contents, the development of sophisticated, natural language–based information-retrieval and search mechanisms.[23] Licklider described, with a contagious amazement, what would become, in the twenty-first century, the Internet at its very best.

In 1962, Licklider left Bolt Beranek and Newman for ARPA, where his many duties included funding behavioral science projects, including Pool's Project ComCom. At ARPA, Licklider also funded research that produced the building blocks of his imagined intergalactic computer network, which came to be called ARPANET. (Paul Baran, at RAND, pioneered the technology known as packet switching, in which data is broken into smaller pieces for faster transfer.) In 1964, not long before Simulmatics began its work in Vietnam, Licklider left ARPA for IBM. Four years later, with nearly all the necessary pieces of ARPANET in place, ARPA awarded a contract to Licklider's old Cambridge research firm, BBN, to build it.[24]

By the time Licklider returned to MIT as a professor of electrical engineering and then director of Project MAC, he'd known Pool for years. In the fall of 1968, Pool wrote to MIT electrical engineering professor and former Kennedy administration science adviser Jerome Wiesner, copying Licklider, to say that ARPA had expressed interest in a major new project. "ARPA has taken its Behavioral Science funds derived from cancelling such projects as mine," Pool told Wiesner, referring to ARPA's termination of its contracts with Simulmatics, "and is proposing to put them into a Behavioral Science Data Analysis Program here."[25]

In New York, Ed Greenfield had failed to sell Simulmatics, but in Cambridge, Pool launched a different sort of salvage operation. As Pool explained to Licklider, this new ARPA-funded project, a behavioral science data analysis program, would require rethinking and expanding MIT's computer facilities to allow for more use of the institute's machines by behavioral scientists and more space for the storage of what Pool referred to as a "data library" that would also store "a set of interactive subsystems for analysis and modelling with that data in an interactive environment."[26] He listed, as the project's first rationale, defense work: "The U.S. Government and the Department of Defense face many problems that are in large part behavioral-science problems, and they need pertinent behavioral-science knowledge for use in solving these problems."[27]

Licklider signed on to what he and Pool at first called Project CAM, short for "computer analysis and modeling," a name they later changed to "Cambridge" because "Project CAM" was easy to confuse with Project MAC.[28] "Project CAM" also stirred bitter memories of Project Camelot, the defense-funded counterinsurgency project, another Pool brainchild, that Robert McNamara had been forced to terminate in 1965. Project Cambridge also, in some minds, raised the specter of the National Data Center, only this time with funding from the Department of Defense. Pool braced for a fight, even as students at MIT were beginning to raise concerns about Pool's earlier work . . . for Simulmatics.

Richard M. Nixon, fifty-five and more furious than ever, accepted the nomination of the Republican Party at its convention in Miami in August 1968. Norman Mailer went to Miami to cover the convention. He found it boring. It was hip to be bored. "Interviewing a candidate is about as intimate as catching him on television," Mailer wrote. The GOP had flown in a pachyderm. Mailer went to the airport, he said, "to cover the elephant."[29]

Nixon knew he could win. He hungered for it. In his acceptance speech, delivered before a sea of delegates wielding blood-red NIXON signs, he held the edges of the podium with both hands as if to stop himself from raising his arms to the sky in defiance of the gods, until he began raising his arms and shaking his fists. The assassinations, the riots, the war: Nixon blamed Democrats. The grief, the humiliation, the fear: Nixon blamed Democrats. "When the strongest nation in the world can be tied down for four years in Vietnam with no end in sight, when the richest nation in the world can't manage its own economy, when the nation with the greatest tradition of the rule of law is plagued by unprecedented lawlessness," he said, "when the president of the United States cannot travel abroad or to any major city at home without fear of a hostile demonstration, then it is time for new leadership for the United States of America." The crowd rose to its feet, stomping like a herd of elephants, and gave him a standing ovation.[30]

The Democrats met three weeks later for their own convention, in Chicago, brokenhearted, divided, and leaderless. Inside the hall, there was chaos, betrayal, and regret. Outside the hall, there was chaos, violence, and bloodshed. The streets of Chicago ran red. Students for a Democratic Society, the National Mobilization Committee to End the War in Vietnam, the Black Panthers, and the Youth International Party had all planned to protest the convention. Some ten thousand demonstrators came to the city. But the mayor of Chicago, Richard Daley, was determined to stop them. He fortified the city with twelve thousand Chicago police officers, six thousand soldiers from the National Guard, six thousand army troops, and a thousand undercover intelligence agents. They arrested hundreds of demonstrators and beat and bullied more. The leaders, who'd also organized the march on the Pentagon in 1967, went to jail: David Dellinger and Jerry Rubin were two of the Chicago Seven, charged with conspiracy and crossing state lines to incite riot. The Yippies held their own nominating convention, in Lincoln Park, and nominated a pig, handing out flyers that said, "VOTE PIG IN 68."[31] The Republicans had a pachyderm; disenchanted Democrats answered with swine.

Lyndon Johnson had withdrawn from the race. Robert F. Kennedy had been killed. Democrats had scrambled. Hubert Humphrey hadn't entered a single primary but nevertheless defeated the antiwar candidate, Eugene McCarthy, mainly through the machinations of party leaders. Fannie Lou Hamer's eloquent protest in 1964 about the immorality and illegality of all-white delegations didn't stop the party from seating them in 1968. That did end, though, in 1968. After the convention, South Dakota senator George McGovern, who'd also unsuccessfully sought the nomination, agreed to co-chair the Commission on Party Structure and Delegate Selection. Its recommendations changed the rules by which the Democratic Party operated: it established requirements for the representation of people of color on state delegations and shifted the balance of power between the primaries and the national nominating conventions, predicting that "instantaneous polls of the entire electorate" conducted by "central computers from every home" would make nominating conventions obsolete and purely ceremonial. The primaries became, effectively, binding. Before 1968, the primaries had hardly mattered; after 1968, the conventions hardly mattered.[32]

Simulmatics couldn't offer its services to the Humphrey campaign, since it had mainly closed shop. But after the Democratic convention, Ithiel de Sola Pool — who may have submitted the "Simultron Project" memo to the Nixon campaign earlier in 1968 — submitted a memo to the Humphrey campaign, offering advice about how he could hold the party together. "Have Humphrey meet at length and 'reason together' with some dramatic newsmaking groups," Pool suggested. For instance, he could "meet in private in Harlem with 8–12 black extremists such as Claude Brown," then get photographed on his way out. "Ditto in Berkeley or Chicago with a group of photogenic hippie students." Pool also recommended that Humphrey assemble a task force on Vietnam, to include Pool, Samuel Popkin, and Daniel Ellsberg. In two days, these men could come up with "their ideal program on how to de-escalate and still win." Or, better yet, "why not pack the Vietnam task force on an airplane" and send them on a much-publicized trip to Vietnam, where Humphrey could join them, appearing at refugee camps and other photogenic whistle-stops.[33]

The Humphrey campaign did not take

Pool's suggestions. Nor did Humphrey hold together a Democratic coalition. American politics had begun to polarize, the New Right pulling the Republican Party toward an anti-government conservatism, the New Left pulling the Democratic Party toward an anti-government radicalism. They shared a common enemy: liberalism. In 1968, Arthur Schlesinger, the quintessential midcentury liberal, found himself attacked when he spoke in public, heckled, howled at, shouted down. "You know what you are?" a man said to him after a speaking engagement. "You're a murderer, a murderer, and a traitor, and a mother fucker. It's against the wall for you. Do you know what's going to happen? You are going to be executed."[34] The worst William F. Buckley Jr. had ever done to Schlesinger had been to ship him a donkey or mock him when Schlesinger declined to appear on Buckley's talk show, *Firing Line*.[35] The New Left threatened him with an actual firing line.

Schlesinger retreated from those skirmishes and backed away from public life. By 1968, he'd also separated from his wife; she stayed in Cambridge in the house on Irving Street, while Schlesinger settled in New York, began teaching at the Graduate Center of the City University of New York, and became rather a man about town. Pool could not avoid those skirmishes. He was still living on Irving Street and still teaching at MIT, a campus that had become a hotbed of student unrest.

Even as late as 1967, MIT had mostly been free of antiwar protest. But no American university took in more funds from the Department of Defense. Of MIT's $214 million 1968 budget, $173.8 came from the federal government, and of that, $111 million came from the Department of Defense.[36] On December 13, 1967, on the floor of the Senate, J. William Fulbright indicted the Johnson administration for dedicating new and ever greater resources to an unwinnable, immoral war and argued that American universities funded by the Department of Defense had abandoned the role of the university in a democracy. "The universities might have formed an effective counterweight to the military-industrial complex by strengthening their emphasis on the traditional values of our democracy," Fulbright said, "but many of our leading institutions have instead joined the monolith, adding greatly to its power and influence." He denounced any university that had made "itself an appendage to the Government," a description that fit no American university better than MIT.[37]

Meanwhile, the Johnson administration and its supporters sought to calm the student protest movement. In the fall of 1968, in advance of the election, McGeorge Bundy, a former dean of Harvard who'd served as

Kennedy's and Johnson's national security adviser, now president of the Ford Foundation, went on a mission to speak on college campuses. On October 12, he spoke at DePauw University.[38] He was scheduled to speak at MIT four days later. He'd been invited by the Political Science Department, where Pool was chair. The department had arranged for Bundy to give a lecture, to be followed by comments from a panel that included Pool and Max Millikan, head of the Center for International Studies. After students objected to the composition of the panel, Pool agreed to add Noam Chomsky.

They were slated to speak at Kresge Auditorium, the largest building at MIT, glass walls topped by a curved metal roof, a building that resembled nothing so much as an indoor shopping mall from *The Jetsons*. The main hall seats some twelve hundred people. Bundy's visit filled Kresge to capacity, with an overflow of students seated in the nearby Student Union. Bundy delivered a version of the address he'd given at DePauw, offering a softening of his earlier positions on the war by advocating a stop to the bombing and a gradual troop withdrawal. Pool and Millikan likely concurred. But then, as the MIT student newspaper reported, "McGeorge Bundy ran head on into the formidable Noam Chomsky." Chomsky denounced the war and called for an immediate withdrawal.

Pool fought back. The battle had a certain gladiatorial quality: "the veracity of not only ideas but evidence on all sides was challenged," as the student reporter delicately put it.[39] The battle lines between Pool and Chomsky had been drawn. Two and a half weeks later, Richard Nixon was elected president of the United States.

Bundy's visit, planned to quiet the student movement, instead ignited it. Chomsky didn't lead the student movement. He wasn't even convinced that the student movement was particularly effective; nor did he agree with its tactics or priorities. But, notwithstanding his tendentiousness, and even because of it, he had become a national leader of the national antiwar movement, a prominent political dissident, and one of the New Left's most visible intellectuals. After the Kresge forum, he began writing an essay that stated his objections to Bundy and, above all, to Pool and, especially, to the Simulmatics Corporation. It appeared in the *New York Review of Books* on January 2, 1969, as "The Menace of Liberal Scholarship."

"The Menace of Liberal Scholarship" is both a fierce and relentless indictment of liberal intellectuals for their complicity in a war of American imperial aggression and a quite personal attack on Ithiel de Sola Pool. Chomsky named Pool seventeen times and discussed two of Simulmatics' projects in

477

Vietnam, projects that, before that day, had been very little in the public eye. (How Chomsky knew so much about Simulmatics is somewhat mysterious, but people who worked in the New York office — the office that had been repeatedly cited by ARPA for lapses of security strictures, including leaving classified reports lying around — told me that Chomsky had often stopped by to visit Urban Studies Division director Sol Chaneles.)[40] In the most widely cited part of the essay, Chomsky quoted Pool as having written, as in fact Pool did write, that in poor countries like Vietnam, "order depends on somehow compelling newly mobilized strata to return to a measure of passivity and defeatism from which they have recently been aroused by the process of modernization." Pool defended his work in Vietnam as part of a campaign not only to counter insurgency and halt revolution but also to accomplish those objectives by reducing people like the Vietnamese to despair and docility.[41]

Pool, meanwhile, continued to seek favor in the Pentagon, allying himself with the incoming administration and offering his continuing services as a behavioral scientist. The day Chomsky's essay appeared, Pool wrote to Henry Kissinger, Richard Nixon's national security adviser. Pool asked Kissinger to help him secure funding to allow Joseph Hoc to continue with his psychological warfare operation in Saigon.[42] (He was unsuccessful.) Days later, Pool traveled to Key West, by military transport, for a meeting of the Defense Science Board.[43] Later that month, possibly under pressure, he resigned as chairman of MIT's Political Science Department.[44]

On January 15, days before Nixon's inauguration, the Department of Defense completed the report Robert McNamara had commissioned in 1967 about the history of U.S. involvement in Vietnam, beginning in 1945. It ran to forty-seven volumes and included four thousand pages of documents and three thousand pages of analysis. It weighed sixty pounds. The Pentagon made fifteen copies: seven remained at the Department of Defense; two were deposited at the National Archives; two were sent to the State Department, one to McNamara, and one to the new secretary of defense. Two copies were made for RAND; Daniel Ellsberg was charged with carrying them to RAND's office in Santa Monica. It is not impossible that Pool read the report.[45]

After the attack by Chomsky, only a handful of Pool's supporters came to his defense. A Political Science Department graduate student proposed organizing an ad hoc seminar on Chomsky's "Menace of Liberal Scholarship," specifically in order to dissect it as "a confused collection of <u>non sequiturs,</u>

slurs, unreliable references cited as fact, and quotes taken out of context and misconstrued" that "completely distorts Pool's position." The graduate students held an organizing meeting, Chomsky indicated that he was willing to participate, but the seminar never came off.[46]

In February, Pool engaged in some back-and-forth with Chomsky in the *New York Review of Books,* to little avail.[47] Schlesinger, reviewing Chomsky's new book, *American Power and the New Mandarins,* wrote that "Chomsky's idea of the responsibility of an intellectual is to forswear reasoned analysis, indulge in moralistic declamation, fabricate evidence when necessary, and shout always at the top of one's voice."[48] But the students who were listening to Noam Chomsky had no use for Arthur Schlesinger.

The antiwar movement on campus had grown; it had also divided between those favoring peaceful protest and calling for an end to defense-funded work and those demanding violent action. The first formed the Science Action Coordinating Committee, or SACC, while those more willing to commit violence joined a militant faction of Students for a Democratic Society, or SDS. The student members of SACC, along with forty-eight MIT faculty members, called for joining a nationwide strike and a teach-in on

March 4, 1969, aimed at defense scientists working on university campuses. Student protesters called MIT "the Second Pentagon."[49] Both groups, in leveling their charges, almost always singled out Pool, who, unlike many of his colleagues, had the courage to engage students in debate. That meant that reporters almost always singled him out, too. In April, a *Boston Globe* columnist called for people "to flush out the names of university professors doing government research on campus and with military-industrial affiliations." One name the reporter already knew was that of Ithiel de Sola Pool, a man he described as "a large shareholder (and founder) in Simulmatics Corp."[50] Confronted by MIT students in April, Pool refused to list his government contracts and consulting obligations.[51]

Because Chomsky had named him, and because Pool didn't retreat, Pool found himself in considerable peril, his very safety threatened. Someone planted a bomb in the bathroom of the Political Science Department. Someone threw Molotov cocktails at Pool's house on Irving Street. Eleven-year-old Adam, home alone after school one day, answered the door to find his neighbor Daniel Patrick Moynihan on the doorstep; given the new threats on Ithiel's life, Moynihan had come to bring Adam over to his house, where he'd be safe.[52]

The attacks on Pool, as well as on Simulmatics, soon extended to Project Cambridge. In May, students announced a planned protest of the new project. "I hope we will have substantial department representation at the rally about Project CAM on Kresge lawn, Tuesday at noon," Pool wrote to his colleagues in the Political Science Department, asking for faculty support.[53] Another memo, filed in Pool's papers, is unsigned, but it appears to have been a draft of remarks Pool intended to deliver to his colleagues, urging them to join him in engaging with the students, which was one of Pool's commitments. "The Department of Defense is not the KKK," the memo reads. "It is an agency of the government of the United States, and it has a perfectly legitimate interest in certain kinds of social science. Yet, these days, for obvious and painful reasons, association with DoD offends some people. Shall we ignore their sense of offense, on the grounds of academic freedom, or shall we somehow accept it? For those who are offended, misguided or no, are members of our community, and to ignore their anger, whatever the substantive decision regarding the project, is not only to invite trouble — there are, after all, issues worth standing up for — but also to exclude them from the community."[54]

Pool's commitment to debate, though, didn't suit the times. The protesters continued to attack both him and Project Cambridge, which, in student pamphlets and leaflets, they invariably associated with Simulmatics. "Pool employs professors and graduate students to make secret pacification studies for the US in South Vietnam," one indictment read. As for his latest initiative, "Behind all the glop, Project CAM is specific. Funds will go to back Pool's work."[55] That said, students weren't alone in objecting to Project Cambridge. Distinguished MIT faculty members objected, too, including four of Licklider's colleagues in the School of Engineering: they objected to its affiliation with the Department of Defense.[56]

The national press picked up the story in May.[57] A feature story about the feud between Pool and Chomsky appeared in the *New York Times Magazine*.[58] A *Washington Post* syndicate piece on "secret research" on college campuses called Pool "one of the most successful of these Defense-endowed intellectuals," associating him with Project CAM, which it linked to Project Camelot, and describing him as "a co-founder as well as a major stockholder and director of Simulmatics, a thriving private research and consulting firm which has done extensive work in South Vietnam."[59] This coverage did not help efforts to find a buyer for Simulmatics.

Even as indictments of Simulmatics were rolling off the presses, Simulmatics, so far from thriving, neared bankruptcy. "At this moment we have a substantial tax loss carry-forward," Pool wrote to a financial consultant, begging for advice.[60] Pool hired a Harvard Business School–trained consultant named George Leyland to write up a new business plan. Leyland had worked for the Census Bureau, and also for Honeywell, in its Data Processing Division.[61] Pool hoped Leyland would become Simulmatics' new president. "Sales promotion activities in South East Asia are the only current activities of the company," Leyland reported, recommending that even if new contracts came in, these should be separated off into another company and Simulmatics should concentrate on "social science oriented computer applications."[62]

That summer, when the campus emptied and quieted, Pool began distributing Leyland's report to potential investors.[63] He tried to enlist support for restructuring the company.[64] But by the middle of August, he'd begun informing inquirers that "the Cambridge office of Simulmatics is no longer functioning."[65] The Saigon office had closed. The New York office had closed. The Cambridge office had closed. What was left? When MIT students came back to campus in September 1969, lugging duffel bags, and got agitated about Simulmatics all over again, they were protesting a company that existed only on paper, punch cards, and magnetic reels.

More than half a million American troops were stationed in Vietnam in 1969. Nearly twelve thousand were killed in action that year. Americans were tired of the war and they were scared of the war and they were angry about the war — a war the United States would not get out of for four more long years. That fall, Daniel Ellsberg and Anthony Russo smuggled the Pentagon Papers out of the offices of RAND, in Santa Monica. They'd decided to leak the report to the public. Ellsberg left RAND and accepted a position at MIT, as a senior research associate at the Center for International Studies, housed in the Hermann Building, the same building as the Political Science Department. He and his wife moved into a house next door to Sam Popkin and his wife. Popkin, who finished his PhD in 1969, accepted a position as an assistant professor of government at Harvard.

"I don't want to talk anymore, so you people aren't going to be disturbed by talk," the MIT student government president told incoming freshman at the beginning of the fall semester. "You're going to be disturbed by action."[66] In September 1969, at an all-day teach-in, Licklider patiently defended Project Cambridge. He had an extraordinary capacity to keep calm. He pointed out that students had been invited to the Project Cambridge planning meetings but none had ever shown up. He issued the invitation all over again.[67] But he was overshadowed by the day's other speakers: Noam Chomsky and the Boston University political scientist and activist Howard Zinn.[68]

In October, about 150 students demonstrated at MIT's Center for International Studies, where both Ellsberg and Pool had offices. Demonstrators handed out leaflets that read, "Meet your local war criminal; come to the CIS." Ellsberg had by now begun speaking against the war, allying himself with Chomsky and Zinn; secretly he was also engaged in the very dangerous work of trying to get the Pentagon Papers into the hands of the chair of the Senate Foreign Relations Committee, J. William Fulbright. But not all the student organizers, engaged in a Stalinist-style purge of intellectuals with whom they disagreed, were convinced that a man from RAND could be trusted. One leaflet argued that Ellsberg, despite his "conversion" to the antiwar cause, "is still suspect." But, they said, "People's Enemy No. 1 is Ithiel de Sola Pool." Forty students confronted Pool in his office foyer, demanding answers about Project Cambridge and Project ComCom.[69] Mainly, though, they indicted Simulmatics: "In 1959 Pool started Simulmatics, a corporation to sell the software he developed to the government. <u>Its major contract was development of the strategic hamlet program in Vietnam</u>."[70]

The student groups made demands, but they didn't all make the same demands. MIT SDS demanded the immediate termination of seven research projects, including Project ComCom and Project Cambridge.[71] Some groups called for Pool and other members of the faculty to be fired, comparing them to German scientists who did research for the Nazis, although, as one flyer pointed out, other students had asked, "So what if they're fired? If their work is so important, won't they just go to the RAND Corp. (or Simulmatics)?"[72]

When students protested outside his door, Pool invited them into his office to talk, but when asked to participate in a public debate, without assurance about the rules, he refused. "I will not appear at a mock trial by Storm Troopers," he said.[73] He'd begun to find it impossible to remain calm. Maybe he could understand the attacks on Simulmatics. But the attacks on Project Cambridge — on no other grounds than its ARPA funding and Pool's involvement — drove him over the edge. "We're not here to enslave people," Pool told a reporter. "But if people think the ultimate plan of the American Government is to enslave the world, they might think for a moment about the possibility of using Project

Cambridge to improve the goodness of life."[74]

MIT students began, that fall, to attack not only Pool and Simulmatics and Project Cambridge but also, by association, ARPANET. In the fall of 1969, while college campuses thundered with the sound of protests, ARPANET went live for the first time, establishing a connection between computers at Stanford and UCLA.[75] "What hath God wrought?" Samuel Morse had tapped out in the first message sent by telegraph in 1844. At ten-thirty p.m. on October 29, 1969, a single message traveled from Los Angeles to Menlo Park, California: "LO," the first two letters of the word "LOGIN," before the system crashed. What would become the Internet had dawned, with very little sense of it having been wrought by God.

This demonstration didn't attract attention, except in technical circles. But MIT was a technical circle. And, possibly because ARPANET was understood to be a brainchild of Licklider's, some student protesters began to conflate ARPANET with Project Cambridge, as a de facto National Data Center.

"The whole computer set-up and the ARPA computer network will enable the government, for the first time, to consult relevant survey data rapidly enough to be used in policy decisions," one SDS publication warned. "The so-called basic research to be supported by Project CAM will deal with questions like why do peasant movements or student groups become revolutionary. The results of this research will similarly be used to suppress progressive movements." The Octoputer, they called it, adopting the name of an actual computer, a time-sharing system manufactured by RCA.[76]

Pool and Licklider had invited Harvard to jointly sponsor Project Cambridge, because it made sense as a multi-campus initiative but also because they hoped Harvard's affiliation would dilute the appearance that the project was yet another MIT defense project. But Harvard Yard had witnessed plenty of its own unrest. In the spring of 1969, the day after the Harvard-Radcliffe SDS pinned to the door of the president's house a list of demands, including the end of ROTC on campus, Harvard students occupied University Hall after forcibly carrying a dean outside and chaining the doors shut. The university called in city and state police. With a battering ram, the officers forced their way into University Hall and removed the students, beating many with clubs, and arrested more than two hundred. Undergraduate Chris Wallace, son of the CBS newsman Mike Wallace (and himself later a Fox News host), reported on the scene for the student radio station; he, too, was carried away by the police. Given the chance to make a phone call, he called the radio station and filed a report, signing off, "This is Chris Wallace in custody."[77]

By the fall of 1969, the state of affairs at Harvard made its involvement in Project Cambridge far stickier than it had been a year before. Harvard therefore convened a faculty committee to consider the question of whether or not to join with MIT in the project. The committee solicited letters from members of the university community, to be held in confidence. The responses were evenly split, the community bitterly divided. The proposal drew strong support, mainly on the back of the argument that Harvard's social sciences would be left behind if the university declined to participate. But those who opposed participation were not limited to student groups like the Harvard-Radcliffe SDS. A PhD political science candidate wrote the committee about "limitations as to the price an academic community can afford to pay in the pursuit of knowledge."[78] An eminent cognitive psychologist called the proposal a "hasty mess."[79] A philosopher asked the committee to evaluate the proposal alongside the ethical commitments of Kant and Rawls.[80] An engineering graduate student declared, in the statement most commonly made by those who opposed the initiative, "I believe that the purposes of this project are immoral."[81]

But no one objected to Harvard's participation more eloquently than the well-known political sociologist Barrington Moore, who vigorously opposed Project Cambridge in a twelve-page letter to the committee. Computers, Moore argued, had been chiefly used by social scientists who were blinded by "the illusion of technical omnipotence." Far more carefully than Chomsky, Moore made the case against government-funded behavioral science. "In Vietnam the combination of liberal paternalism with this illusion of technical omnipotence has encountered a defeat that has exposed not only the futility of this combination but also its cruelty and moral bankruptcy," Moore wrote. "What was intended to bring freedom, abundance, and happiness to the world has turned quite literally to ashes, bringing in place of happiness death, terror, destruction and disease."[82] Moore wanted to know: By what possible measure had these people earned the trust of their colleagues for the support of future work?

Even as the faculty committee was readying for a meeting, Harvard students marched into the dean's office, demanding "the total dismemberment of Project Cambridge." Outside the building, they chanted, "Stop Cam!"[83] David Bruck, an undergraduate who later became a defense attorney, wrote a scathing attack on the project in the student newspaper, the *Harvard Crimson*. "It was not to be an information gathering project as Camelot had been, but would center instead on developing new ways of using and interpreting behavioral science data," he wrote, depicting the project as a cynical attempt by Pool to squeeze yet more money out of ARPA for already discredited work. Pool and Licklider had looked to Harvard out of a need to cloak this act of intellectual prostitution because, Bruck wrote, "M.I.T. is the Defense Department's house whore."[84]

Pool wrote a lengthy response to Bruck's article, in which he compared the student rebellion to McCarthyism and pointed out that Bruck had not even attempted to interview him. The *Crimson* declined to publish his letter.[85] Nor did the efforts of MIT's Project Cambridge advisory committee to allay suspicion quiet the protests.[86] By late fall, those protests had risen to charges of war crimes.

On November 4, 1969, during a long-planned, coordinated demonstration at MIT known as the "November Actions," protesters occupied the president's office for three hours and, in a corridor outside, held a mock trial of Pool and three of his colleagues for war crimes.[87] They'd advertised the trial with posters pasted all over campus, featuring photographs of the four men.

WANTED
FOR SERVING U.S. IMPERIALISM,
FOR PLANNING AND DEVELOPING
COUNTERINSURGENCY TECHNIQUES
AT THE
CENTER FOR INTERNATIONAL STUDIES
ABOLISH THE CIS

Outside the Hermann Building, in a slashing rain, they directed their ire at Pool and his colleague Lucian Pye, chanting, "We won't die for Pool and Pye." About a hundred faculty and staff members and graduate students staged a counterprotest, wearing blue arm bands on their raincoats and carrying a banner they'd stitched together from blue sheets, urging an end to the war and celebrating the right to disagree.[88] They were defending what remained of liberalism.

But Pool, in the tumult and the terror, would no longer defend liberalism, instead following a path taken by so many former Trotskyites turned Cold Warriors, toward neoconservatism. He left liberalism behind. He left Simulmatics behind. He left Project Cambridge and Project ComCom behind. After 1969, he dedicated himself to writing about technologies of communication, and the communications revolution, and its implications for political life. He wrote the founding political theory of the Internet. He became, to technological utopians, a prophet.

"The propagandists of the SDS have found it convenient to make the name Pool a codeword," he once complained, exhausted. They'd done the same with the Simulmatics Corporation, made it a codeword, used it as a buzzword, a cudgel. In the spring of 1970, antiwar activists marched from Boston Common to the last known address of Simulmatics in Cambridge, 930 Massachusetts Avenue, a tiny little house halfway between Harvard and MIT. The office was closed. The office was empty. There was no office.[89]

And then, like everyone else in America, they forgot all about the Simulmatics Corporation. They forgot what it had once done, and what it meant, and what it had left behind, and how it had invented the future.

CHAPTER 14
THE MOOD CORPORATION

"What about the computers? Is that real data you're running through the system or is it just practice stuff?"

"You watch," he said.

— Don DeLillo, *White Noise,* 1985

Simulmatics underwent bankruptcy proceedings in a courthouse in Manhattan on August 26, 1970, a day that happened to mark the fiftieth anniversary of the Nineteenth Amendment, guaranteeing women the right to vote. To honor the anniversary, fifty thousand people, mostly women, marched through the city in the Women's Strike for Equality. In miniskirts and bell-bottoms and sheath dresses, flowered and paisley, they hooked arms and raised clenched fists and carried signs that read, W♀MEN UNITE! and EVE WAS FRAMED and ERA YES. They marched for equal rights and equal pay, for lesbian and gay rights, and for an end to the war in Vietnam. Organizers handed out questionnaires: "Have you ever resented it, even a little, that

495

almost all the important political decisions are made by men?"[1] Naomi Spatz had loaned money to Ed Greenfield; she had to go to court to be registered as a Simulmatics creditor. She did that, and then she walked out of the courthouse and joined the march.[2]

Simulmatics died. The fantasy of predicting human behavior by way of machines did not. Instead, it took new forms, forms that depended on forgetting that Simulmatics had ever existed.

By the time Simulmatics declared bankruptcy, the automated computer simulation of human behavior had fallen into disrepute, a casualty of McNamara's simulated Vietnam, and of the disaster of urban renewal, its systems gone awry, its predictions proven wrong. In what one historian has called "the twilight of simulation," universities stopped dedicating funding to systems analysis and simulation, journals stopped publishing, labs closed. MIT shut down its Urban Systems Lab, once headed by Ithiel de Sola Pool, in 1974. For a while it seemed that simulation would live on only in the form of computer games; the first version of *SimCity* appeared in 1989.[3]

The people who'd worked for Simulmatics scattered. Peter Shulman, the onetime deputy director of Simulmatics' Urban Studies Division, went to Vietnam with RAND in 1967, but when he came back to the States the next year, he left New York City for the country-side and became a dairy farmer. He moved around for a long time before settling on a place he called Moon Shadow Farm, where he installed an outdoor war game, a simulation spread over thirty acres and involving "over 60,000 hand sculpted soldiers and more than 5,000 scale vehicles." They did battle on wooded hills, year after year after year, in an endless, haunted make-believe war.[4]

In 1968, Thomas Morgan, Simulmatics' onetime PR man, served as a press aide to the fervent antiwar candidate Eugene McCarthy. After McCarthy lost, Morgan became press secretary to the mayor of New York, John V. Lindsay. He later became editor of the *Village Voice* and then president of WNYC, a broadcasting conglomerate.[5] Alex Bernstein became a psychoanalyst. Father Joseph Hoc left a monastery outside Saigon for a rectory in Dorchester, Massachusetts, bringing along one of his brothers and some of his nieces and nephews. Ann Penner Winston and Gus Winston had a daughter together and then divorced. After the Tet Offensive, Kim Le and her fiancé, an American military adviser and former Rhodes scholar named Mike Cook, decided the war was hopeless and left Vietnam for the United States.[6] Cuc Thu Duong tried to get out of the country then, too: she wrote to Ithiel de

Sola Pool to ask him for a job, but in 1968 he said he couldn't help her.[7] She didn't manage to get out of Vietnam with her three children until just before the fall of Saigon in 1975. Maureen Shea supported Cuc's petition for political asylum, which was eventually granted by the Immigration and Naturalization Service; she settled in New Jersey.[8] In 1968, Shea moved to Wisconsin to help out on the campaign, for state treasurer, of Les Aspin, an MIT-trained military analyst she'd met in Saigon while working for Simulmatics. They married in 1969 and divorced ten years later. In 1993, Aspin became Bill Clinton's secretary of defense.

Minnow McPhee, her divorce finalized in 1970, remarried in Colorado in 1976. Ten years later she published a book of advice for new parents. She called it *Praise and Patience*.[9] Jim Coleman moved from Johns Hopkins to the University of Chicago. At the height of the fury over the Coleman Report, during the forced busing debates in the early 1970s, leading members of the American Sociological Association tried to revoke his membership; after the fury's ebb, he was elected the association's president. In 1970 Al de Grazia founded an alternative university, the University of the New World, in the Swiss Alps. When Pool came to speak as a guest lecturer, students protested his visit. A Swiss television journalist investigated the university for fraud.[10] In 1976, Daniel Patrick Moynihan won election to the Senate, as a Democrat from New York. He was succeeded, in 2001, by Hillary Clinton. Newton Minow, after serving in the Kennedy administration as chairman of the Federal Communications Commission, returned to law practice in Chicago at Sidley Austin. In 1988, Minow hired as a summer intern a young law school student named Barack Obama. Bill Clinton awarded Arthur Schlesinger Jr. the National Humanities Medal in 1998; Schlesinger died in 2007. Barack Obama awarded Newton Minow the Presidential Medal of Freedom in 2016, the month that Donald Trump, following a campaign aided by Facebook News and Cambridge Analytica, defeated Hillary Clinton to become the forty-fifth president of the United States.

After Patty Greenfield's death in 1968, her children went to friends' houses and boarding schools in western Massachusetts. Their father mailed them packages, fancy boxes of opulent gifts. Jennifer, the littlest, who'd lost her mother when she was only four, lived for a while with her elementary school teacher. When she was six and seven, her father would arrange for her to take a Greyhound bus to New York, where he lavished her with affection. He took her for lunch in the Grand Ballroom at the Plaza, like the children's book character Eloise. He took her shopping at Saks Fifth Avenue. He kept his suit pockets full of cash, wads of money. They toured apartments he could not possibly afford and studied brochures for vacations to Europe they would never take.[11]

He dreamed more dreams. He dreamed of understanding other people's feelings. He got an idea for another company. "I'm going to do this again," he'd say. It'd be easier this time. There was so much more data. Nothing could stop anyone from simply taking it, scooping up data like shovelfuls of sand from a beach. He died of a heart attack in 1983, only fifty-six years old, this dream unrealized, but he'd planned to collect so much data about so many things that he'd be able to tell how people were feeling and then to sell that information to other companies: a tracking service.[12] He called it the Mood Corporation.[13]

No one ever really publicly debated what Simulmatics had done. It had been too small a player, too weak a company, too clumsy a corporation. But a national debate did follow, a debate that buried the history of Simulmatics under its weight, like the wreckage after a storm.

Early in 1971, a Senate investigation demonstrated that the U.S. Army had collected data and conducted surveillance on civil rights figures, antiwar activists, and political dissidents and had stored those records in computers, treating American citizens as if they were foreign combatants. In a report entitled *Federal Data Banks, Computers, and the Bill of Rights,* a Senate subcommittee concluded that "the men who ran the domestic war room kept records not unlike those maintained by their counterparts in the computerized war rooms in Saigon."[14] Those war rooms in Saigon, if not the computers, became the subject of far more scrutiny just months later when the *New York Times* began publishing the McNamara-commissioned study of U.S. involvement in Vietnam from 1945 to 1967 that Daniel Ellsberg had smuggled out of his office at RAND in 1969. (Pool, on hearing that it was Ellsberg who'd leaked the Pentagon Papers, said, "Doesn't he realize he'll never get a clearance again?" "I was expecting to go to prison for the rest of my life," Ellsberg wryly said later, "and Ithiel wanted to know whether I understood that I'd never get another dollar from the federal government.")[15]

The McNamara report chronicled the lies and blunders of one administration after another, from Truman through Johnson. It did not indict the Nixon administration, but Nixon nevertheless resolved both to halt publication and to punish the leakers. On June 15, 1971, two days after the *Times* printed its first excerpt from the Pentagon

501

Papers, the U.S. Department of Justice filed suit in the U.S. District Court for the Southern District of New York seeking a temporary restraining order; the court granted the request. Days later, the *Washington Post* began publishing excerpts, after which Nixon's Department of Justice sought a similar restraining order in the U.S. District Court for the District of Columbia. The judge denied the request, a decision upheld the next day by the U.S. Court of Appeals for the D.C. Circuit. On June 23, the U.S. Court of Appeals for the Second Circuit remanded the *Times* case to district court. Finally, on June 30, the U.S. Supreme Court ruled, 6–3, that the U.S. government could not halt American newspapers from publishing the Pentagon Papers.

Thwarted in its attempt to halt the publication of the Pentagon Papers, the Nixon Justice Department redoubled its effort to punish the leakers. A grand jury in Los Angeles indicted Ellsberg and Anthony Russo on charges of theft and unauthorized possession of documents. A second grand jury was impaneled in Boston to identify people who had assisted Ellsberg in securing, photocopying, and releasing the papers. In particular, the Boston investigation sought knowledge of who, in Massachusetts, had helped get the papers from Ellsberg to *New York Times* reporter Neil Sheehan. But to a lot of people, the Boston grand jury looked like an attempt to hassle antiwar activists, "an uncalled for inquisition of Ellsberg's friends and associates."[16] Those friends included Simulmatics' Samuel Popkin.

Popkin, twenty-nine, had gotten friendly with Ellsberg in 1970 when, as neighbors in Cambridge, they and their wives spent a lot of time together. Popkin had been in Hong Kong when the Pentagon Papers were published, but on July 12, after he returned to the United States, the FBI questioned him in Cambridge.[17] Popkin told them he'd learned about the leaking of the release of the report only by reading about it in the newspaper. But the FBI also investigated reporters who went to Cambridge that summer. On July 21, *Detroit Free Press* reporter Saul Friedman, working on a profile of Ellsberg, stayed at the Sheraton Commander Hotel, on Cambridge Common, and spent nine minutes on the phone with Ithiel de Sola Pool. The next morning, Friedman called Ellsberg. The FBI acquired Friedman's phone records on August 8.[18] And the FBI's interest in Popkin increased when agents discovered that the Popkins and the Ellsbergs used the same maid, a Chilean woman, whom the FBI questioned repeatedly, not least because she told an agent that "chief among her duties at the Ellsbergs' was tidying up Daniel Ellsberg's office when he slept."[19]

On August 11, a Special Investigations Unit whose formation Nixon had ordered immediately after the release of the Pentagon Papers reported to the president that Popkin was on a list of people to be subpoenaed by the grand jury in Boston, a list that included Noam Chomsky and *New York Times* reporter David Halberstam.[20] Most of those subpoenaed by the Boston grand jury did not cooperate. Halberstam pleaded the First Amendment. On August 19, 1971, the prosecutor asked Popkin, "What is your opinion as to persons you believe possessed a copy of the Pentagon Papers in Massachusetts prior to June 13, 1971?" Popkin refused to answer.

Critics of the Nixon Justice Department's persecution of dissenters viewed the Boston and Los Angeles investigations as two arms of the same inquiry, denounced both as unconstitutional, and understood them as of a piece with the U.S. Army's domestic surveillance program. "The grand jury probe, which continues at high speed in Boston and in Los Angeles, is a direct threat to the anti-war movement and to the constitutional rights which have traditionally allowed such movements to exist," a reporter for the *Harvard Crimson* wrote in September 1971. "These events are not occurring in a vacuum," the *Crimson* writer continued. "For the past two and a half years, the Administration's cavalier views of civil liberties have combined with recent decisions by the federal judiciary to turn the grand jury system into a powerful political weapon in the government's battle with the American left."[21]

Popkin, who had long since become an opponent of the war, fought to protect his sources and he also fought for scholarly independence from the federal government. In defying the grand jury, he argued that to answer questions about his knowledge of Ellsberg and the Pentagon Papers would be to risk jeopardizing people he'd interviewed in Vietnam, beginning in the summer of 1966, including top military brass who'd spoken to him in confidence. He refused to betray them. Called twice more, he twice more refused to answer, citing a "scholar's privilege" to protect the sources of his research, akin to the protection claimed by journalists as a "reporter's privilege."

Popkin's plight had very little to do with Simulmatics, other than the fact that it was working for Simulmatics that had taken him to Vietnam in the first place, but, especially after the tumult on the campuses of Harvard and MIT in 1968 and 1969, Popkin's dilemma called into sharp relief the question of where the line is to be drawn between the university and the government. The Harvard Faculty Council voted unanimously to support Popkin's right to refuse to speak to the grand jury. Twenty-four political and social scientists from Harvard and Yale faculty submitted affidavits in his support.[22]

Pool by now publicly supported the Nixon administration and, before the election of 1972, actively advanced the Nixon campaign. He told a reporter he'd been forced to leave the Democratic Party. "It is the party that has turned around, not me," he said.[23] In 1950, when Pool and Nixon were both much younger men, Nixon had come to Pool's aid, using his influence to help Pool get a security clearance. They'd been Cold Warriors together, sharing foreign policy objectives even as they disagreed about a domestic agenda. Pool had worked for every Democratic presidential candidate since Adlai Stevenson's campaign in 1952. But the Vietnam War had realigned the parties. In 1968, Pool had wavered. In 1972, he threw his weight behind Nixon.

But Nixon, running for reelection, had grown more paranoid and more vengeful than ever. He'd also become dangerously reckless. On June 17, 1972, five men broke into rooms in the hotel that served as the headquarters of the Democratic National Committee: the offices of Lawrence O'Brien (whom Simulmatics had reported to during the 1960 Kennedy campaign). These five men, who came to be known as the "plumbers," were trying to bug the DNC — or, rather, to repair bugs they'd installed during an earlier break-in, at the end of May. After bugging O'Brien's office, they were supposed to go next to the offices of the presumed Democratic nominee, South Dakota senator George McGovern, but they never got there. Instead, they were caught by a security guard and arrested by the police. Quite who they were, and who'd hired them, was not immediately clear.

Three days after the arrests at the Watergate, Daniel Patrick Moynihan, who worked in the Nixon administration as an assistant to the president on domestic affairs, alerted Nixon aide H. R. Haldeman that "an amazing number of people around Harvard and MIT are extremely upset at the prospect of McGovern's nomination" and named "one particular" individual that the Nixon campaign ought to reach out to: Ithiel de Sola Pool.[24] By October, Pool had been named an honorary vice chair of Democrats for Nixon.[25] He joined Scholars for the President, with fifty-four other academics, including Irving Kristol, Milton Friedman, and Robert Bork, and signed a letter published in the *New York Times* on October 15, supporting the Committee for the Re-election of the President.[26]

While Pool avowed Nixon, Popkin's persecution at the hands of the Nixon Justice Department became a national political cause. As the *New York Times* reported on November 1, the grand jury had become entangled "in an almost impenetrable thicket of legal objections raised by a group of doggedly recalcitrant witnesses." Popkin did appear before the grand jury but, as the *Times* reported, "the jury of 23 mostly dour, middle-aged Bostonians heard hardly a word of testimony."[27] Cited for contempt, Popkin appealed to the U.S. Supreme Court, which, in *United States v. Popkin,* refused to hear the case.[28] He was next scheduled to appear before the grand jury in November 1972, just after the election.

To the public, the June 17 break-in at the Watergate Hotel looked like little more than a bungled burglary until October 10, 1972, when the *Washington Post* published the first of a series of stories reported by Bob Woodward and Carl Bernstein: the break-in, the FBI now confirmed, had been part of "a massive campaign of political spying and sabotage conducted on behalf of President Nixon's reelection and directed by officials of the White House and the Committee for the Re-election of the President."[29]

Nevertheless, on November 7, Nixon was reelected, winning more than 60 percent of the popular vote, eviscerating McGovern, who lost even his home state of South Dakota and won only a single state: Massachusetts. "What ever happened to Watergate?" an ABC News correspondent asked Nixon aide John Ehrlichman during the jubilation at the

Nixon campaign on Election Night. Ehrlichman laughed and laughed. "I don't know," he said. "Apparently nothing!"[30]

Two weeks later, on November 21, a judge in Boston sent Popkin to prison for contempt of court and sentenced him to serve for the life of the grand jury in Norfolk County Jail in Dedham, Massachusetts.[31] *Times* columnist Tom Wicker wrote a piece called "Liberty in Shackles" to call attention to Popkin's plight, lamenting that "the Government has demonstrated no purpose of any legitimate kind in taking a teacher off to prison in handcuffs."[32] Columnist Anthony Lewis suggested that "if the spirit of liberty survives in the United States, we shall be grateful to Samuel Popkin."[33] Newspapers published photographs of Popkin behind bars. The *Times* reported that he was "believed to be the first American scholar . . . jailed for protecting sources of information."[34] Seventy professors, mostly from MIT and Harvard, including Ithiel de Sola Pool, Daniel Patrick Moynihan, and two more of Popkin's former Simulmatics colleagues, Robert Abelson and James Coleman, signed a letter characterized as a "brief statement of moral support," indicating their "respect and admiration for his willingness to face imprisonment."[35]

And then, suddenly, with no explanation to the public but apparently in order to avoid the appearance of entanglement with Ellsberg's trial in Los Angeles, the Boston grand jury was discharged and Popkin was released.[36] Not long afterward, Ellsberg's first trial in Los Angeles was declared a mistrial. Eventually, charges against both Ellsberg and Russo were dropped after it was revealed that burglars hired by the White House had been caught breaking into the office of Ellsberg's psychiatrist. Popkin never knew quite why he'd been the subject of so much grand jury attention, and historians have never been able to figure it out, either.[37]

The 1972 election obscured a turning point in the history of technology: the beginning of personal computing and the unveiling of what would become the Internet. "Ready or not, computers are coming to the people," Stewart Brand predicted in *Rolling Stone* in December 1972.[38] Woodward and Bernstein broke the Watergate story. Stewart Brand broke the story of the coming computer revolution. Hardly anyone else noticed, at least not in 1972, a year when all eyes turned to what had happened in the Watergate Hotel, but very few paid attention to what had gone on in another Washington hotel, the Hilton.

On October 24, 1972, at the first ever meeting of the International Conference on Computer Communication, in the ballroom of the Washington, D.C., Hilton, ARPANET, the network first imagined by J.C.R. Licklider, the network that would become the Internet, was demonstrated to the public for the first time. Anyone who wanted to could log on to the network from any one of twenty-nine different terminals. It was possible, within that room, to play chess with a computer at MIT's Artificial Intelligence Lab, to get psychiatric advice from a program at the Cambridge firm of Bolt Beranek and Newman, or to ask questions of a terminal at UCLA named TIMMY. Each program demonstrated the reliability, speed, and capacity of the network: from a terminal in Washington, it was possible to talk to computers all over the country.[39]

No major news outlet reported on the demonstration of ARPANET, which would turn out to be one of the most important events in human history, as important as the first demonstration of the printing press. At the time, outside the worlds of computer science and engineering, its significance was difficult to see. And between Watergate and the election, reporters had their hands full.

Brand, though, wasn't interested in Watergate or even the election. He was interested in the coming countercultural computer revolution. Born in Illinois in 1938, Brand had graduated from Stanford in 1960 and, after serving in the army, had joined a movement known as the New Communalism, which was powerfully influenced by the eccentric visionary Buckminster Fuller, the same man who, for his friend Frank Safford, Patty Greenfield's father, had built the geodesic dome, where the scientists of Simulmatics had met in 1961. Brand carried Fuller's technological utopianism into his vision for what would one day become the Internet.

In the 1960s, Fuller's iconic geodesic domes — miniature Spaceship Earths — were favored structures on communes. Fuller was the great-nephew of the radical writer, transcendentalist, and social utopian Margaret Fuller, and his futurism borrowed from transcendentalism and attached to it a mid-twentieth-century vision of technological utopia, an imagined escape from the dour and deathly machines of the Cold War and into a new era in which machines would be designed, and put to use, for human fulfillment, a realization of bliss. "If man is to continue as a successful pattern-complex function in universal evolution," Fuller wrote in 1960, "it will be because the next decades will have witnessed the artist-scientist's spontaneous seizure of the prime design responsibility and his successful conversion of the total capability of tool-augmented man from killingry to advanced livingry." For Stewart Brand, an LSD advocate and one of Ken Kesey's Merry Pranksters, the commune answered the atomization of Cold War

America. But communal living required tools, a taking back of the machine. In 1968, from his base in Menlo Park, California, Brand launched the *Whole Earth Catalog,* with the motto "access to tools." ("The insights of Buckminster Fuller are what initiated this catalog," Brand wrote in its inaugural issue.)[40]

In 1972, when ARPANET made its debut, Brand celebrated the liberation of the computer from big business. "That's good news, maybe the best since psychedelics," he wrote in *Rolling Stone.* While students at MIT protested the Vietnam War, people around Stanford rebuilt the machine as something that could be homegrown and emancipatory, a tool for personal liberation and collective transformation. They formed outfits like the Homebrew Computer Club and the People's Computer Company.[41] Brand, in *Rolling Stone,* called ARPA "an astonishingly enlightened research program from the very top of the Defense Department." The essay, a fawning profile of Stanford's Artificial Intelligence Lab, with photographs by Annie Leibovitz, celebrated the zaniness of the lab's computer scientists and the playfulness of their new online game, a war simulation called Spacewar. "The setting and decor at AI is Modern Mad Scientist," Brand wrote, "long hallways and cubicles and large windowless rooms, brutal fluorescent light, enormous machines humming and clattering, robots on wheels, scurrying arcane technicians." The computer scientists there were antiwar and anti-Nixon, as the posters on the wall made clear. But, mainly, they were anti-establishment. They were hackers, freaks, geeks. They'd linked their lab up to ARPANET, but Brand wasn't down on ARPA or its Cold War, missile-defense origins: "Poetically enough it owes its origin to real spacewar," he wrote, and its early history was "full of freedom and weirdness." ARPA was groovy. Giant institutionally owned computers, so far, had been used mainly, as Brand wrote, "to answer in detail questions that begin, 'What if?' " but their real usefulness had yet to be seen, or even glimpsed. He was sure of two things: computers were coming to the people and the revolution was at hand.[42]

The New Left's rage against the machine — Do Not Fold, Bend, Spindle, or Mutilate — ended in 1972. Hackers became the new New Communalists, and the coming network of networks their new commune. With the rise of the personal computer in the 1970s and 1980s, the machine became the New Left's salvation, the coming Internet an engine of personal liberation, a computer version of Buckminster Fuller's idea of the world as Spaceship Earth, a free, universal, online commune, a Whole Earth experiment.[43] In one of the stranger ironies of this history, Ithiel de Sola Pool became this new movement's idol and prophet.

Few observers could have detected signs of this particular realignment. Nixon took his reelection as a repudiation of liberalism, the death knell for the Great Society, and a turning point for a new, anti-government conservatism. In his Second Inaugural Address, on January 20, 1973, he promised to end the war in Vietnam and to end the rule of experts. "In trusting too much in government," he said, "we have asked of it more than it can deliver." Two days later, Lyndon B. Johnson, at home on his ranch in Texas, suffered a heart attack; he died, some said, of a broken heart.

But if conservatives argued against trusting too much in government, so, too, did liberals. As the Watergate scandal unraveled, the investigation, down to the strange case of Samuel Popkin, demonstrated the extent of Nixon's willingness to use the federal government and, in particular, the powers of the executive to spy on American citizens and hobble and punish his political enemies. Nixon resigned on August 9, 1974. Among the many legacies of Watergate, the debate that had begun with the proposed National Data Center in 1966 raged on, animated by a fear of data collection that was almost exclusively directed at the federal government, not at corporations. Only months after Nixon's resignation, Congress passed the Privacy Act, which opened with an indictment: "Increasing use of computers and sophisticated information technology, while essential to the efficient operations of the Government, has greatly magnified the harm to individual privacy that can occur." It required the Department of Justice to establish the Data Integrity Board. It made scant provisions for protecting private data from corporations.[44]

Not long afterward, fear of government-run computers and government-held data led to a misguided attempt to expose ARPANET, which had never been secret. Early in 1975, an ambitious young NBC reporter named Ford Rowan began an investigation of his own into the Nixon White House. In June 1975, in a series of three explosive reports on *NBC Nightly News,* Rowan claimed to have uncovered yet another conspiracy and another cover-up.

"NBC News has learned that the government has built a secret electronic intelligence network that gives the White House, the CIA, and the Defense Department instant access to computer files on millions of Americans," Rowan told Americans. "The secret computer network was made possible by dramatic breakthroughs in the technique of hooking different makes and models of computers together . . . so they can talk to one another and share information." Americans sat up in their chairs. Members of Congress began fielding calls from constituents. Rowan went on: "If you pay taxes, or use a credit card, if you drive a car, or have ever served in the military, if you've ever been arrested, or even investigated by a police agency, if you've had major medical expenses or contributed to a national political party, there is information on you somewhere in some computer."

Rowan tied this "secret computer network" to the long-abandoned National Data Center. "Congress has always been afraid that computers, if all linked together, could turn the government into 'big brother.' . . . In 1968 it killed a proposal for a national data bank. . . . But NBC News has learned that while Congress was voting down plans for big computer link-ups, the Defense Department was developing exactly that capability: the technology to connect virtually every computer . . . is now in operation."[45]

Hardly anyone knows about this secret network, Rowan said. But, actually, ARPANET was not and had never been a classified project. No one had been much interested when it had been demonstrated at the Washington Hilton in October 1972, a demonstration open to the public. And while it was true that not many people knew about it, that was only because, aside from Stewart Brand, not many people had cared.

Nevertheless, the Senate, still in a Watergate mood, opened hearings into this supposedly secret government network. Testifying before a Senate committee, the deputy assistant secretary of defense acknowledged that the U.S. Army had indeed conducted surveillance on and kept data files on Americans involved in "civil disturbances" — political protests. But that had nothing to do with ARPANET, which, as he pointed out, had been "discussed extensively in open literature since 1969," with the idea that it would become "commercially available" both in the United States and around the world: "We are at a loss to imagine why anyone would try to hide outdated data on civil disturbances over 5 years old in an unclassified computer network widely available to subscribers in the academic community." He professed himself perplexed. "Let me emphasize that it is not a 'secret' network, that it is used for scientific research purposes, that it contains no sociological or intelligence data on personalities, and that it is a marvel in many ways," he said. "It does not fit the Orwellian mold attributed to it."[46]

The Senate committee also called Jerome Wiesner, who had been a science and technology adviser to the Kennedy administration and was in 1975 the president of MIT. (Wiesner had also been called to testify in 1971, at the hearings about federal data banks and the Bill of Rights.) Wiesner discounted Rowan's story as misinformed, but he still took the Senate investigation seriously: he expressed more concern than anyone else at the hearings about the issues Rowan had raised. Wiesner warned of the danger of "information tyranny," given the lack of safeguards on the collection and analysis of data, both by public agencies and by private interests. He urged Congress, as he had done in 1971, to establish a federal watchdog authority, possibly through the FCC, "to review regularly the public and private information gathering and processing activities within the country. The agency should have the authority to examine the nature and extent of such activities and should report its findings to the Congress and the public."[47] The FCC did not assume this oversight authority, partly because Ithiel de Sola Pool spent much of the remainder of his career arguing that it shouldn't.

The MIT student protests over Simulmatics and Project Cambridge in 1968 and 1969 had made it almost impossible for Pool to continue in the fields of computer modeling and simulation.[48] As he moved into new research, much of Pool's early work was either forgotten or credited to other people. He spent much of the rest of his life, at great length and over many years, making the case that new forms of communication should be allowed to develop without government regulation.

In the 1970s and early 1980s, Pool wrote, influentially and brilliantly, about emerging technologies: cable television, e-mail, ARPANET, microcomputers, personal computers, social networks and the fast-developing Internet, and the international network of networks. He'd been making spot-on predictions since the 1960s. "In the 21st Century, the sort of critic who now attacks conformity in society may be complaining of an atomized society," Pool had written in 1968. "Modern technology, he'll assert, has destroyed our common cultural base and has left us living in a little world of his own."[49] With each passing year, his predictions only got more accurate. "New technology is allowing the editors of mass media to tailor their contents to the interest of small specialized audiences," he wrote in the early 1980s, two decades before other observers would first notice this trend. Networking computer networks, he predicted, would produce "communities without boundaries."[50] He also advocated the abolition of the FCC's Fairness Doctrine (critiqued by conservatives as unduly suppressing conservative political views). And he argued against government regulation of what would become the Internet.

"Communications policy in the United States most seriously lost its way" in the 1920s, Pool argued in his award-winning 1983 book, *Technologies of Freedom: On Free Speech in an Electronic Age.* The 1927 Radio Act established the federal government regulation of radio transmissions; Congress extended that principle in the 1934 Communications Act, which established the Federal Communications Commission. Pool derided this body of legislation as emerging out of a misguided concern about corporate monopoly and representing the overreach of the New Deal. Its era, he said, had come to an end. In the new world, there will be networks upon networks upon networks, all interconnected, Pool told his readers. The world will shrink. Processes will be distributed; searching will be conducted by artificial intelligence that will learn by doing. The regulatory regimes of older technologies — the common carrier services of radio, telephone, and television — had been an error. The new regime could correct this error by deregulating what had already been invented and not regulating what was yet to be imagined. "The future of communications will be radically different from its past because of such artificial intelligence," he wrote. "If media become 'demassified' to serve individual wants, it will not be by throwing upon lazy readers the arduous task of searching vast information bases, but by programming computers heuristically to give particular readers more of what they chose the last time."[51]

Few people read Pool's words more avidly than Stewart Brand. "With each passing year the value of this 1983 book becomes more evident," he wrote.[52] Pool died at the age of sixty-six in the Orwellian year of 1984, the year Apple launched its first Macintosh, the year MIT was establishing a new lab, the Media Lab. Two years later, Brand moved to Cambridge to take a job at the Media Lab, a six-story, $45 million building designed by I. M. Pei and named after Jerome Wiesner, a building that represented nothing so much as a newer version of Buckminster Fuller's geodesic dome. Brand didn't so much conduct research at the Media Lab as promote its agenda, as in his best-selling 1987 book, *The Media Lab: Inventing the Future at M.I.T.* The entire book, Brand said, bore the influence of Ithiel de Sola Pool, especially his *Technologies of Freedom.* "His book was the single most helpful text in preparing the book you're reading and the one I would most recommend for following up issues raised here," Brand wrote. "His interpretations of what's really going on with the new communications technologies are the best in print." Brand cited Pool's work on page after page after page, treating him as the Media

Lab's founding father, which he was.[53]

"A global computer is taking shape, and we're all connected to it," Brand wrote.[54] It would be a new commune. The MIT Media Lab's founders encouraged its faculty to spend their time in lucrative consulting arrangements, Simulmatics-style, all aimed at inventing the future, a future that Stewart Brand pictured as a post-national global world of harmony, a world of people carrying computers in their pockets or implanted in their heads. Brand made a cult of Media Lab figures, the visionary scientist Marvin Minsky, the dead prophet Ithiel de Sola Pool. Corporations funded almost all of the Media Lab's research. In this world, the future of the Media Lab's invention, technologies of communication — from electronic publishing and e-mail to "personal television" and the "personal newspaper" — solve all problems: social problems, economic problems, political problems.

Brand had long since left life on an actual commune behind, in favor of corporate life — newly dubbed, in Silicon Valley, "entrepreneurship." In 1987, he co-founded the Global Business Network, encouraging online entrepreneurship. In 1990, the Department of Defense turned ARPANET over to the National Science Foundation's NSFNET network, which was decommissioned in 1995 when the private sector took it over. Libertarians had made common cause with conservatives, arguing for opening the Internet up to commercial traffic as a libertarian and even anarchist free market utopia, outside of any possible federal government agency or regulation. In 1994 *Wired* magazine published "The Merry Pranksters Go to Washington." The New Communalists, in the form of men like Brand, joined forces with the New Right, led by Newt Gingrich. An unregulated Internet would be the least noticed element of Gingrich's Contract with America.[55]

"The real legacy of the sixties generation is the computer revolution," Brand announced in *Time* magazine. "The counterculture's scorn for centralized authority provided the philosophical foundations of not only the leaderless Internet but also the entire personal-computer revolution," he insisted.[56] But the real philosophical foundation for the argument that these emerging technologies should be free of all government regulation came less from the counterculture than from Pool's *Technologies of Freedom* and from his *Technologies Without Boundaries,* posthumously published in 1990. Pool's arguments in these two books served as the backbone for the Newt Gingrich–affiliated Progress and Freedom Foundation's 1994 *Magna Carta for the Knowledge Age.* "The meaning of freedom, structures of self-government, definition of property, nature of competition, conditions for cooperation, sense of community and nature of progress will each be redefined for the Knowledge Age," the new Magna Carta read, in lines that could have been written by Pool.[57] Pool's work equally informed the 1996 "Declaration of the Independence of Cyberspace," drafted by another leading counterculture figure, John Perry Barlow, a former lyricist for the Grateful Dead. "We are creating a world where anyone, anywhere may express his or her beliefs, no matter how singular, without fear of being coerced into silence or conformity," Barlow wrote.[58] And, most lastingly, Pool's argument for treating the emerging Internet as a "technology of freedom" formed the underlying logic of the 1996 Telecommunications Act, under whose terms earlier campaigns to treat the Internet the way the federal government treated radio — as a regulated public utility — were abandoned.

The new Internet followed no rules but many mantras. Content must be free. Media solves all problems. Data drives predictions. Soon, the Internet began to host social media companies. With the founding of Facebook, Pool's theory of social networks at last found the perfect algorithmic expression, a twenty-first-century version of the Mood Corporation.

Years before, so many long years before, while trying to perfect a mathematical model,

Pool had sent around the country a list of names, asking people he knew if they knew any of the same people he knew: If you ran into "Edward L. Greenfield, Public Relations, N.Y.C. Formerly Univ. of Chicago, Yale Law School," on the street, would you recognize him? Would you say hello? Would Ed Greenfield know you, too? "Check here if you know this person."[59] Was he an acquaintance? Was he a friend? How many degrees of separation is the population of the entire planet from Edward L. Greenfield?

Years before, so many years before, Ed Greenfield had pulled together the best and the brightest: scientists of human behavior, ad men, computer men. He wanted to build a machine to predict human behavior. Greenfield must have believed they'd failed. He'd lost everything. But Simulmatics hadn't failed. The automated simulation of human behavior became the human condition.

Maybe Bill McPhee, the man who'd invented the People Machine and then walked away from Simulmatics in 1961, understood better than anyone the nature and tragedy of its success. Maybe he'd even come to understand that the Simulmatics scientists' bargain had been Faustian. In Colorado, he settled in a cabin high in the Rocky Mountains, where he lived alone, often snowbound. In 1984, he retired from the University of Colorado and moved to New Mexico. He went on and off his psychiatric medication. He suffered, too, from diabetes and, a lifelong smoker, from emphysema.[60]

The Internet opened to commercial traffic, and in 1996, a year during which ten thousand websites were launched every day, Bob Dole became the first U.S. presidential candidate to host a website. Two years later, the year Google was founded, Bill McPhee died at the age of seventy-seven. Sitting in front of his computer, he shot himself in the head.[61]

EPILOGUE:
META DATA

So we made our own computer out of macaroni pieces
And it did our thinking while we lived our lives.
— Regina Spektor, "The Calculation," 2009

The Simulmatics Corporation is a relic of its time, an artifact of the Eisenhower-Kennedy-Nixon Cold War, a product of the Madison Avenue of Ralston Purina and Colgate-Palmolive and Young & Rubicam and BBDO, a casualty of midcentury American liberalism. But the People Machine was also hobbled by its time, by the technological limitations of its day: the 1960s threw sand in its gears. Data was scarce. Models were weak. Computers were slow. The machine faltered, and the men who built it could not repair it: the company's behavioral scientists had very little business sense, its chief mathematician struggled with insanity, its computer scientist fell behind the latest research, its president drank too much, and nearly all

528

of their marriages were falling apart. *They treat their wives like dirt,* said Minnow McPhee. The machine sputtered, sparks flying, smoke rising, and ground to a halt, its lights blinking, wildly, desperately, before going dark.

Simulmatics failed, but not before its scientists built a very early version of the machine in which humanity would in the early twenty-first century find itself trapped, a machine that applies the science of psychological warfare to the affairs of ordinary life, a machine that manipulates opinion, exploits attention, commodifies information, divides voters, fractures communities, alienates individuals, and undermines democracy. "What does it take for people to recognize a dystopia?" the virtual reality pioneer Jaron Lanier asked in 2019, anguished, heartbroken.[1] Long before the age of quarantine and social distancing, Simulmatics helped atomize the world.

It had begun sixty years before, with the best of intentions. In 1959, hoping to build a better America, Simulmatics pioneered the use of computer-run simulation, pattern detection and prediction in American political campaigning, segmenting the electorate into voter types and issues into clusters in order to advise candidates about strategies for voter-targeted issues. In 1961, Simulmatics introduced what-if simulation to the advertising industry, targeting segmented consumers with custom-fit messages. In 1962, Simulmatics became the first data firm to provide real-time computing to an American newspaper for the purpose of analyzing election results. In 1963, Simulmatics simulated the entire economy of a developing nation, with an eye toward halting the advance of socialism and inoculating post-colonial nations from the temptations of Communist revolution. Beginning in 1965, Simulmatics conducted psychological research in Vietnam as part of a larger project of waging a war by way of computer-run data analysis and modeling. In 1967 and 1968, on the streets of American cities, Simulmatics attempted to build a race-riot prediction machine. In 1969, after mad assassins murdered Martin Luther King Jr. and Robert F. Kennedy, after Patty Greenfield fell off a balcony and Ithiel de Sola Pool's students called him a war criminal, the People Machine crashed, and Simulmatics filed for bankruptcy.

"The Company proposes to engage principally in estimating probable human behavior by the use of computer technology," Ed Greenfield had promised investors. By the early twenty-first century, the mission of Simulmatics had become the mission of many corporations, from manufacturers to banks to predictive policing consultants. Collect data. Write code. Detect patterns. Target ads.

Predict behavior. Direct action. Encourage consumption. Influence elections. It became impossible to read a newspaper or open a refrigerator or buy shampoo or cast a vote or sign a petition or go to the dentist without having had one's probable human behavior estimated by the use of computer technology. If, then, long ago, Simulmatics had not undertaken this work, it would have been done by someone else. But if, then, someone else had done it, it might have been done differently.

Or maybe the government would have accepted its obligation to monitor emerging technologies. Instead, the government abdicated that obligation, a responsibility to establish rules, safeguards, and standards relating to the collection and use of personal data for the purpose of directing human behavior. Plainly, all of this might have gone differently. Plenty of people believed at the time that a people machine was entirely and utterly amoral. "My own opinion is that such a thing (a) cannot work, (b) is immoral, (c) should be declared illegal," Newton Minow had written to Arthur Schlesinger in 1959. "Please advise."[2] *Please advise.* "Some critics say this is dehumanizing and an affront to human dignity," Ithiel de Sola Pool admitted in 1962. He waved that criticism aside. "Our assumption is that knowledge is a good thing. Why should politicians operate in the dark?

If there are *two* People Machines working against each other in a political campaign — that would be progress."[3] That came to pass. It was not progress.

Prophecy is ancient. Estimating probable outcomes is a feature of human curiosity. And using computers to make predictions is a very sensible idea. In the physical and natural sciences, the estimation of probable outcomes by the use of computer technology has produced vast new bodies of knowledge and saved and improved countless lives. In engineering, computer-aided pattern detection and prediction has improved the safety of buildings, factories, and vehicles. In environmental science, it is essential in reckoning with catastrophic climate change. In astronomy, it has made possible the seeing of new stars. In medical research, pharmacology, and public health, it helps researchers find cures for diseases, avert epidemics, and defeat pandemics. Nothing could be more vital.

But the study of the human condition is not the same as the study of the spread of viruses and the density of clouds and the movement of the stars.[4] Human nature does not follow laws like the law of gravity, and to believe that it does is to take an oath to a new religion.[5] Predestination can be a dangerous gospel. The profit-motivated collection and use of data about human behavior, unregulated by any governmental body, has wreaked havoc on human societies, especially on the spheres in which Simulmatics engaged: politics, advertising, journalism, counterinsurgency, and race relations. Its rise also marked the near abandonment of humanistic knowledge. Ithiel de Sola Pool declared the social sciences — the predictive, computational behavioral sciences — to be "the New Humanities of the Twentieth Century." The future was everything, the past nothing: a void, the humanities obsolete. *Apollo* reached the moon. Icarus reached the sun, and his wings did not melt. Instead, the light blinded him.

"The best minds of my generation are thinking about how to make people click ads," one of Facebook's first employees howled in 2011. "That sucks."[6] And it did suck. But it had sucked before. The best minds of another generation had tried to simulate the human mind to sell shampoo and dog food and to win the hearts and minds of rice farmers in Vietnam.

In the 1950s, a flood of money into universities attempted to make the study of human behavior into a science. The Vietnam War exposed the moral bankruptcy of much of this research and elicited an assault from the New Left. But the New Left's assault on liberal intellectuals in the 1960s had much the same cultish and anti-humanist quality that behavioral science had in the 1950s. On top of the ruins of the theory it had demolished, the New Left erected its own theory of knowledge: the theory was that all knowledge is biased. The New Right agreed. And then, while the world reeled from the ensuing epistemological chaos, start-up whiz kids, venture capitalists, and Silicon Valley entrepreneurs held out the promise of newly certain knowledge: big data, machine learning, algorithmic truth. Unlike their Cold War forefathers, their interest wasn't national security; their interest was profit. They built a new people machine, bigger, sleeker, faster, and seemingly unstoppable. They built it out of old bits and pieces made by Simulmatics: content analysis, the simulation of human behavior, targeted messages, social networks, IF/THEN statements. Unlike the scientists of Simulmatics, many of the people who built the new machine did not begin with the best of intentions. They boasted that, at best, they had no bad intentions. "Don't be evil," the motto of Google, marked the limit of a swaggering, devil-may-care ethical ambition; doing good did not come into it.[7]

Incubated decades before, beneath a honeycombed, geodesic dome in Wading River, this work found a place, too, in universities. In the 2010s, a flood of money into universities attempted to make the study of data a science, with data science initiatives, data science programs, data science degrees, data science centers.[8] Much academic research that fell under the label "data science" produced excellent and invaluable work, across many fields of inquiry, findings that would not have been possible with computational discovery.[9] And no field should be judged by its worst practitioners. Still, the shadiest data science, like the shadiest behavioral science, grew in influence by way of self-mystification, exaggerated claims, and all-around chicanery, including fast-changing, razzle-dazzle buzzwords, from "big data" to "data analytics." Calling something "AI," "data science," and "predictive" became a way to raise huge amounts of venture capital funding; a credulous press aided in hyping those claims, and a broken federal government failed to exercise even the least oversight. The most egregious social-outcome products to emerge from this industry had a great deal in common with the projects undertaken by Simulmatics, decades before: they aimed to predict the behavior of women, children, people of color, and the poor, with tools that claimed to be able to predict criminal behavior, job performance, criminal recidivism, school performance, the likelihood that at-risk children will fail, and the likelihood that immigrants will become terrorists.[10]

Few rules governed this field, especially

outside the university; its leaders touted its anarchy as a measure of its creativity. In Silicon Valley, nearly all of the leaders of companies selling snake oil were men, the great-grandsons of the scientists of Simulmatics, but they believed themselves to be orphans, parentless, fatherless, sui generis self-made geniuses.[11] They made no room in their world for women, or family, or knowledge other than the calculations of computers. Places like MIT's corporate-funded Media Lab cultivated a "hacker ethic," which meant, in many quarters, no ethics at all. In 2016, the director of the Media Lab accepted $1.7 million from convicted felon Jeffrey Epstein, after he'd registered as a sex offender and pleaded guilty to procuring an underage girl for sex (Epstein helped the lab pull in another $7.5 million from other donors), and announced a "Disobedience Award" to celebrate "responsible, ethical disobedience," making of heedless audacity a fetish.[12]

MIT's Media Lab served as a convenient scapegoat, a distraction from a broader ethical aimlessness not only in Silicon Valley but on college campuses. Much of university life had by the 2010s followed the model of the Media Lab, collapsing the boundaries between corporate commissions, academic inquiry, and hucksterism; at its worst, behavioral data science's self-mystification was meant to boggle the mind, daunt critics, and entice corporate sponsors and venture capitalists. "Data science is in its infancy," wrote one MIT computer scientist in 2015. "Few individuals or organizations understand the potential of and the paradigm shift associated with Data science, let alone understand it conceptually."[13] The more mystification, the wealthier the donors. The number of data science programs stretched into the hundreds, even though little consensus had been reached on the meaning or purpose of "data science." New disciplines and methods take time to find their way; that's all to the good. But the fledgling field of data science claimed to have triumphed over all other ways of knowing. In a time of global crisis, this made it feel difficult to know anything at all.

Behavioral data science presented itself as if it had sprung out of nowhere or as if, like Athena, it had sprung from the head of Zeus. The method Ed Greenfield dubbed "simulmatics" in 1959 was rebranded a half century later as "predictive analytics," a field with a market size of $4.6 billion in 2017, expected to grow to $12.4 billion by 2022.[14] It was as if Simulmatics' scientists, first called the "What-If Men" in 1961, had never existed, as if they represented not the past but the future. "Data without what-if modeling may be the database community's past," according to a 2011 journal article, "but data with what-if modeling must be its future."[15] A

2018 encyclopedia defined "what-if analysis" as "a data-intensive simulation," describing it as "a relatively recent discipline."[16] What if, what if, what if: What if the future forgets its past?

Behavioral data science didn't spring from the head of Zeus. Cambridge Analytica used Facebook data in an attempt to influence the 2016 U.S. presidential election in a way that Bill McPhee and Ed Greenfield and Ithiel de Sola Pool and Joseph Hoc would have understood, even if they could only dream about its gargantuan quantity of data or the ability to run simulations in real time, dynamically. "The bulk of our resources went into targeting those whose minds we thought we could change," former Cambridge Analytica executive Brittany Kaiser explained in 2019. "We called them 'the persuadables.' . . . We bombarded them through blogs, websites, articles, videos, ads, every platform you can imagine. Until they saw the world the way we wanted them to. Until they voted for our candidate. It's like a boomerang: you send your data out, it gets analyzed, and it comes back at you as targeted messaging to change your behavior."[17]

Cambridge Analytica's Facebook "likes"? That was Ithiel de Sola Pool's "cross-cutting pressures." Cambridge Analytica's boomerang model? That was Bill McPhee's "three-stage model of communication." Faster, better, fancier, pricier, but the same hucksterism, and as for the claims of its daunting efficacy, the same flimflam. Donald Trump's 2016 presidential campaign didn't need Cambridge Analytica. Facebook, alone, could target specific voters with custom-made messages. Commentators accused the Trump campaign of using a "weaponized AI propaganda machine," describing a new and "nearly impenetrable voter manipulation machine."[18] New? Hardly. Simulmatics invented that machine in 1959.

In twenty-first-century Silicon Valley, the meaninglessness of the past and the uselessness of history became articles of faith, gleefully performed arrogance. "The only thing that matters is the future," said the Google and Uber self-driving car designer Anthony Levandowski in 2018. "I don't even know why we study history. It's entertaining, I guess — the dinosaurs and the Neanderthals and the Industrial Revolution and stuff like that. But what already happened doesn't really matter. You don't need to know history to build on what they made. In technology, all that matters is tomorrow."[19]

This cockeyed idea isn't an original idea; it's a creaky, bankrupt Cold War idea, an exhausted and discredited idea. The invention of the future has a history, decades old, dilapidated. Simulmatics is its cautionary tale, a timeworn fable, a story of yesterday.

Because tomorrow is not all that matters. Nor is technology, or the next president, or the best dog food. What matters is what remains, endures, and cures.

ACKNOWLEDGMENTS

Harvard professors keep clear, on the whole, of things like Simulmatics.
— Christopher Rand, *Cambridge, U.S.A.,*
1964

I came across the story of Simulmatics when I was working on a story for *The New Yorker* in 2015 and I went to MIT to look at the papers of Ithiel de Sola Pool. I opened box after box and began to think there might be a book in those boxes, which led me to still more boxes, all over the country. That means that I could not have written this book without the expert assistance of archivists and librarians at many institutions. Particular thanks to Nora Murphy at the MIT Institute Archives and Special Collections; Joan Gearin at the National Archives, Boston; Sarah Pratt at the Howard Gotlieb Center, Boston University; Melinda Kent at Harvard Law School; and Fred Burchsted at Harvard Library. Thanks, as well, to everyone who

541

showed me around Vietnam, including Nguyen Duy Thang and Duy Nguyen.

Family members provided me with archival material, too. Adam de Sola Pool and Jonathan Robert Pool shared with me their father's family letters. Sarah Neidhardt sent me her extraordinary collection of the correspondence of her maternal grandmother, Minnow McPhee. Maureen Shea sent me her letters from Vietnam. Anne-Marie de Grazia e-mailed me dozens of scanned Simulmatics letters. I am grateful, as well, to everyone I interviewed, including Elizabeth Bernstein Rand and Max Bernstein; Michael Burdick; Thomas Coleman; Victoria de Grazia; Cuc Thu Duong; Daniel Ellsberg; Frances FitzGerald; Jennifer, Susan, and Ann Greenfield; Ward Just; Kim Le Cook; Wendy, John, and Sarah McPhee; Newton Minow; Kate Tarlow Morgan; Adam de Sola Pool, Jeremy Pool, and Jonathan Robert Pool; Samuel Popkin; Edwin Safford; Maureen Shea; Naomi Spatz; and Ann Penner Winston.

On matters big and small, I consulted with a number of scholars, not only historians and other humanists but also scientists who work with data (and who tolerated my skepticism about behavioral science and data science with incredible grace): Francine Berman, Mercè Crosas, Alyssa Goodman, Elizabeth Hinton, Gary King, Alexandra Lahav, David Lazer, Rebecca Lemov, Andrew Lippman,

Khalil Muhammad, Julie Reuben, Todd Rogers, Peter Simonson, Hu-Tam Ho Tai, Sharon Weinberger, and Cliff Weinstein. Thanks to all, and particular thanks to Martha Minow, Beverly Gage, and Rogers Smith, for supporting this project.

Funding for this research came from a Dean's Competitive Fund for Promising Scholarship at Harvard and from a fellowship from the Radcliffe Institute for Advanced Study, where I also presented this work to a wonderful community of fellows. My thanks to anonymous readers of grant applications for their candid feedback. Harvard deans Claudine Gay and Nina Zipser granted me permission to see previously closed files in the Harvard University Archives. At the Yale Media Freedom and Information Access Clinic, Charles Crain, John Langford, David Schulz, Jacob Schriner-Briggs, Rachel Cheong, Jonathan Albano, Noah Kaufman, Ellis Liang, and Jessica Baker prepared and argued my case, attempting to unseal the records of the Pentagon Papers 1971 Boston grand jury. Thanks, too, to everyone who submitted declarations in support of my petition, and thanks as well to David McCraw, Bruce Craig, Julian Zelizer, and Ann Marie Lipinski.

My students Luke Minton, Jordan Virtue, and Angel Mata provided invaluable and ingenious research assistance. Thanks as well to Gemma Collins and Clio Griffin. Guy Fedorkow, out of his own insatiable curiosity, tackled Simulmatics' original data set and also reviewed the manuscript and, with Carl Claunch and the Computer History Museum, scanned and read thousands of ancient punch cards that I found in Pool's papers, in an attempt to rerun Simulmatics' 1960 election study.

Francine Berman, Henry Finder, Elizabeth Hinton, Jane Kamensky, Fred Logevall, and Martha Minow each read drafts. I owe them more than I can say. My gratitude, always, to Tina Bennett. The extraordinary Emily Gogolak and Julie Tate checked facts. Rebecca Karamehmedovic and Melissa Flamson sorted out permissions. Bonnie Thompson caught errors like fish in a net. Thanks to them, and to everyone at Liveright, particulary Gabriel Kachuck and Don Rifkin. Above all, thanks to Robert Weil.

Thank you, too, to Adrianna Alty, Elise Broach, Deb Favreau, Jane Kamensky, Elisabeth Kanner, Lisa Lovett, Sophie McKibben, Liz McNerney, Benjamin Naddaff-Hafrey, Dan Penrice, Bruce Schulman, Rachel Seidman, Ramie Targoff, Sue Vargo, and Denise Webb. Love and thanks to Doris and Paul, and to Gideon, Simon, and Oliver, always, and to Timothy Leek, no simulation but the real thing. If, then, there had been no him, I cannot begin to predict . . .

Is there any reason why the book should not be titled "Simulmatics" — a word like Cybernetics that should become generic?

— Thomas B. Morgan to
Ithiel de Sola Pool, 1961

NOTES

Abbreviations Used in the Notes

People
AB Alex Bernstein
ADG Alfred de Grazia
AS Arthur Schlesinger Jr.
DPM Daniel Patrick Moynihan
EB Eugene Burdick
ELG Edward L. Greenfield
IP Ithiel de Sola Pool
JC James Coleman
JFK John F. Kennedy
JH Joseph Hoc
LBJ Lyndon B. Johnson
MLK Martin Luther King Jr.
MMc Minnow McPhee
MS Maureen Shea
NM Newton Minow
RFK Robert F. Kennedy
RN Richard Nixon
SP Samuel Popkin
TBM Thomas B. Morgan

WMc William McPhee

Manuscript Collections

BASR Records Bureau of Applied Social Research Records, Rare Book and Manuscript Library, Columbia University Library

Burdick Collection Eugene Burdick Collection, Boston University Libraries, Howard Gotlieb Archival Research Center

CASBS Records Center for Advanced Study in the Behavioral Sciences Records, Department of Special Collections at University Archives, Stanford University

Coleman Papers James Coleman Papers, Special Collections Research Center, University of Chicago Library

Grazian Archive Digital archive of the writings of Al de Grazia at grazian-archive.com

Greenfield Papers Family Papers of Edward L. Greenfield and Patricia Greenfield, collection of the Greenfield family

Emery-McPhee Papers Papers of the Emery and McPhee families, collection of Sarah Neidhardt

Johnson Library Lyndon B. Johnson Presidential Library

Kennedy Library John F. Kennedy Presidential Library and Museum

Lasswell Papers Harold Dwight Lasswell Papers, Manuscripts and Archives, Yale University

Moynihan Papers Daniel P. Moynihan Papers, Manuscripts Division, Library of Congress

Nixon Library Richard Nixon Presidential Library and Museum

NYT Records New York Times Company Records, Manuscripts and Archives Division, New York Public Library

O'Brien Papers Lawrence F. O'Brien Personal Papers, Kennedy Library

Pool Family Papers Ithiel de Sola Pool family papers, collection of Adam de Sola Pool

Pool Papers Ithiel de Sola Pool Papers, MIT Institute Archives and Special Collections

Project Agile Records National Archives RG 330, Records Pertaining to Project Agile, 1962–1968

Shea Letters Letters of Maureen Shea, collection of Maureen Shea

Stevenson Papers Adlai E. Stevenson Papers, Public Policy Papers, Department of Rare Books and Special Collections, Princeton University Library

Wayman Papers Dorothy Wayman Papers, College of the Holy Cross Archives and Special Collections

Folder names are given only for folders that lack numbers.

Prologue: What If?

1. The dome, although never finished, was built by Buckminster Fuller for Frank Saf-

ford; Safford later gave it to his son, Edwin (b. 1936). Edwin Safford, interview with the author, April 23, 2018.

2. The Simulmatics Corporation has been the subject of very little scholarly inquiry outside of the scholarship produced by the Simulmatics scientists themselves. The company's history has never been the subject of a book, and all published discussions of the firm confine their attention to one or another of its projects rather than to the firm's origins, ambitions, scope, and demise. Simulmatics' work in the 1960 election is treated in Sasha Issenberg, *The Victory Lab: The Secret Science of Winning Campaigns* (New York: Broadway Books, 2013), 119–23, and I also discuss that work in "Politics and the New Machine," *New Yorker,* November 16, 2015. Simulmatics' work in Vietnam is discussed in Joy Rohde, "The Last Stand of the Psychocultural Cold Warriors: Military Contract Research in Vietnam," *Journal of the History of the Behavioral Sciences* 47 (2011): 232–50, and Sharon Weinberger, *The Imagineers: The Untold History of DARPA, the Pentagon Agency That Changed the World* (New York: Knopf, 2017), ch. 10, "Blame It On the Sorcerers." The firm's work in Vietnam is also briefly mentioned in Yasha Levine, *Surveillance Valley: The Secret Military History of*

the Internet (New York: PublicAffairs, 2018), and in Joy Rohde, *Armed with Expertise: The Militarization of American Social Research During the Cold War* (Ithaca: Cornell University Press, 2013). Its work for the Kerner Commission is discussed in Thomas J. Hrach, *The Riot Report and the News: How the Kerner Commission Changed Media Coverage of Black America* (Amherst: University of Massachusetts Press, 2016), ch. 6: "Simulmatics Produces a Contradictory Analysis."

3. William M. Freeman, "Life Is Imitated for Research," *New York Times* (hereafter *NYT*), August 27, 1961.

4. Adam de Sola Pool, interview with the author, May 19, 2018. MMc to Eleanor Emery, August 14, 1961, Emery-McPhee Papers.

5. ELG, "Statement to Simulmatics Stockholders, September 20, 1966," printed brochure, Pool Papers, Box 177, no folder.

6. TBM, "The People-Machine," *Harper's*, June 1961.

7. EB, *The 480* (New York: McGraw-Hill, 1964). Daniel F. Galouye, *Simulacron-3* (1964; repr., Rockville, MD: Phoenix Pick, 2011).

8. Jean Baudrillard, *Simulacra and Simulation* (1981; Ann Arbor: University of Michigan Press, 1994), 18.

Chapter 1: Madly for Adlai

1. That wise man was Paul Baran of RAND. See his RAND obituary at https://www.rand.org/news/press/2011/03/28/index1.html.
2. Abigail Casey, Registrar's Office, Wabash College, e-mail to the author, July 23, 2018. Hugo Vasquez, Office of the Registrar, University of Chicago, e-mail to the author, August 8, 2018. (Owing the University of Chicago a library book: Naomi Spatz, interview with the author, May 24, 2018.) Michael Frost, Public Services, University Archives, Yale University, e-mail to the author, July 31 and August 2, 2018.
3. Harold Lasswell to the Yale Club of New York City, April 17, 1968, Lasswell Papers, Box 40, Folder 547.
4. Ann Greenfield, interview with the author, June 9, 2018. Edwin Safford, interview with the author, April 23, 2018. The films made by Edwin Denby starring the Safford children, are in the possession of Jennifer Greenfield. Some of them have been publicly screened, including *The Dogwood Maiden,* a copy of which is held by the Museum of Modern Art. On Wright's summer at Wading River in 1947, see Toru Kiuchi, Yoshinobu Hakutani, *Richard Wright: A Documented Chronology, 1908–1960* (Jefferson, NC: McFarland, 2014), 227.

5. Edwin Safford, interview with the author, April 23, 2018. Susan Greenfield, interview with the author, July 27, 2018.

6. Harris Wofford to MLK, April 1, 1960, *The Papers of Martin Luther King, Jr.,* ed. Clayborne Carson et al. (Berkeley: University of California Press, 2005), 5: 404–5.

7. Robert L. Heilbroner, "Public Relations: The Invisible Sell," *Harper's,* June 1957.

8. Vance Packard, *The Hidden Persuaders* (New York: McKay, 1957).

9. My full bibliography for this research can be found at https://scholar.harvard.edu/ files/jlepore/files/Bibliography.pdf.

10. Campaign Procedures, Whitaker & Baxter Campaigns, Inc. Records, California State Archives, Office of the Secretary of State, Sacramento, Box 9, Folder 27. A very long typescript titled "AMA's Plan of Battle: An Outline of Strategy and Policies in the Campaign Against Compulsory Health Insurance" and identified as written by W&B, Directors of the National Education campaign of the AMA, February 12, 1949, p. 1.

11. "Truman Blames A.M.A. for Defeat of Security Bill," *Boston Daily Globe,* May 22, 1952.

12. Carey McWilliams, "Government by Whitaker and Baxter," *Nation,* May 5, 1951, 368.

13. Jill Lepore, "How to Steal an Election,"

New Yorker, June 27, 2016. And see the bibliography for that essay at https://scholar .harvard.edu/files/jlepore/files/lepore _conventions_bibliography.pdf.

14. Longines Chronoscope with Senator Estes Kefauver (February 11, 1952), National Archives.

15. Adlai E. Stevenson, Speech Accepting the Democratic Presidential Nomination, July 26, 1952.

16. James Madison, *The Federalist Papers,* no. 10 (1787).

17. Citizens for Eisenhower, "I Like Ike for President," television commercial, 1952. Citizens for Eisenhower, "The Man from Abilene," television commercial, 1952.

18. David Haven Blake, *Liking Ike: Eisenhower, Advertising, and the Rise of Celebrity Politics* (New York: Oxford University Press, 2016), 71–74, 106, 109.

19. Edward Hunter, *Brain-Washing in Red China: The Calculated Destruction of Men's Minds* (New York: Vanguard, 1951).

20. On the diffusion of this idea through popular culture, especially in film, see Marcia Holmes, "Brainwashing the Cybernetic Spectator: *The Ipcress File,* 1960s Cinematic Spectacle and the Sciences of Mind," *History of the Human Sciences* 30 (2017): 3–24.

21. Blake, *Liking Ike,* 107.

22. "How to Insure an Eisenhower Victory in November," Rosser Reeves Papers, Wisconsin Historical Society Archives, Box 19, Folder 12.

23. Jill Lepore, narr., *The Last Archive,* podcast, Pushkin Industries, 2020, Season 1, Episode 5.

24. George Ball, *The Past Has Another Pattern: Memoirs* (New York: Norton, 1982), 125.

25. Joseph McCarthy, Address on Communism and the Candidacy of Adlai Stevenson, October 27, 1952. Sam Tanenhaus, "Who Stopped McCarthy?" *Atlantic,* April 2017.

26. Ball, *The Past Has Another Pattern,* 127–28.

27. Larry J. Sabato, *The Rise of Political Consultants: New Ways of Winning Elections* (New York: Basic Books, 1981), 112, 113, 117, 114.

28. Noel L. Griese, "Rosser Reeves and the 1952 Eisenhower TV Spot Blitz," *Journal of Advertising* 4 (1975): 30.

29. "Stevenson Bids Eggheads Unite," *New York Herald Tribune,* March 23, 1954.

30. "National Affairs: The Third Brother," *Time,* March 31, 1958.

31. Ira Chinoy, "Battle of the Brains: Election-Night Forecasting at the Dawn of the Computer Age" (PhD diss., Johns

Hopkins University, 2010).

32. At first, IP was merely sharing research he had already conducted. Jack Shea to Thomas L. Finletter, February 28, 1956, encloses a "first draft of a study by de Sola Pool on impressions of Eisenhower and Stevenson through TV and radio media in the 1952 campaign. This was made available to me by Ed Greenfield on a confidential basis." Stevenson Papers, Box 2, Folder 9.

Chapter 2: Impossible Man

1. Robert Henderson to EB, December 15, 1948, Burdick Collection, Box 91, Folder *The New Yorker.*
2. Wallace Stegner, "Eugene Burdick," *Book-of-the-Month Club News,* May 1956.
3. Ernest Freeberg, *Democracy's Prisoner: Eugene V. Debs, the Great War, and the Right to Dissent* (Cambridge, MA: Harvard University Press, 2008), 5.
4. Tity de Vries, "A Year at the Center: Experiences and Effects of the First International Study Group of Fellows at the Center for Advanced Study in the Behavioral Sciences, Palo Alto, ca. 1954–1955," *Scholarly Environments* (2004): 169–79.
5. EB, *The Ninth Wave* (Boston: Houghton Mifflin, 1956), 14.
6. Ibid., 143.

7. Ibid., 66.

8. Carl Sandburg, *The People, Yes* (New York: Harcourt, Brace, 1936).

9. A brief biography can be found in Clark Kerr, recommendation for EB, 1954, CASBS Records, Box 45, Folder 8. The best contemporary magazine portrait of him is Jerry Adams, "Eugene Burdick: Writer, Teacher and Iconoclast," *San Francisco Examiner,* [1962]. Burdick's birth record, his World War II draft card, and his marriage license are on Ancestry, but see also EB, "Navy War Service Record," undated, Burdick Collection, Box 98, Folder "U.S. Navy Records." And, for a more recent assessment, see especially Chris Smith, "Intellectual Action Hero: The Political Fictions of Eugene Burdick," *California Magazine,* Summer 2010. My thanks to Smith for sharing his research materials with me.

10. Margaret O'Mara, *The Code: Silicon Valley and the Remaking of America* (New York: Penguin Press, 2019), 14–18, 24. And see Christophe Lécuyer, *Making Silicon Valley: Innovation and the Growth of High Tech, 1930–1970* (Cambridge, MA: MIT Press, 2006).

11. Stegner, "Eugene Burdick," 4–5. EB, "Rest Camp on Maui," *Harper's,* July 1946, 81–90.

12. Stegner, "Eugene Burdick," 4–5.

13. Adams, "Eugene Burdick."

14. EB, "Burdick with a Baedeker," *Isis,* October 12, 1949, 14–15. And see "American Is Not Impressed with Oxford," *Madera* [CA] *Tribune,* October 15, 1949.

15. Adams, "Eugene Burdick."

16. EB, Political Science 212B: Contemporary Political Theory, Spring 1952, syllabus, University of California, Berkeley, Burdick Collection, Box 78, Folder 9.

17. EB, *The 480,* vii.

18. Leo Rosten, "Harold Lasswell: A Memoir," *Saturday Review,* April 15, 1967, CASBS Records, Box 45, Folder 22.

19. Naomi Spatz, interview with the author, May 24, 2018.

20. For a concise statement, see Harold Lasswell, "The Structure and Function of Communication in Society," in *The Communication of Ideas,* ed. Lyman Bryson (New York: Harper & Row, 1948): 37–51.

21. Gabriel A. Almond, "Harold Dwight Lasswell, 1902–1978: A Biographical Memoir" (Washington, DC: National Academy of Sciences, 1987).

22. Christopher Simpson, *Science of Coercion: Communication Research and Psychological Warfare, 1945–1960* (New York: Oxford University Press, 1994), 11, 23–24. Timothy Glander, *Origins of Mass Com-*

munications Research During the American Cold War (Mahwah, NJ: Lawrence Erlbaum Associates, 2000), 27.

23. Paul F. Lazarsfeld, Bernard Berelson, and Hazel Gaudet, *The People's Choice: How the Voter Makes Up His Mind in a Presidential Campaign,* 2nd ed. (1944, repr., New York: Columbia University Press, 1948), 1.

24. Bernard R. Berelson, Paul F. Lazarsfeld, and William N. McPhee, *Voting: A Study of Opinion Formation in a Presidential Campaign* (University of Chicago Press, 1954). The voting study began with a series of questions posed to voters: "As you read or listen to the news these days, do you find you are paying a great deal of attention to things about the election, only a little attention, or no attention at all?" And: "Do you think that people like you have a lot of influence on how the government runs things, some influence, or not much influence at all?" "The 1948 Voting Study First Questionnaire, June 1948," BASR Records, Columbia University, Box 17, Folder "Elmira: Instructions, Plans, Etc." See also "Specifications for Third Wave of 1948 Voting Study," BASR Records, Columbia University, Box 17, Folder "Elmira: Instructions, Plans, Etc."

25. H. Rowan Gaither, chairman, *Report of the Study for the Ford Foundation on Policy*

and Program (Detroit: Ford Foundation, November 1949), 95. H. Rowan Gaither, quoted in Dwight Macdonald, *The Ford Foundation: The Men and the Millions* (New York: Reynal, 1956), 80.

26. Daniel Bessner, *Democracy in Exile: Hans Speier and the Rise of the Defense Intellectual* (Ithaca, NY: Cornell University Press, 2018), 180–81. Mark Solovey, *Shaky Foundations: The Politics–Patronage–Social Science Nexus in Cold War America* (New Brunswick, NJ: Rutgers University Press, 2013), 112–19. Between 1952 and 1962, Ford Foundation programs "granted $82 million for social or behavioral science research" (105).

27. For a brilliant treatment of the history of the future, see Jenny Andersson, *The Future of the World: Futurology, Futurists, and the Struggle for the Post–Cold War Imagination* (New York: Oxford University Press, 2018).

28. Ibid., 2–26.

29. Rebecca S. Lowen, *Creating the Cold War University: The Transformation of Stanford* (Berkeley: University of California Press, 1997), 193–95. Solovey, *Shaky Foundations,* 129–31; the quotation is from p. 130. And see especially Bessner, *Democracy in Exile,* 181–95.

30. De Vries, "A Year at the Center," 172–73.

31. EB, *The 480,* 72.

32. It appears that the confessions, as well as many of the procedures and plans at the center that first year, were Burdick's idea. See EB et al. to Ralph Tyler, July 31, 1954, CASBS Records, Box 45, Folder 8.

33. De Vries, "Year at the Center," 173–75

34. Kenneth E. Boulding, *The Image* (Ann Arbor: University of Michigan Press, 1954).

35. EB to William A. Becker, November 15, 1955, Burdick Collection, Box 17, Folder "Presidential Papers, 1956." In the middle of his fellowship year, Burdick drafted "A Proposal for an Institute for Integrated Studies in the Social Sciences" for the University of California, hoping to replicate, by the bay, what had been built on the hill. EB, "A Proposal for an Institute for Integrated Studies in the Social Sciences [1955]," Burdick Collection, Box 78, Folder "Center for Integrated Studies."

36. IP contributed an essay called "TV and the Image of the Candidate." EB to IP, March 14, 1956, Burdick Collection, Box 57, Folder "Voting Continuities Correspondence 1956."

37. EB, "Political Theory and the Voting Studies," in Eugene Burdick and Arthur J. Brodbeck, eds., *American Voting Behavior* (Glencoe, IL: Free Press, 1959), 136–49.

38. [Kenneth Boulding], "Folk Songs for the Center," undated typescript, Burdick Collection, Box 114, Folder "Center An-

nouncements, Tax, Etc."

39. EB to Edith Haggard, Curtis Brown, April 1, 1955, Burdick Collection, Box 91, Folder "Curtis Brown."

40. EB, *The Ninth Wave,* 45, 90.

41. Maurice Isserman and Michael Kazin, *America Divided: The Civil War of the 1960s* (New York: Oxford University Press, 2008), 57. Joe Kennedy Sr. is quoted in David Burner, *John F. Kennedy and a New Generation* (Glenview, IL: Scott Foresman, 1988), 10.

42. RN defenders insist that RN only ever said this privately. Frank Gannon, "The Pink Lady Revisited," post at nixonfounda tion.org, November 22, 2009.

43. EB, *The Ninth Wave,* 214–15.

44. Ibid., 216.

45. Ibid., 218–19.

46. Oscar Godbout, " 'The Ninth Wave' Bought for Film," *NYT,* July 12, 1956. The book was finished and was supposed to appear in the spring of 1956, but publication was delayed until the fall for better market placement. See Burdick to Paul Brooks, December 21, 1955, and Burdick to Wallace Stegner, December 30, 1955, Wallace Stegner Papers, Special Collections, J. Willard Marriott Library, University of Utah, Box 12, Folder 51.

47. Hedda Hopper, " 'Ninth Wave' Aimed at Brando, Sinatra," *Los Angeles Times,* July

12, 1956.

48. Hedda Hopper, "Sinatra Will Costar in 'Solo' with June," *Los Angeles Times,* October 16, 1956.

49. ELG & Co. to Thomas K. Finletter, November 18, 1955, Stevenson Papers, Box 2, Folder 9.

50. AS, "Adlai Ewing Stevenson II," *American National Biography,* published in print in 1999; published online in February 2000.

51. Josephine Baskin Minow and Newton N. Minow, *As Our Parents Planted for Us, So Shall We Plant for Our Children: A Family Memoir* (Chicago: J. B. Minow, 1999), 71–72.

52. Adlai E. Stevenson, *The New America,* ed. Seymour E. Harris, John Bartlow Martin, and Arthur Schlesinger Jr. (New Yorker: Harper & Brothers, 1957), xiv.

53. For an interesting vantage, see Ball, *The Past Has Another Pattern,* 136.

54. Earl Mazo, "Stevenson Favors Going Slow on Desegregation," *New York Herald Tribune,* February 8, 1956. Seymour Korman, "Adlai Loses Adherents After Both Speak," *Chicago Tribune,* February 6, 1956.

55. Isserman and Kazin, *America Divided,* 41.

56. Stevenson, *The New America,* 5.

57. Ibid., 6.

58. Packard, *Hidden Persuaders,* 197.

59. Blake, *Liking Ike,* 141–42.

60. ELG & Co. to the New York Committee for Stevenson-Kefauver, September 11, 1956, Stevenson Papers, Box 2, Folder 9. He also had an operation in Oregon: see ELG & Co., New York–San Francisco, "A Report on the Oregon Primary Campaign," 1956, Stevenson Papers, Box 2, Folder 9.

61. Packard, *Hidden Persuaders*, 6, 21–22, 29.

62. Aldous Huxley, *Brave New World Revisited* (New York: Harper & Brothers, 1958), 59, 69–71.

63. EB circulated the Planning Papers fairly widely within the tight circles of the campaign leadership and voting behavior experts, where they were avidly read and much discussed. Lasswell congratulated him on his "famous memorandum." Harold Lasswell to EB, November 23, 1955, Burdick Collection, Box 57, Folder "Voting Continuities Volume Correspondence, 1955."

64. The landmark study is Angus Campbell, Philip E. Converse, Warren E. Miller, Donald E. Stokes, *The American Voter* (Chicago: University of Chicago Press, 1960). The most influential summary and analysis of this research is Philip E. Converse, "The Nature of Belief Systems in Mass Publics," in *Ideology and Discontent,* ed. David E. Apter (New York: Free Press of Glencoe, 1964), 207–60. And see Eugene Burdick,

"The Perfect President," *This Week,* January 1, 1956. EB to William A. Becker, November 15, 1955, Burdick Collection, Box 117, Folder "Presidential Papers, 1956."

65. EB, "How You'll Vote in '56," *This Week,* May 13, 1956.

66. EB's notes on interviewing Joseph Houghtelling, 1956, Pool Papers, Box 187, Folder "Stevenson 1956 Campaign."

67. EB [with Harold Lasswell and Paul Lazarsfeld], Planning Paper #1, Stevenson Papers, Box 15, Folder 112. EB, incidentally, had also asked Stevenson to contribute an essay to his volume of essays on voting behavior. EB to William Blair Jr., September 1, 1955, Stevenson Papers, Box 15, Folder 11.

68. EB [and four others], "The Pattern of Governor Stevenson's Speeches: Analysis and Recommendations," Planning Paper #3 [undated but 1955], Burdick Collection, Box 53, Folder 4. Stevenson's advisers replied with expressions of concern. "I see no objection to your going ahead with your project as long as we are completely disassociated from it," wrote one. "I really would prefer not having it be known that we were consulted at all!" Another urged EB not to circulate the paper, lest it get into the hands of a reporter and embarrass Stevenson. William Blair to EB, September 23, 1955,

Stevenson Papers, Box 15, Folder 11. "I think it dangerous to circulate your observations on Mamie, Truman and Ike" was the reply. "A copy may be made and reach unfriendly hands, or an indiscreet remark may reach one of the parties, or a reporter learn of it and Stevenson and the campaign are seriously embarrassed." William Rivkin to Eugene Burdick, July 13, 1955, Stevenson Papers, Box 15, Folder 11.

69. EB to Dear _____, undated but before August 15, 1955, Burdick Collection, Box 117, Folder "Presidential Papers, 1956."

Chapter 3: The Quiet American

1. IP's voice can be heard at http:// americanarchive.org/catalog/cpb-aacip_15-1615f3mm.

2. IP, Statement to the Industrial Employment Review Board, December 7, 1950, Pool Family Papers.

3. MIT SDS, "Wanted for War Crimes, Ithiel de Sola Pool," 1968, MIT Institute Archives and Special Collections, Science Action Coordinating Committee Records, Box 2, Folder 76.

4. Bessner, *Democracy in Exile,* 141–46.

5. Allan A. Needell, " 'Truth Is Our Weapon': Project TROY, Political Warfare, and Government-Academic Relations in the

National Security State," *Diplomatic History* 17 (1993): 399–420.

6. Joseph M. Goldsen (RAND), to IP, April 28, 1948, Pool Family Papers.

7. Bessner, *Democracy in Exile,* 148–49.

8. Francis M. Wray (Industrial Employment Review Board) to IP, November 3, 1950, Pool Family Papers.

9. IP, Statement to the Industrial Employment Review Board, December 7, 1950, Pool Family Papers.

10. Marc D. Angel, "Rabbi Dr. David de Sola Pool: Sephardic Visionary and Activist," https://www.jewishideas.org/article/rabbi-dr -david-de-sola-pool-sephardic-visionary -and-activist.

11. "Tamar de Sola Pool, 90, Author and Former Head of Hadassah," *NYT,* June 2, 1981. Tamar was both a scholar of languages and a devoted mother. When her baby turned one, she made a list of everything he'd learned to say, from "butter please" to "I see you." Tamar de Sola Pool, "Ithiel's Vocabulary," December 23, 1918, Pool Papers, Box 180, Folder "Ithiel's Childhood."

12. Adam de Sola Pool, interview with the author, May 19, 2018.

13. IP, Statement to the Industrial Employment Review Board, December 7, 1950, Pool Family Papers.

14. Francis M. Wray (Industrial Employment

Review Board) to IP, November 3, 1950, Pool Family Papers.

15. IP, "Faust," Pool Papers, Box 180, Folder "Ithiel's Childhood."

16. IP to David and Tamar de Sola Pool, August 18, 1936, Pool Papers, Box 180, Folder "Letters from Mexico 1936."

17. IP's exams from 1939 can be found in the Ithiel de Sola Pool Papers, Special Collections Research Center, University of Chicago Library, Box 2.

18. IP, Statement to the Industrial Employment Review Board, December 7, 1950.

19. Harold D. Lasswell and Nathan Leites, eds., *Language of Politics: Studies in Quantitative Semantics* (New York: George W. Stewart, 1949), ch. 3.

20. N. C. Leites and IP, "Communist Propaganda in Reaction to Frustration," December 1, 1942, Pool Papers, Box 186, in a plastic bag instead of a folder.

21. "I have recently been considerably more retiring on matters of politics than in the past, which may in part explain the confusion on my views." IP, "To Whom It May Concern," April 18, 1942, Pool Family Papers. For a related essay, see IP, "How to Prevent a Third World War [1944]," Pool Papers, Box 175, Folder "How to Prevent a Third World War [1944]."

22. IP kept in close touch with Lasswell during these years. See their correspondence in

the Lasswell Papers, Box 77, Folder 975. IP also took an extended medical leave of absence, beginning in the summer of 1945. Tricia McEldowney, Archivist & Special Collections Librarian, Warren Hunting Smith Library, Hobart and William Smith Colleges, e-mail to the author, February 11, 2019.

23. IP, Statement to the Industrial Employment Review Board, December 7, 1950.

24. Harold D. Lasswell, Daniel Lerner, and IP, "The Comparative Study of Symbols: An Introduction," *Hoover Institute Studies,* Series C, no. 1 (January 1952). For a related RADIR project — a content analysis of constitutions — see IP and Harriet Alden, "A Comparative Study of World Constitutions," undated but apparently c. 1953; the study was never completed (Lasswell to IP, April 7, 1953), Lasswell Papers, Box 77, Folder 975.

25. *Exposé of the Communist Party of Western Pennsylvania: Hearings Before the House Committee on Un-American Activities, Eighty-First Congress, Second Session, Appendix, March 24 and 25, 1950* (Washington, DC: GPO, 1950).

26. IP, Statement to the Industrial Employment Review Board, December 7, 1950.

27. Ralph de Toledano to RN, undated but mid-1950, Nixon Library, Pre-Presidential

Papers, 1946–1963, General Correspondence, 1946–1963.

28. IP, Statement to the Industrial Employment Review Board, December 7, 1950.

29. J. Tenney Mason (Industrial Review Board) to IP, December 21, 1950, Pool Family Papers.

30. IP to RN, December 29, 1950, Nixon Library, Pre-Presidential Papers, 1946–1963, General Correspondence, 1946–1963.

31. Saul Bellow, *A Theft* (New York: Penguin, 1989), 56.

32. John L. Boies, *Buying for Armageddon: Business, Society, and Military Spending Since the Cuban Missile Crisis* (New Brunswick, NJ: Rutgers University Press, 1994), 1. O'Mara, *The Code,* 23–24.

33. O'Mara, *The Code,* 25.

34. He went on to contribute an essay to Burdick's anthology *American Voting Behavior.* IP to EB, November 4, 1952, Burdick Collection, Box 117, Folder 58.

35. Lowen, *Creating the Cold War University,* 204–8.

36. Solovey, *Shaky Foundations,* 119–27.

37. William F. Buckley Jr., "Our Mission Statement," *National Review,* November 19, 1955, https://www.nationalreview.com/1955/11/our-mission-statement-william-f-buckley-jr/.

38. Donald L. M. Blackmer, *The MIT Center*

for International Studies: The Founding Years, 1951–1969 (Cambridge, MA: MIT Press, 2002), chs. 1 and 2; the quotation is from p. 57. For more on the Ford funding, see IP to Lasswell, March 15, 1956, Lasswell Papers, Box 77, Folder 975. Max F. Millikan to Ithiel de Sola Pool, January 29, 1953, Pool Papers, Box 180, Folder "Letters Inviting Ithiel to MIT." Lerner and IP soon worked together at MIT on a project on the Middle East, in collaboration with the Bureau of Applied Social Research at Columbia. IP to Lasswell, April 6, 1955, Lasswell Papers, Box 77, Folder 975.

39. It also marooned the only behavioral scientist in Stanford's political science department, Alfred de Grazia. In 1953, ADG wrote to the university's president to protest the loss of Pool and Lerner. Two years later, ADG was denied tenure. Lowen, *Creating the Cold War University*, 209, 282. ADG's most noted work was *Public and Republic: Political Representation in America* (New York: Knopf, 1951).

40. O'Mara, *The Code*, 35.

41. Adam de Sola Pool, interview with the author, May 19, 2018.

42. Jean MacKenzie de Sola Pool, "The Story of My Life," pp. 27–29, unpublished manuscript, based on interviews conducted by Hattie Belin and adapted by Dorothy

Danker, 1998, Schlesinger Library, Radcliffe Institute for Advanced Study, Harvard University.

43. "Have I told you that I am getting married in March to Jean MacKenzie, a psychologist on Moe Stein's creativity project?" IP to Lasswell, January 25, 1956, Lasswell Papers, Box 77, Folder 975.

44. Bellow had known IP fairly well in the 1930s (Saul Bellow to Jean Pool, February 22, 1988), but Jean appears to have been ambivalent about the portrait of him he offered in *A Theft* (Jean Pool to Saul Bellow, March 22, 1989). Both letters are in the Pool Papers, Box 180, Folder "Saul Bellow."

45. Bellow, *A Theft,* 7, 18, 8–9.

46. Ibid., 17–18.

47. IP to RN, December 29, 1950, Nixon Library, Pre-Presidential Papers, 1946–1963, General Correspondence, 1946–1963.

48. FDR, Memorandum to Cordell Hull, January 24, 1944, as quoted in Isserman and Kazin, *America Divided,* 70.

49. Fredrik Logevall, *Embers of War: The Fall of an Empire and the Making of America's Vietnam* (New York: Random House, 2012). James Carter, *Inventing Vietnam: The United States and State Building, 1954–1968* (New York: Cambridge University Press, 2008).

50. Richard West, "Graham Greene and 'The

Quiet American,' " *New York Review of Books,* May 16, 1991; Clive Christie, *"The Quiet American" and "The Ugly American": Western Literary Perspectives on Indo-China in a Decade of Transition, 1950–1960,* Occasional Paper No. 10 (Canterbury: Centre of South-East Asian Studies, University of Kent, 1989), 28–29, 44–45. James Gibney, "The Ugly American," *NYT,* January 15, 2006. Max Boot, *The Road Not Taken: Edward Lansdale and the American Tragedy in Vietnam* (New York: Liveright, 2018). And see Jonathan Nashel, *Edward Lansdale's Cold War* (Amherst: University of Massachusetts Press, 2005), ch. 6.

51. Graham Greene, *The Quiet American* (1955; repr., New York: Bantam, 1974), 4, 21.

52. ELG to IP, December 1, 1955, Pool Papers, Box 187, Folder "Stevenson 1956 Campaign."

53. IP to Manfred Kochen, November 30, 1956, Pool Papers, Box 59, Folder "Contact Nets Diary."

54. IP, Questionnaire, undated, Pool Papers, Box 59, Folder "Contact Nets Diary."

55. That IP "thinks that any two Americans, of any sort, can be linked by a chain of six acquaintanceships or less" was reported in *The New Yorker* in 1964, in an essay later published as a book by Christopher Rand:

Cambridge, U.S.A.: Hub of a New World (New York: Oxford University Press, 1964), 140. And see IP and Manfred Kochen, "Contacts and Influence," *Social Networks* 1 (1978–79): 5–51.

56. ELG to Jack Shea, April 26 and June 6, 1956, Stevenson Papers, Box 2, Folder 9.

57. IP, "The Use of Feedback Technique by the Stevenson Volunteers," Pool Papers, Box 187, Folder "Stevenson 1956 Campaign." Rules: "<u>Never, never, call this a poll</u>" (because it wasn't a random sample; they were interviewing registered Democrats). Also: "This is not the place to go into details on the social science research on which this plan is based. Suffice it to say that a great many studies have shown that giving a person a chance to participate plays a great role in teaching him. People like to be listened to. They are more favorable to someone who does listen to their views."

58. "Major Needs of the Stevenson Campaign in the Los Angeles Area," Pool Papers, Box 57, Folder "Stevenson Campaign in California [May 13, 1956]."

59. IP, EB, Warren Miller, and Dwaine Marvick, "The Stevenson Campaign in California: A Research Report from Edward L. Greenfield & Co. [May 1956]," Stevenson Papers, Box 299, Folder 5.

60. Totton J. Anderson, "The 1956 Election in California," *Western Political Quarterly*

10 (1957): 102–16. The California campaign material can be found in the Stevenson Papers, Box 249, Folder 1.

61. Gore Vidal, *The Best Man: A Play About Politics* (Boston: Little, Brown, 1960), 7, 9, 77, 94.

62. AS, *A Thousand Days: John F. Kennedy in the White House* (Boston: Houghton Mifflin, 1965), 8.

63. JFK, "Nominating Adlai Stevenson for President of the United States," Democratic National Convention, August 16, 1956.

64. AS, *Thousand Days,* 8–9. Ball, *The Past Has Another Pattern,* 140–41.

65. Stevenson, *The New America,* 4.

66. Richard Aldous, *Schlesinger: The Imperial Historian* (New York: Norton, 2017), 176.

67. Ball, *The Past Has Another Pattern,* 142–45.

68. Stevenson campaign, "The Man from Libertyville," television commercial, 1956.

69. Eisenhower campaign, "I Like Ike for President," television commercial, 1956.

70. Russel Windes Jr., "Speech-Making of Adlai E. Stevenson in the 1956 Presidential Campaign" (PhD diss., Northwestern University, 1959), 174.

71. "The Stevenson Campaign in California, II: A Research Report from Edward L. Greenfield and Company, New York-San Francisco . . . Under the direction and

compiled by the Social Science Research and Political Analysis Divisions of Edward J. Greenfield & Co., New York," including "Field Reports From IP, EB, Warren Miller, Dwaine Marvick, with a special report from Lewis Dexter and Davis Bobrow." Pool Papers, Box 57, Folder "Stevenson Campaign in California [May 13, 1956]."

72. Robert Blanchard, "House Leader Martin Here, Raps Stevenson," *Los Angeles Times,* October 9, 1956.

73. Anderson, "The 1956 Election in California."

74. This was not the only idea on which IP and ELG collaborated in 1958. They also attempted to start an international editors' panel. The advisory board included WMc. IP to Lasswell, July 15, 1958, Lasswell Papers, Box 77, Folder 975. IP appears to have mailed Lasswell a proposal for Project Macroscope at the same time. See "Planning Paper: Voting Behavior Machine," undated but 1958, same folder.

Chapter 4: Artificial Intelligence

1. [Andy Logan and Brendan Gill], "Runner-Up," *New Yorker,* November 29, 1958.

2. Ibid.

3. Naomi Spatz, interview with the author, May 24, 2018. Elizabeth Bernstein Rand, interview with the author, August 20, 2018.

4. "Don't Pity the Liberal Arts Grads — They're Bossing the Engineers," *Hartford Courant,* February 13, 1962.

5. T. R. Kennedy Jr., "Electronic Computer Flashes Answers, May Speed Engineering," *NYT,* February 15, 1946.

6. See my discussion in *These Truths,* 557–65.

7. Norbert Wiener, *Cybernetics* (New York: Wiley, 1948).

8. Clive Thompson, "The Secret History of Women in Coding," *NYT,* February 13, 2019.

9. Grace Hopper, "The Education of a Computer [1952]," *Annals of the History of Computing* 9 (1988): 272.

10. Louis Hyman, *Temp: How American Work, American Business, and the American Dream Became Temporary* (New York: Viking, 2018), 127.

11. Joy Lisi Rankin, *A People's History of Computing in the United States* (Cambridge, MA: Harvard University Press, 2018), 14.

12. Robert Jungk, *Tomorrow Is Already Here: Scenes from a Man-Made World* (London: Rupert Hart-Davis, 1954), 178–82, 187.

13. O'Mara, *The Code,* 11. Emerson W. Pugh, *Building IBM: Shaping an Industry and Its Technology* (Cambridge, MA: MIT Press, 1995), chs. 12 and 13. See also Chinoy, "Battle of the Brains."

14. Rankin, *A People's History of Computing,* 13.

15. Pugh, *Building IBM,* 194–96.

16. *Fortran: Automatic Coding System for the IBM 704* (New York: IBM, 1956).

17. John McCarthy, Marvin Minsky, Nathaniel Rochester, and Claude Shannon, "A Proposal for the Dartmouth Summer Research Project on Artificial Intelligence," August 31, 1955, Rauner Special Collections Library, Dartmouth College.

18. A ProQuest Historical Newspapers search conducted in 2019 located 745 uses in 1954 and 4,300 in 1955.

19. "Rand Declares New Computer Is Step to 'Automatic Factory,' " *New York Herald Tribune,* February 5, 1953. When the air force announced plans to build a flight simulation laboratory in Ohio, the *Boston Globe* explained, not getting it quite right, that the purpose of the lab was "to test make-believe craft," and a writer for the *Christian Science Monitor* wondered whether, in the end, they would "have to fly the real rocket at all." "AF Planning Largest Flight Simulation Lab," *Austin American,* August 21, 1955. "Air Force Orders Laboratory to Test Make-Believe Craft," *Boston Daily Globe,* August 21, 1955. William H. Stringer, "Here Comes (There Goes) Uni-

vac!" *Christian Science Monitor,* June 26, 1955.

20. Eva C. Freeman, ed., *MIT Lincoln Laboratory: Technology in the National Interest* (Boston: Nimrod Press, 1995), ch. 2.

21. Engineering and Research Corp. Advertisement, *Philadelphia Inquirer,* April 13, 1952. O'Mara, *The Code,* 24.

22. Edgar Allan Poe, "Maelzel's Chess Player," *Southern Literary Messenger,* April 1836.

23. AB and Michael de V. Roberts, "Computer v. Chess-Player," *Scientific American,* June 1958, 96–107. And see AB et al., "A Chess Playing Program for the IBM 704," *Proceedings of the May 6–8, 1958, Western Joint Computer Conference,* 157–59.

24. Claude E. Shannon, "A Chess Playing Machine," *Scientific American,* February 1950.

25. [Logan and Gill], "Runner-Up."

26. Taylor Branch, *The King Years: Historic Moments in the Civil Rights Movement* (New York: Simon & Schuster, 2013), ch. 1.

27. Harvard Sitkoff, *The Struggle for Black Equality* (New York: Hill & Wang, 2008), 50.

28. Elizabeth Bernstein Rand, interview with the author, August 20, 2018.

29. McCarthy et al., "A Proposal for the Dartmouth Summer Research Project on

Artificial Intelligence." The best discussion of the Dartmouth Project is Grace Solomonoff, "Ray Solomonoff and the Dartmouth Summer Research Project in Artificial Intelligence, 1956," http://raysolomonoff.com/dartmouth/dartray.pdf. With thanks to Grace Solomonoff. On Bernstein, see also Pamela McCorduck, *Machines Who Think: A Personal Inquiry into the History and Prospects of Artificial Intelligence* (Boca Raton, FL: CRC Press, 2004), 180–82.

30. AB and Michael de V Roberts, "Computer v. Chess-Player," *Scientific American,* June 1958, 96–107.

31. [Logan and Gill], "Runner-Up."

32. AB and Michael de V. Roberts, "Computer v. Chess-Player," 96–107. And see AB et al., "A Chess Playing Program for the IBM 704," 157–59.

33. Bruce J. Schulman, *Lyndon B. Johnson and American Liberalism: A Brief Biography with Documents* (Boston: Bedford Books, 1995), 53–54; Dan T. Carter, *The Politics of Rage: George Wallace, the Origins of the New Conservatism, and the Transformation of American Politics* (1995; repr., Baton Rouge: Louisiana State University Press, 2000), 96–97.

34. Michael J. Klarman, *Brown v. Board of Education and the Civil Rights Movement*

(New York: Oxford University Press, 2007), 187–91. Michael D. Davis and Hunter R. Clark, *Thurgood Marshall: Warrior at the Bar, Rebel on the Bench* (New York: Birch Lane Press, 1992), 458.

35. Douglas Brinkley, *American Moonshot: John F. Kennedy and the Great Space Race* (New York: HarperCollins, 2019), 118–19, 210.

36. George Reedy, letter to Lyndon Baines Johnson on October 17, 1957, as quoted in Brinkley, *American Moonshot,* 133–35.

37. IP thought a good idea would be to ask IBM to pay Kochen's way. "We would like to explore the possibility of asking IBM to send you, and it should be a reasonable thing for them to do for you will be addressing a group which will be developing considerable work of interest to IBM." IP to Manfred Kochen, September 30, 1957, Pool Papers, Box 59, Folder "Contact Nets Diary."

38. IP to Manfred Kochen, December 9, 1957, Pool Papers, Box 59, Folder "Contact Nets Diary."

39. IP to Ralph Tyler, April 26, 1957, CASBS Records, Box 48, Folder 18.

40. Jean MacKenzie Pool, "The Story of My Life," 40–42. Adam de Sola Pool, e-mail to the author, March 7, 2019.

41. IP, "Acquaintanceship Networks: Project

Proposal," Pool Papers, Box 59, Folder "Drafts of Content Nets Problem, 1958."

42. IP and Kochen received an NSF grant, Research Grant NSF-G7196, on January 21, 1959. On calling this branch of research the study of the "small world," see IP, "Project Proposal" [to the NSF], May 1, 1961, Pool Papers, Box 59, Folder "NSF Proposal Acquaintanceship Nets."

43. IP and Manfred Kochen, "Contacts and Influence," *Social Networks* 1 (1978–79): 5–51. Manfred Kochen, ed., *The Small World* (Norwood, NJ: Ablex Publishers, 1989).

44. Bruce J. Schulman, *From Cotton Belt to Sunbelt: Federal Policy, Economic Development, and the Transformation of the South, 1938–1980* (Durham, NC: Duke University Press, 1994), 147–48.

45. Brinkley, *American Moonshot,* 260–61.

46. American Experience, *Chasing the Moon,* episode 1, directed by Robert Stone, documentary film, 2019.

47. Rachel Carson to Dorothy Freeman, February 1, 1958, in *Always Rachel: The Letters of Rachel Carson and Dorothy Freeman, 1952–1964,* ed. Martha Freeman (Boston: Beacon, 1995), 249.

48. Hannah Arendt, *The Human Condition* (Chicago: University of Chicago Press, 1958), prologue.

Chapter 5: Project Macroscope

1. Sarah Neidhardt, e-mail to the author, March 6, 2019.
2. WMc, Curriculum Vitae, c. 1964, in the possession of his son, John McPhee. John McPhee, "A Life Remembered: Dr. William N. McPhee (1921–1998)," in the possession of John McPhee. Rebecca Syzmanski, "Ailing Ex-Professor Chose to End Life," *Albuquerque* [NM] *Journal,* May 29, 2001; John McPhee, interview with the author, July 24, 2018. The fullest biographical treatment of McPhee is Peter Simonson, "Writing Figures into the Field: William McPhee and the Parts Played by People in Our Histories of Media Research," in *The History of Media and Communication Research: Contested Memories,* ed. Peter W. Park and Jefferson Pooley (New York: Peter Lang, 2008), ch. 12.
3. Eleanor Emery to Miriam Emery, November 17, 1942, Emery-McPhee Papers.
4. Sarah McPhee, interview with the author, July 30, 2018. Wendy McPhee, interview with the author, July 16, 2018. John McPhee, interview with the author, July 24, 2018.
5. WMc, Curriculum Vitae, c. 1964.
6. Betty Friedan, *The Feminine Mystique* (New York: Norton, 1963).
7. Sarah McPhee, interview with the author,

July 30, 2018. Wendy McPhee, interview with the author, July 16, 2018.

8. MMc to Eleanor Emery, May 17, 1951, Emery-McPhee Papers.

9. MMc to Eleanor Emery, Thanksgiving Day, 1952, Emery-McPhee Papers.

10. MMc to Eleanor Emery, April 1, 1953, Emery-McPhee Papers.

11. MMc to Ruth Washburn, June 18, 1953, Emery-McPhee Papers.

12. MMc, "Look at Me Today," in *Watch Me*, with drawings by Heidi Brandt (Colorado Springs, CO: Ruth Washburn Cooperative Nursery School, 1966).

13. Sylvia Plath, "Daddy," *Ariel* (London: Faber & Faber, 1965).

14. E.g., Sylvia Plath to Aurelia Schober Plath, October 8, 1960, in Sylvia Plath, *Letters Home*, selected and edited by Aurelia Schober Plath (New York: Harper & Row, 1975, 396. N.B.: The Plath letter that serves as the epigraph to this chapter is Plath to Aurelia Schober Plath, July 19, 1960, in *The Letters of Sylvia Plath*, ed. Peter K. Steinberg and Karen V. Kukil, vol. 2, *1956–1963* (New York: HarperCollins, 2018), 495.

15. MMc to Eleanor Emery, March 27, 1957, Emery-McPhee Papers.

16. MMc to Eleanor Emery, Thanksgiving Day, 1952, Emery-McPhee Papers.

17. MMc to Eleanor Emery, October 31 or November 7, 1951, Emery-McPhee Papers.

18. MMc to Eleanor Emery, June 18, 1953, Emery-McPhee Papers.

19. MMc to Eleanor Emery, October 9, 1951, Emery-McPhee Papers. JC and WMc collaborated closely in the 1950s, on a number of projects. The intensity of their professional and personal relationship can be glimpsed in a series of letters in the Coleman Papers, Box 53, Folder "McPhee."

20. Thomas Coleman, interview with the author, March 18, 2019.

21. JC broke the relationship off in 1958, in a letter in which he regretted the banality of 1950s-style marriage. JC to Jane Emery, April 29, 1958, Emery-McPhee Papers.

22. Edward Albee, *Who's Afraid of Virginia Woolf?* (New York: Atheneum, 1962), 34.

23. MMc to Eleanor Emery, January 23, 1953, Emery-McPhee Papers.

24. MMc to Charles Francis Emery, October 23, 1953, Emery-McPhee Papers.

25. MMc to Eleanor Emery, March 13, 1956, Emery-McPhee Papers.

26. MMc to Eleanor Emery, March 9 and 15, 1954, Emery-McPhee Papers.

27. Wendy McPhee to Eleanor Emery, April 1954, Emery-McPhee Papers.

28. MMc to Eleanor Emery, July 7, 1954, Emery-McPhee Papers.

29. Jane Emery to WMc, undated but c. 1956, Emery-McPhee Papers.

30. MMc to Eleanor Emery, February 22, 1955, Emery-McPhee Papers.

31. WMc's weight is frequently discussed, mainly because he struggled to control it. E.g., from when he was 219 pounds: "He hasn't weighed that since before Wendy was born so he is making headway." MMc to Eleanor Emery, October 2, 1957, Emery-McPhee Papers.

32. Charles Emery to Jane Emery, November 10, 1955, Emery-McPhee Papers.

33. MMc to Eleanor Emery, April 29, 1951, and see, e.g., Charles Emery to Jane Emery, October 25, 1955, Emery-McPhee Papers.

34. MMc to Eleanor Emery, December 4, 1956, Emery-McPhee Papers.

35. C. Wright Mills to Dr. Paul Lazarsfeld, May 6, 1959, Paul F. Lazarsfeld Papers, Columbia University Archives, as quoted in Glander, *Origins of Mass Communications Research* (Mahwah, NJ: L. Erlbaum, 2000), 211.

36. MMc to Jane Emery, June 21, 1956, Emery-McPhee Papers. The Dartmouth Workshop ran from about June 18 to August 17, 1956. WMc does not appear on any of the (scant and not necessarily comprehensive) lists of participants but may have attended as an observer. Solomonoff, "Ray Solomonoff and the Dartmouth Sum-

mer Research Project in Artificial Intelligence, 1956."

37. MMc to Eleanor Emery, February 2 and February 17, 1956, Emery-McPhee Papers.

38. Miriam McPhee to Jane Emery, November 4, 1955, Emery-McPhee Papers.

39. Charles Emery to WMc, undated but c. 1956, Emery-McPhee Papers.

40. Jane Emery to Charles Emery, August 23, 1957, Emery-McPhee Papers.

41. Simulmatics, *Human Behavior and the Electronic Computer: An Information Brochure* (New York: Simulmatics Corporation, undated, but c. 1962), Pool Papers, Box 67, Folder "Simulmatics: Correspondence," 7.

42. IP, Abelson, and SP, *Candidates, Issues, Strategies* (Cambridge, MA: MIT Press, 1964), 7.

43. Jane Emery to Charles Emery, February 26, 1958, Emery-McPhee Papers.

44. Grace Metalious, *Peyton Place* (New York: Julian Messner, 1956), 1, 17. And see the Introduction, "Open Secrets: Rereading *Peyton Place*," by Ardis Cameron, in *Peyton Place* (1956; repr., Boston: Northeastern University Press, 2011).

45. Jane Emery to Charles Emery, April 16, 1958, Emery-McPhee Papers.

46. Ibid.

47. Jane Emery to Charles Emery, April 21, 1958, Emery-McPhee Papers.

48. Simonson, "Writing Figures," 304.

49. Social Science Research Council, "Simulation of Cognitive Processes: A Report on the Summer Research Training Institute, 1958." *Items* 12, no. 4 (1958): 37–48. Abelson's attendance is also mentioned in his bio in a Simulmatics brochure: Simulmatics Corporation, *Human Behavior and the Electronic Computer.* See another draft, possibly slightly different, in Lasswell Papers, Box 40, Folder 548.

50. IP, Abelson, and SP, *Candidates, Issues, Strategies,* 8, n12.

51. WMc to Wendy, John, and Sarah McPhee, undated but summer 1958 (because from the Hotel Monica, Santa Monica), Emery-McPhee Papers; this portion of the letter is directed at John.

52. Eleanor (Cooie) Emerson to Charles and Jane Emery, March 3, 1959, Emery-McPhee Papers.

53. WMc to JC, December 27, no year (the yellow-sheet letter), and WMc to JC, undated (the white-sheet letter), Coleman Papers, Box 87, Folder "McPhee."

54. McPhee provided the behavioral theory and the mathematical model, but another of Lazarsfeld's graduate students, Robert Smith, did a lot of the original programming. On Smith's collaboration with McPhee, see Robert B. Smith, "Innovations

at Columbia," 2006, typescript, in the possession of Peter Simonson, and Camelia Florela Voinea, *Political Attitudes: Computational and Simulation Modelling* (New York: Wiley, 2016), ch. 3. Smith died in 2017, and I was unable to interview him for this book. More about this work does not appear to have been among his papers. Joanna Handlin Smith, e-mail to the author, July 31, 2018.

55. WMc and Robert B. Smith, "A Model for Analyzing Macro-Dynamics in Voting System," 1959, BASR Records, Box 27, Folder B0395. See also Robert B. Smith and WMc, "A Fully Observable Electorate," 1961, BASR Records, Box 62, Folder B-0725; and WMc, John Ferguson, and Robert B. Smith, "A Model for Simulating Voting Systems," in John M. Dutton, ed., *Computer Simulation of Human Behavior* (New York: Wiley, 1971), 469–81, but note that this paper is simply "an edited composite" of earlier papers by WMc and Smith (1962) and WMc (1961).

56. MMc to Eleanor Emery, October 30, 1958, Emery-McPhee Papers.

57. An early version of the proposal is "Project Macroscope," Pool Papers, Box 93, Folder "McPhee." This draft appears to be written by ELG with edits by IP. And for the recruitment of Abelson to this project,

see the correspondence between IP and Abelson, 1958–59, also in this folder.

58. Elizabeth Bernstein Rand, interview with the author, August 20, 2018.
59. [ELG], "Project Macroscope" [1959], Stevenson Papers, Box 38, Folder 7.
60. [ELG], "Voting Behavior Machine," Stevenson Papers, Box 38, Folder 7.
61. [ELG], Memo, "Use of Machine Computers in Political Survey Research," Stevenson Papers, Box 38, Folder 7.
62. ELG to NM, March 18, 1959, Stevenson Papers, Box 38, Folder 7.
63. David Marcus, "The Power Historian," *Nation,* October 30, 2017.
64. NM to AS, March 25, 1959, Stevenson Papers, Box 38, Folder 7.
65. AS to NM, April 17, 1959, Stevenson Papers, Box 38, Folder 7.

Chapter 6: The IBM President

1. The Simulmatics Corporation, Common Stock, offered by Russell & Saxe, Inc., New York, May 15, 1961, Pool Papers, Box 142, Folder "Enquiries."
2. ELG to NM, March 18, 1959, Stevenson Papers, Box 38, Folder 7. Lasswell, in contrast to NM, wrote IP, "I am delighted to learn that the machine project is going through." Lasswell to IP, May 5, 1959, Lasswell Papers, Box 77, Folder 975.

3. "Chicago-N.Y. Air Crash: Fear 58 of 73 Aboard Die; Find 10 Survivors," *Chicago Daily Tribune,* February 4, 1959. MMc to Jane Emery, February 6, 1959, Emery-McPhee Papers.

4. EB to Sonia Lilienthal (McGraw-Hill), May 18, 1964, Burdick Collection, Box 35, Folder "The 480 — Correspondence."

5. Christie, *"The Quiet American" and "The Ugly American,"* 38. Norton promoted the book very heavily, a campaign that was largely designed by Burdick himself and was later subject to much criticism. See especially, Joseph Buttinger, "Fact and Fiction on Foreign Aid: A Critique of 'The Ugly American,' " *Dissent,* Summer 1959, and Burdick's reply, "Imagination and Dreams in Foreign Aid," August 10, 1959, Burdick Collection, Box 19, Folder "Ugly American — Dissent Episode."

6. On that friendship, see RN to William Lederer, October 10, 1957, and December 19, 1961, William Lederer Papers, Special Collections and University Archives, University of Massachusetts, Amherst, Series 2.

7. AS, *Thousand Days,* 63.

8. EB was deeply involved with the controversy over the film adaptation of *The Ugly American.* See, e.g., EB to H. N. Swanson, undated, Burdick Collection, Box 17, Folder "Ugly American Correspondence,

Lederer to Burdick," and EB, "Why Can't the Movie Be More Like the Book?," undated typescript, Burdick Collection, Box 23, Folder "From Novel to Movie." Note that this folder contains audio recordings — Dictaphone vinyl — of the original novel.

9. EB, Political Science 194A, Syllabus, Burdick Collection, Box 65, Folder "Political Science 194A."

10. EB, *The 480,* ii.

11. Quoted in Sean Wilentz's foreword to Arthur Schlesinger Jr., *"The Politics of Hope" and "The Bitter Heritage": American Liberalism in the 1960s* (Princeton, NJ: Princeton University Press, 2008).

12. Adlai Stevenson, *The Papers of Adlai E. Stevenson: Continuing Education and the Unfinished Business of American Society, 1957–1961,* ed. Walter Johnson (Boston: Little, Brown, 1977), 385, and quoted in John Bartlow Martin, *Adlai Stevenson and the World: The Life of Adlai E. Stevenson* (New York: Doubleday, 1977), 471.

13. Wendy McPhee, interview with the author, July 16, 2018.

14. Mary McGrory, "The Uneasy Politician," in *Candidates 1960: Behind the Headlines in the Presidential Race,* ed. Eric Sevareid (New York: Basic Books, 1959), 216, 254, 219.

15. AS, *Thousand Days,* 9–10.

16. TBM, "The People-Machine." And for a progress report on the state of the work, see "Project on the Compilation and Detailed Analysis of Political Survey Data: Progress Report," October 15, 1959, Burdick Collection, Box 34, no folder.

17. IP to Lawrence O'Brien, July 16, 1960, O'Brien Papers, Box 16, Folder 10.

18. Theodore H. White, *The Making of the President 1960* (1961; repr., New York: Pocket Books, 1962), 58.

19. The Simulmatics Corporation, Common Stock, offered by Russell & Saxe, Inc., New York, May 15, 1961, Pool Papers, Box 142, Folder "Enquiries."

20. IP, Abelson, and SP, Candidates, Issues, Strategies, 15–16.

21. Ibid., 17.

22. TBM, "The People-Machine."

23. Sarah Igo, *The Averaged American: Surveys, Citizens, and the Making of a Mass Public* (Cambridge, MA: Harvard University Press, 2009), 137.

24. Simulmatics noted, with regard to its conclusion, "The validity of this conclusion is sometimes denied on the grounds that many Negroes are non-voters. If that were true, obviously their weight in the electorate would be proportionally less than their numbers would indicate. However, it is not true that Northern Negroes stay away from

the polls. This belief is based upon national public opinion poll figures which combine figures from both North and South to produce enough cases for separate analyses. The Simulmatics procedure of combining both local and national polls in a single treatment permits us for the first time to make separate comparisons of non-voting. Between middle class Negroes and whites in Northern cities; between lower status Negroes and whites in Northern cities." Simulmatics Corporation, "Negro Voters in Northern Cities," May 15, 1960, Kennedy Library, Democratic National Committee Papers, Box 212 (another copy can be found in the O'Brien Papers).

25. Austin Ranney, "The 1960 Democratic Convention: Los Angeles and Before," in *Inside Politics: The National Conventions, 1960,* ed. Paul Tillett (Dobbs Ferry, NY: Eagleton Institute of Politics, 1962), 7.

26. Gary A. Donaldson, *The First Modern Campaign: Kennedy, Nixon, and the Election of 1960* (Lanham, NJ: Rowman & Littlefield, 2007), 38.

27. AS, *Thousand Days,* 17.

28. On the defection, see especially Aldous, *Schlesinger,* ch. 11.

29. White, *Making of the President,* 38.

30. Donaldson, *The First Modern Campaign,* 42.

31. For a critique, see, e.g., W. J. Rorabaugh, *The Real Making of the President: Kennedy, Nixon, and the 1960 Election* (Lawrence: University Press of Kansas, 2009).

32. Isserman and Kazin, *America Divided,* 33.

33. AS, *Thousand Days,* 23.

34. Ranney, "The 1960 Democratic Convention," 10.

35. Simulmatics Corporation, "Negro Voters in Northern Cities," May 15, 1960, Kennedy Library, Democratic National Committee Papers, Box 212.

36. IP to Lawrence O'Brien, July 16, 1960, O'Brien Papers, Box 16, Folder 10.

37. Ball, *The Past Has Another Pattern,* 158.

38. AS in *"Let Us Begin Anew": An Oral History of the Kennedy Presidency,* by Gerald S. Strober and Deborah H. Strober (New York: HarperCollins, 1993), 9.

39. TMB, "Madly for Adlai," *American Heritage,* August–September 1984.

40. Donaldson, *The First Modern Campaign,* 43–44.

41. Harold Taylor, "Stevenson Named as Candidate by Group of His Supporters," Press Release, June 9, 1960, and "Petition to Delegates to the DNC on Behalf of the Nomination of Adlai Stevenson" (June 13, 1960), 1960, Presidential Campaign Series, Stevenson Papers. "An Important Message to All Liberals" (June 17, 1960),

Theodore C. Sorensen Personal Papers, Box 23, Campaign Files, Kennedy Library. Schlesinger, *Thousand Days*, 28–29. And see Donaldson, *The First Modern Campaign*, 44.

42. AS, *Thousand Days*, 28.
43. According to AS, NM "stayed with Stevenson" but "had no illusions as to what was going to happen." AS in Strober and Strober, *"Let Us Begin Anew,"* 8–9.
44. Ibid., 9.
45. TBM, "Madly for Adlai."
46. IP to Theodore Sorensen, Pool Papers, Box 93, Folder "Negro Vote in Kennedy 1960 Election."
47. Ranney, "The 1960 Democratic Convention," 10–11. Donaldson, *The First Modern Campaign*, 73–74.
48. White, *Making of the President*, 181.
49. TBM, "Madly for Adlai."
50. Donaldson, *The First Modern Campaign*, 75.
51. AS, *Journals, 1952–2000* (New York: Penguin Press, 2007), 71 (July 9, 1960).
52. TBM, "Madly for Adlai."
53. IP, Abelson, and SP, *Candidates, Issues, Strategies*, 23.
54. Harris Wofford to MLK, April 1, 1960, *The Papers of Martin Luther King, Jr.,* 5: 404–5.
55. Ranney, "The 1960 Democratic Convention," 13.

56. "1960 Democratic Party Platform," The American Presidency Project, https://www.presidency.ucsb.edu/documents/1960-democratic-party-platform.

57. Donaldson, *The First Modern Campaign,* 78–79.

58. AS, *Journals,* 73 (July 13, 1960). Newton Minow, e-mail to the author, November 11, 2019.

59. Martin, *Adlai Stevenson and the World,* 522–23. And retold in Thomas Oliphant, *The Road to Camelot: Inside JFK's Five-Year Campaign* (New York: Simon & Schuster, 2017), 244–45.

60. AS, *Thousand Days,* 34.

61. White, *Making of the President,* 195.

62. TBM, "Madly for Adlai." AS, *Journals,* 72 (July 13, 1960).

63. TBM, "Madly for Adlai."

64. White, *Making of the President,* 197.

65. TBM, "Madly for Adlai."

66. White, *Making of the President,* 198–99.

67. TBM, "Madly for Adlai."

68. H. L. Mencken, "Post-Mortem," July 14, 1924, in *On Politics: A Carnival of Buncombe* (Baltimore: Johns Hopkins University Press, 1956), 79.

69. White, *Making of the President,* 198–99.

70. Ibid., 200–201. TBM, "Madly for Adlai."

71. AS, Thousand Days, 49.

72. Lepore, *These Truths,* 570–71.

73. AS, *Journals,* 77–79 (July 15, 1960).

74. AS, *Kennedy or Nixon: Does It Make Any Difference?* (New York: Macmillan, 1960). Aldous, *Schlesinger,* 213–15.

75. Aldous, *Schlesinger,* 216.

76. IP to RFK, July 16, 1960, O'Brien Papers, Box 16, Folder 10.

77. IP, "The Simulmatics Project," fifteen-page typescript, June or July 1960.

78. IP to Lawrence O'Brien, July 16, 1960, O'Brien Papers, Box 16, Folder 10.

79. George Belknap to Ralph Dungan, August 19, 1960, O'Brien Papers, Box 16, Folder 10.

80. The Simulmatics Corporation, "The Simulmatics Corporation," ninety-one-page type-script, August 1960, O'Brien Papers, Box 16, Folder 11. Emphasis in the original.

81. IP, Abelson, and SP, *Candidates, Issues, Strategies,* 18–19.

82. Jeremy de Sola Pool, e-mail to the author, March 5, 2019.

83. Edwin Safford, letter to the author, April 24, 2018.

84. The corporation reported having paid $22,214.09 for "computer time" in 1960, and another $2,393.65 for "computer materials." Simulmatics stock offering, May 15, 1961.

85. "When they were presented, there was also a briefing session with him and his top staff." IP, Abelson, and SP, *Candidates, Is-*

sues, Strategies, 23.

86. Simulmatics, "Kennedy Before Labor Day," August 25, 1960, 17, 22, 23, in O'Brien Papers, Box 16, Folder 11 (another copy, also in the Kennedy Library, can be found in the Democratic National Committee Papers, Box 212).

87. Ibid., 25, emphasis in original. But note that this page of the report was missing from the first copy and was only later submitted to O'Brien, so it is filed in a different folder: O'Brien Papers, Box 16, Folder 10.

88. Ibid., 26, 33.

89. Ibid., 25a. Again, note that this page of the report was missing from the first copy and was only later submitted to O'Brien, so it is filed in a different folder: O'Brien Papers, Box 16, Folder 10. Simulmatics, "Kennedy, Nixon, and Foreign Affairs," Democratic National Committee Records, Kennedy Library, RFK Pre-Administration Papers, Political Files: 1960 Campaign and Transition Series, Box 212.

90. IP, Abelson, and SP, *Candidates, Issues, Strategies,* 22.

91. E.g., "Their subsequent actions were almost perfectly consistent with the Simulmatics recommendations," one historian has observed. Edmund F. Kallina Jr., *Kennedy v. Nixon: The Presidential Election of*

1960 (Gainesville: University Press of Florida, 2010), 104–5.

92. Donaldson, *The First Modern Campaign,* 155–56.

93. This is according to Sargent Shriver, as related in Strober and Strober, *"Let Us Begin Anew,"* 35.

94. "Address to the Houston Ministers Conference, 12 September 1960," Kennedy Library, https://www.jfklibrary.org/asset -viewer/archives/IFP/1960/IFP-140/IFP -140.

95. MMc to Eleanor Emery, September 13, 1960, Emery-McPhee Papers.

96. Simulmatics Corporation, "Nixon Before Labor Day," August 25, 1960, p. 2, Pool Papers, Box 187.

97. "The People Machine," *Newsweek,* April 2, 1962.

98. White, *Making of the President,* 18.

99. IBM, *The Fastest Reported Election* (New York, 1960), in Pool Papers, Box 141, Folder "POQ Comments."

100. Ibid.

101. IBM, *The Fastest Reported Election.*

102. White, *Making of the President,* 14.

103. IBM, *The Fastest Reported Election.*

104. IBM, *The Fastest Reported Election.*

105. Rorabaugh, *The Real Making of the President,* 5.

106. Earl Mazo and Stephen Hess, *President*

Nixon: A Political Portrait (New York: Harper & Row, 1967), 249. Edmund F. Kallina Jr., *Courthouse to the White House: Chicago and the Presidential Election of 1960* (Orlando: University of Central Florida Press, 1988), chs. 5–7.
107. "Had African-Americans cast their votes in 1960 the same as they had in 1956, Kennedy would most likely have lost Illinois, NJ, MI, SC, and DE" (Donaldson, *The First Modern Campaign,* 155–56).

Chapter 7: Billion-Dollar Brain

1. TBM, "The People-Machine."
2. Roscoe Drummond, "People Predictor Revealed as Kennedy's Secret Weapon," *New York Herald Tribune,* December 19, 1960. Greenfield gleefully sent clippings to Lasswell, January 11, 1961, Lasswell Papers, Box 39, Folder 544.
3. Editorial, "The People Machine," *Chicago Sun-Times,* December 28, 1960. For other representative stories, see "Kennedy's 'Thinking Machine,' " *Montgomery* [AL] *Advertiser,* December 24, 1960; Tom Donnelly, "Damned Internal, These Machines," *Washington* [DC] *News,* December 21, 1960; "Of 'Machine' Politics," *Arkansas Gazette* (Little Rock), December 21, 1960; "And Now, a 'People Machine,' " *Charlotte* [NC] *Observer,* December 22, 1960; and

"The Test," *Intelligencer* (Wheeling, WV), December 20, 1960.

4. Editorial, "No Dissent for People Machine," *World* (Coos Bay, OR), January 12, 1961.

5. TBM, "The People-Machine."

6. TBM, *Self-Creations: 13 Impersonalities* (New York: Holt, Rinehart & Winston, 1965), 3.

7. Ibid., 1.

8. TBM, "Madly for Adlai."

9. TBM, *Self-Creations,* 3–15. Kate Tarlow Morgan, interview with the author, July 2, 2018. Marc Weingarten, *Who's Afraid of Tom Wolfe? How New Journalism Rewrote the World* (London: Aurum, 2005), 44–46. Douglas Martin, "Thomas B. Morgan, Writer, Editor and Lindsay Press Aide, Dies at 87," *NYT,* June 18, 2014.

10. TBM, "The People-Machine."

11. James E. Doyle to NM, May 6, 1959, Stevenson Papers, Box 38, Folder 7. Wisconsin lawyer James E. Doyle served as the executive director of the 1960 National Stevenson for President Committee. On Doyle, see James E. Doyle Oral History Interview, January 15, 1966, interviewed by Charles T. Morrissey, Kennedy Library.

12. I have been unable to find audio of the CBS radio story, but it is described and quoted in Robert Abelson to IP, December 22, 1960, Pool Papers, Box 96, Folder

"Simulmatics: Abelson, Robert."

13. "Did IBM Computer Draft Strategy Send Kennedy to White House?," *Press & Sun-Bulletin* (Binghamton, NY), December 19, 1960.

14. TBM, "Madly for Adlai."

15. Labor's Committee for Kennedy and Johnson, *If Automation Takes Over Your Job . . . Who Will You Want in the White House?,* brochure, 1960, Burdick Collection, Box 53, Folder 4.

16. "Top Aides Deny It: 'Brain' Assist Seen in Kennedy Campaign," *Los Angeles Times,* December 19, 1960.

17. See the UPI story as it appears in listings generated by Newspapers.com. That headline is from *Albuquerque Journal,* December 21, 1960.

18. "Did IBM Computer Draft Strategy Send Kennedy to White House?," *Press & Sun-Bulletin* (Binghamton, NY), December 19, 1960. Salinger, in calling Simulmatics' product a "People Predictor," was employing the usage begun by Roscoe Drummond, initially in an article in the *New York Herald Tribune* and continuing in "The People Predictor," *Christian Science Monitor,* December 28, 1960, and widely picked up elsewhere, e.g., "The 'People Predictor' Gives Politicians One More Advantage over the Electorate," *Charleston* [WV] *Mail,*

December 23, 1960.

19. Robert Abelson to IP, November 3, 1960; IP to Robert Abelson, November 10, 1960; IP, "Book Outline," Pool Papers, Box 96, Folder "Simulmatics: Abelson, Robert."

20. Ian Forman, "Political Weather Map Put Kennedy Up There," *Boston Globe,* November 13, 1960.

21. IP, "Book Outline," Pool Papers, Box 96, Folder "Simulmatics: Abelson, Robert."

22. "Books — Authors," *NYT,* September 30, 1960. "Work in Progress," *Philadelphia Inquirer,* September 11, 1960.

23. IP, "Book Outline," Pool Papers, Box 96, Folder "Simulmatics: Abelson, Robert."

24. TBM to IP, March 2, 1961, Pool Papers, Box 142, untitled folder.

25. Editorial, "Delphi Revisited," *St. Louis Post-Dispatch,* January 4, 1961.

26. Editorial, "Politics' Ultimate Weapon?" *Cincinnati Enquirer,* December 22, 1960.

27. Victor Lasky, " 'People Machine' Helped Kennedy," *Indianapolis Star,* September 4, 1963. And see Victor Lasky, *JFK: The Man and the Myth* (New York: Macmillan, 1963), 428–30.

28. "Presses Stopped on Anti-J.F.K. Bestseller," *Des Moines Register,* November 25, 1963, as typed into notes kept by EB, Burdick Collection, Box 35, Folder "General Notes: 'The 480.' "

29. "He [WMc] said that Kennedy forces had certainly seen the Simulmatics results which had been produced for the Democratic Advisory Committee but he was not at all convinced that they had anything more than a marginal effect, if that, on the campaign." John F. Maloney to William Lederer, September 7, 1961, William Lederer Papers, Special Collections and University Archives, University of Massachusetts, Amherst.

30. Robert Abelson to IP, December 22, 1960, and [Robert Abelson], "PRESS RELEASE: CLARIFYING THE ROLE OF THE SIMULMATICS CORP. PROJECT IN THE PRESIDENTIAL CAMPAIGN," [December 22, 1960], Pool Papers, Box 96, Folder "Simulmatics: Abelson, Robert."

31. Aldous, *Schlesinger,* 5, 219–20.

32. Dwight D. Eisenhower, Farewell Address, January 17, 1961.

33. AS, *Thousand Days,* 164–65.

34. Ibid., 3. Robert Frost, "Dedication," 1961, in Alan Wirzbicki, "The Poem Robert Frost Wanted to Read at John F. Kennedy's Inauguration," *Boston Globe,* January 20, 2011.

35. The Simulmatics Corporation, *Human Behavior and the Electronic Computer.*

36. MMc to Eleanor Emery, January 15, 1961, Emery-McPhee Papers. The firm's

attorneys advised: "For the protection of the Company and the individuals involved, it would be good if each of the controlling persons of the Company were advised that he may not trade in Simulmatics stock. In essence, the prohibition goes to the sale of stock." Donald H. Rivkin, Memo, May 22, 1961, Pool Papers, Box 142, Folder "Publicity."

37. MMc to Eleanor Emery, January 25, 1961, Emery-McPhee Papers. For more on the stock option discussions with the scientists, see JC to ELG, January 17, 1961, and enclosures of notes, Coleman Papers, Box 65, Folder "Simulmatics Corp., 1961." This folder also includes JC's formal contract with Simulmatics.

38. Synopsis of Results of Simulmatics Conference, Hotel Barclay, February 5–6, 1961, Lasswell Papers, Box 39, Folder 144.

39. TBM to Abelson, JC, ELG, WMc, TBM, IP, and others, Call Report, April 6, 1961, Pool Papers, Box 142, untitled folder.

40. AB et al. to Abelson, JC, ELG, WMc, TBM, IP, and others, Call Report, April 6, 1961, and AB and TBM to Abelson, JC, ELG, WMc, IP, and others, Call Report, May 5, 1961, Pool Papers, Box 142, untitled folder.

41. Aldous, *Schlesinger,* 224, 227.

42. ELG to NM, November 20, 1960, and "Possible Government Uses of Simulation,"

November 1960, Stevenson Papers, Box 38, Folder 7. AB prepared a fuller set of proposals. See AB, "Simulmatics Models for Government and Industry," undated but 1961, and also "Clients and Models: Preliminary Survey of Opportunities in Governmental and Industrial Areas for The Simulmatics Corporation," STRICTLY CONFIDENTIAL, Prepared for Meeting of The Corporation, February 5–6, 1961, Coleman Papers, Box 67, Folder "Simulmatics Corporation, 1961."

43. Simulmatics Corp., "Simulmatics Physicians Drug Acceptance Model," March 17, 1961, Coleman Papers, Box 65, Folder "Simulmatics Corp., 1962." Addenda, Box 87, Folder "Simulmatics." See also JC to James F. Dodd (J. B. Roerig & Company), January 18, 1962, Coleman Papers, Box 65, Folder "Simulmatics Corp., 1962."

44. IP to Carl Barnes (USDA), Draft Proposal, June 9, 1961. And see William W. Cochrane to IP, August 30, 1961, Pool Papers, Box 142, Folder "Agr."

45. "McNamara Reported Ordering Study on Prevention of Accidental A-War," *Washington Post,* March 25, 1961, Pool Papers, Box 142, untitled folder.

46. TBM to ELG, Memo, June 23, 1961, Pool Papers, Box 142, untitled folder.

47. E.g., ELG and IP to Abelson, JC, WMc, TBM, IP, and others, Call Report, June 27,

1961. Pool Papers, Box 142, untitled folder.

48. Pitches to these clients — from call reports to formal proposals — can be found scattered through the Pool Papers but mainly in Box 142. Other pitches and even some reports can be found in the Coleman Papers, especially Boxes 53, 64, and 65. For more on the beer brand-switching project, see JC to Charles Ramond (Advertising Research Foundation), September 24, 1963, Coleman Papers, Box 53, Folder "Simulmatics, 1960–63." On Ralston Purina, see James Tyson (Simulmatics) to Arthur S. Pearson (Ralston Purina Company), September 27, 1963, with enclosures, and additional correspondence regarding that account, same folder. Papers concerning the Nescafé account are in this same folder of the Coleman Papers, but a full report of JC's "mathematical analysis of Nescafe" can be found in Box 64, Folder "Nescafe, 1962–3." Nevertheless, most records of the actual accounts Simulmatics had with advertising clients, or with any other clients, are not in either the Pool Papers or the Coleman Papers, because neither man ran the business end of the office. Those records would have been in the New York office, and they have not survived.

49. IP to Richard Casey (Benton & Bowles), Memo, July 6, 1961, Pool Papers, Box 142, Folder "Media Mix — Early Notes."

50. Sidney C. Furst, "Simulmatics White Paper: Part One — Sales Approach," April 17, 1961, Pool Papers, Box 142, untitled folder.

51. Documents relating to the proposed merger can be found in the Coleman Papers, Box 64, Folder "Simulmatics, 1962."

52. Memo to Abelson, AB, JC, ELG, IP, and others, October 29, 1962, Pool Papers, Box 142, untitled folder. Also: "As Ithiel pointed out in his memo, we have run into the problem of agencies resisting the Media Mix because of the cost of doing a run for each of a large number of their clients." James Tyson (Simulmatics) to AB, ELG, IP, and others, Memo, December 5, 1962, same folder. (Ithiel's memo is IP to ELG and others, Memo, November 21, 1962, same folder.)

53. Sidney C. Furst, "Simulmatics White Paper: Part One — Sales Approach," April 17, 1961, Pool Papers, Box 142, untitled folder.

54. IP to Abelson, AB, JC, Furst, ELG, WMc, and TBM, Memo, April 24, 1961, Pool Papers, Box 142, Folder "Media Mix — Early Notes."

55. Basil Boothroyd, "Computer v. Consumer," *Punch,* September 19, 1962, 423–24.

56. Simulmatics stock offering, May 15,

1961, Pool Papers, Box 142, Folder "Enquiries."

57. MMc to Eleanor Emery, May 12, 1961, Emery-McPhee Papers.

58. "Book Notes," *New York Herald Tribune,* May 9, 1961.

59. See, e.g., Orville Prescott, "Books of The Times," *NYT,* July 5, 1961.

60. On the distance between White's JFK and the reality, see Robert Dallek, foreword to Theodore H. White, *The Making of the President* (1960; repr., New York: Harper-Perennial, 2009).

61. "Recent Hot Issues Cool in Quiet Trade," *Los Angeles Time*s, May 24, 1961.

62. Editors, "The People-Machine," *Harper's,* [September?] 1961.

63. Sylvia Porter, "Getting In on New Stock Issue 'Deals' Risky Business; Rigging Faces Probe," *Santa Fe New Mexican,* December 24, 1961 (this column appeared in newspapers all over the country). On Porter's influence, see Glenn Fowler, "Sylvia Porter, Financial Columnist, Is Dead at 77," *NYT,* June 7, 1991.

64. Pool, for instance, asked the advertising agency Benton & Bowles to send him "data about people's distribution of their time over activities during the course of the day"; "data which tabulates media use according to activity"; and "data on frequency with which people change the television program

they are watching." IP to Frank Stanton (Benton & Bowles, Inc.), September 5, 1961, Pool Papers, Box 142, untitled folder.

65. Simulmatics Conference, June 2, 3, and 4, 1961, Revised Agenda; and ELG to The Simulmatics Corporation, Memo, June 6, 1961, Box 142, untitled folder. Simulmatics would soon afterward apparently turn down an opportunity to work for a Republican candidate. Newton Steers Jr., to TBM, October 23, 1961, Coleman Papers, Box 64, Folder "Simulmatics, 1961."

66. IP to Abelson, AB, JC, Furst, ELG, WMc, and TBM, Memo, May 16, 1961, Pool Papers, Box 142, Folder "Media Mix — Early Notes."

67. Simulmatics Conference, June 2, 3, and 4, 1961, Revised Agenda; and ELG to The Simulmatics Corporation, Memo, June 6, 1961, Box 142, untitled folder.

68. MMc to Eleanor Emery, March 27, 1961; May 26, 1961; and July 18, 1961, Emery-McPhee Papers.

69. MMc to Miriam Washburn Adams, June 21, 1961, Emery-McPhee Papers.

70. MMc to Eleanor Emery, October 24, 1960, Emery-McPhee Papers.

71. Wendy McPhee, interview with the author, July 16, 2018.

72. Edwin Safford, letter to the author, April 24, 2018.

73. Paul Lazarsfeld to IP, August 18, 1961,

Pool Papers, Box 141, Folder "POQ Comments."

74. Susan Greenfield, interview with the author, July 27, 2018.

75. Wendy McPhee, interview with the author, July 16, 2018.

76. Sarah McPhee to Miriam Washburn Adams, June 21, 1961, Emery-McPhee Papers. MMc to Eleanor Emery, July 13, 1961, Emery-McPhee Papers. Sarah McPhee, interview with the author, July 30, 2018.

77. Jeremy de Sola Pool, interview with the author, May 23, 2018.

78. ELG to IP (at Wading River): "What is the most daring claim you can make for our theory?" July 22, 1961, Pool Papers, Box 142, Folder "Publicity."

79. MMc to Eleanor Emery, July 20, 1961, Emery-McPhee Papers. Kate Tarlow Morgan, interview with the author, July 2, 2018. MMc to Eleanor Emery, August 14, 1961, Emery-McPhee Papers.

80. Sarah McPhee, interview with the author, July 30, 2018.

81. Wendy McPhee, interview with the author, July 16, 2018. John McPhee, interview with the author, July 24, 2018.

82. Ted Cott to ELG, July 19, 1961; Edward P. Seymour (Simulmatics) to IP (at Wading River), August 15, 1961; WMc, "Preliminary Codes for Television Program File (Media Mix), August 21, 1961, Pool Pa-

pers, Box 142, untitled folder.

83. JC to IP, AB, WMc, and Abelson, Memo, August 4, 1961, Pool Papers, Box 142, Folder "Media Mix — Early Notes." On attention, see, e.g., Tim Wu, *The Attention Merchants: The Epic Scramble to Get Inside Our Heads* (New York: Knopf, 2016).

84. Susan Greenfield, interview with the author, July 27, 2018.

85. MMc to Eleanor Emery, July 20, 1961, Emery-McPhee Papers. Kate Tarlow Morgan, interview with the author, July 2, 2018. MMc to Eleanor Emery, August 14, 1961, Emery-McPhee Papers.

86. MMc to Eleanor Emery, March 27, 1961, and August 22, 1961, Emery-McPhee Papers.

87. IP to Jerome Feniger, August 22, 1961, Pool Papers, Box 142, Folder "Cigarettes."

88. IP to ELG and AB, Memo, October 23, 1961, Pool Papers, Box 142, Folder "Media Mix — Early Notes."

89. WMc to Wendy McPhee, postmarked May 27, 1962, Emery-McPhee Papers.

90. ELG, "To our Shareholders," August 21, 1961, Pool Papers, Box 142, Folder "Publicity."

91. TBM, "Computers and Behavioral Science Are Combined in New Applications: Predict Revolution in Business Research," Press Release, August 25, 1961, Pool Papers, Box 142, Folder "Publicity."

92. William M. Freeman, "Advertising: Life Is Imitated for Research," *NYT,* August 27, 1961.

93. "The People Machine," *Newsweek,* April 2, 1962.

94. *Louis Sapir Weekly Newsletter,* August 31, 1961, emphasis in original. "For anyone interested in an offbeat RANK speculation where the risks are tremendous, but the future exciting, we suggest you look closely at SIMULMATICS CORP. which is traded Over-the-Counter and presented quoted at 5–5 1/2. This is a new company that is attempting to introduce a new idea to business and government by using human behavioral factors along with other data in electronic computers. The areas in which they can apply their technique are many and varied, for example: urban renewal, traffic patterns, new product introduction and advertising selection and marketing strategy. The officers of the company are the social scientists' who's who of the academic world, representing Johns Hopkins, Columbia, Yale, Harvard, and MIT. SIMULMATICS already has one contract from the Department of Health Education and Welfare and presently is negotiating an additional $800,000 of business. SIMULMATICS represents an opportunity for risk takers to possibly make substantial gains in the years ahead." Pool Papers, Box 142,

Folder "Enquiries." See also other stories generated by the press release, including "'Simulmatics' Backtracks Behavior," *Christian Science Monitor,* September 9, 1961.

95. Susan Greenfield, interview with the author, July 27, 2018.
96. Adam de Sola Pool, interview with the author, May 19, 2018.

Chapter 8: Fail-Safe

1. On the building, see David W. Dunlap, "Copy!," *NYT,* June 10, 2007.
2. "The IBM 1401," http://ibm-1401.info/1401GuidePosterV9.html. Computer History Museum, *IBM 1401 System 50th Anniversary,* https://www.youtube.com/watch?v=FVsX7aHNENo. Computer History Archives Project, *1970's IBM Vintage Computer Promotional Film — Historical Educational,* https://www.youtube.com/watch?v=wIjgZhAjQS4.
3. A landmark came in 1973 with the publication of Philip Meyer, *Precision Journalism: A Reporter's Introduction to Social Science Methods* (Bloomington: Indiana University Press, 1973).
4. Chinoy, "Battle of the Brains," 82–86, 106.
5. Jill Abramson, *Merchants of Truth: The Business of News and the Fight for Facts* (New

615

York: Simon & Schuster, 2019), 4.

6. David Starr, "The Quiet Revolution," *Bulletin of the American Society of Newspaper Editors,* April 1963, 2, quoted in Matthew Pressman, *On Press: The Liberal Values That Shaped the News* (Cambridge, MA: Harvard University Press, 2018), 35.

7. Ibid., 6, 26–27.

8. Harold Faber to Mr. Turner Catledge, Memo, November 21, 1962, NYT Records, General Files, Box 123, Folder "Simulmatics Reports." Before approaching the *Times,* Simulmatics had proposed joining forces with IBM for Election Night television coverage. See JC to David Holzman (IBM), August 1, 1961, Coleman Papers, Box 67, Folder "Simulmatics Corporation, 1961." N.B.: Records relating to Simulmatics' work for the *New York Times* can be found in several places in the NYT Records: General Records, Box 123, Folder "Elections 1962"; Clifton Daniel Papers, Box 33, Folder "Simulmatics Project"; Daniel Papers, Box 33, Folder "Simulmatics Reports"; Daniel Papers, Box 33, Folder "Simulmatics Articles"; and General Records, Box 25, Folder "Simulmatics Project."

9. JC, WMc, and IP to Joseph Herzberg, Harold Faber, and Chester M. Lewis, Memo, October 31, 1961, NYT Records, Daniel Papers, Box 33, Folder "Simulmat-

ics Project."

10. "The Times to Speed Fall Election Data," *NYT,* June 1, 1962.

11. IP to Leslie Kish (Survey Research Center, Ann Arbor), October 15, 1961, Pool Papers, Box 142, Folder "Media Mix: Sampling Points."

12. Edward Seymour (Simulmatics) to Abelson, ELG, and IP, Call Report, November 1, 1961, Pool Papers, Box 143, Folder "Media Mix: Publicity." He reported having spoken with Dr. Clark Wilson of BBDO. I "told him frankly I wondered why he did not have a particularly favorable attitude towards Simulmatics. In reply, he said that when he heard about Simulmatics last spring he sought out Dr. Pool, and that he felt as a result of their meeting that Simulmatics did not understand the practical problems involved in building models for use in the advertising business. Also, he felt that Simulmatics did not have in its employ at that time a mathematician, which would be very necessary in the building of a computer model. He felt that they could not wait and have, therefore, been building their own model."

13. Charles Mangel, "BBDO Applies Computer Process to Selecting Media: Will Give Principle to All Agencies," Press Release, November 13, 1961, Pool Papers, Box 142, Folder "Media Mix: Publicity." And see

TBM to AB, JC, ELG, IP, and others, Memo, November 17, 1961, Pool Papers, Box 143, Folder "Media Mix: Publicity." TBM discusses the upcoming meeting with Ramond at the foundation: "For this meeting, Dr. Charles Ramond has invited David Lerner of BBD&O who will introduce Milt Godfrey of C-E-I-R to talk about linear programming and Ben Lipstein of Benton and Bowles who will introduce Alex Bernstein of Simulmatics to talk about simulation."

14. ELG, "Simulmatics Media-Mix I: First General Description," November 22, 1961, Pool Papers, Box 143, Folder "Media Mix: Publicity."

15. "The Simulmatics Corporation," brochure, c. 1965, p. 6, Pool Papers, Box 67, Folder "Simulmatics Correspondence."

16. IP, "Simulation: How It Can Help the Marketer," American Marketing Association, March 15, 1962, Pool Papers, Box 67, Folder "Simulmatics Correspondence." On attendance, see Ray Berland to IP, March 27, 1962, Pool Papers, Box 142, Folder "3/6 1962."

17. ELG, "Simulmatics Media-Mix I: General Description," February 1962, p. 2, Pool Papers, Box 67, Folder "Simulmatics Correspondence."

18. Feniger writes, after reading a draft of the Media-Mix brochure, "I would expand at

greater length on the difference between your plan and linear programming since you will obviously be putting your service on the market at the same time as CEIR is selling their plan." Jerome Feniger to ELG, March 19, 1962, Pool Papers, Box 142, Folder "3/6 1962."

19. "Y&R, BBDO Unleash Media Computerization," *Advertising Age,* October 1, 1962, Pool Papers, Box 142, Folder "NAEA meeting Chicago, 1963, Jan. 21." And see Lee Mcguigan, "Selling the American People: Data, Technology, and the Calculated Transformation of Advertising" (PhD diss., University of Pennsylvania, 2018).

20. ELG, IP, and Edward Seymour to Abelson, AB, JC, WMc, TBM, and others, Call Report, June 27, 1961, Pool Papers, Box 142, untitled folder. For Ramond's very useful account of this era in the history of advertising, see Charles Ramond, *Advertising Research: The State of the Art* (New York: Association of National Advertisers, 1976), 30, 70–72. For more on Ramond, see John Pope, "Charles Ramond, Who Forecast Currency Values and Measured Advertising's Impact, Dies at 83," *Times-Picayune* (New Orleans), August 4, 2014. Note that Ramond worked for Simulmatics in Saigon in 1966 and 1967.

21. "Incidentally, they are going to install

their own computer within the year." Edward Seymour (Simulmatics) to Abelson, ELG, and IP, Memo, November 1, 1961, Pool Papers, Box 142, Folder "Media Mix — Publicity."

22. Bernard P. Gallagher, "The Great Computer Hoax," *Gallagher Report* (trade newsletter), November 26, 1962, Pool Papers, Box 142, untitled folder.

23. The Simulmatics Corporation, Newsletter, no. 5, June 11, 1962, Pool Papers, Box 67, Folder "Simulmatics Correspondence." See also "New York Times Breakdown," Coleman Papers, Box 48, Folder "Simulmatics 1962," and "New York Times Project — 1962, General Cost Breakdown," Coleman Papers, Box 53, Folder "Simulmatics, 1960–63."

24. "New York Times Breakdown," Coleman Papers, Box 48, Folder "Simulmatics 1962," and "New York Times Project — 1962, General Cost Breakdown," Coleman Papers, Box 53, Folder "Simulmatics, 1960–63."

25. For preliminary versions of the plan, see Simulmatics Corporation, Memo, May 1, 1962, Coleman Papers, Box 48, Folder "NYTimes, 1961–62," and JC, Memo, June 15, 1962, Coleman Papers, Box 48, Folder "New York Times Paper, 1962." For the final report, see JC, Ernest Heau, Robert Peabody, and Leo Rigsby, "Computers and

Election Analysis: The *New York Times* Project," *Public Opinion Quarterly* 28 (1964): 418–46. On the office, see Leo Rigsby to James Coleman and Robert Peabody, Memo, June 1, 1962, Coleman Papers, Box 48, Folder "NYTimes, 1961–62."

26. Dennis Hevesi, "Harold Faber, Longtime Reporter and Editor for The Times, Dies at 90," *NYT,* January 18, 2010.
27. JC et al., "Computers and Election Analysis," 429.
28. "The IBM 1401," http://ibm-1401.info/ 1401GuidePosterV9.html.
29. Computer History Archives Project, *1970's IBM Vintage Computer Promotional Film — Historical Educational,* https:// www.youtube.com/watch?v=wIjgZhAjQS4.
30. JC et al., "Computers and Election Analysis," 429.
31. Ray Josephs Public Relations, "Preliminary Version of Basic Simulmatics New York Times Story," Press Release, November 1, 1962, NYT Records, Daniel Papers, Box 33, Folder "Simulmatics Project." For more on the proposed analytical work, see JC to Hal Faber, Memo, September 16, 1962, Coleman Papers, Box 48, Folder "Regression, 1962."
32. A. Kuhn, "The New York Newspaper Strike," *International Communication Gazette,* May 1, 1964; James F. Tracy, " 'Labor's Monkey Wrench': Newsweekly Cover-

age of the 1962–63 New York Newspaper Strike," *Canadian Journal of Communication* 31 (2006): 541–60.

33. Documents can be found at "13 Days in October," Kennedy Library, online exhibit, https://microsites.jfklibrary.org/cmc/oct16/index.html.

34. Martin, *Adlai Stevenson and the World,* 719–21.

35. Clifton Fadiman, Review, *Book-of-the-Month Club News,* October 1962, in Burdick Collection, Box 40, Folder "F.S."

36. Linus Pauling, Comments on *Fail-Safe,* Burdick Collection, Box 40, Folder "FS."

37. Orville Prescott, "Books of The Times," *NYT,* October 24, 1962.

38. Jack Raymond, "Pentagon Backs 'Fail-Safe' Setup," *NYT,* October 21, 1962.

39. Eugene Burdick and Joseph Buttinger, "*The Ugly American:* Imagination and Dreams in Foreign Aid," *Dissent* (Winter 1960).

40. J. William Fulbright, "The Ugly American," *Congressional Record,* September 7, 1959, pp. 16820–24. And see Lederer and Burdick's many letters to Fulbright, along with the Norton ad in the *Times,* which appeared on November 28, 1959, all in Burdick Collection, Box 19, Folder "Ugly American — Dissent Episode."

41. Eric Swenson (Norton) to EB and William J. Lederer, August 8, 1962, Burdick

Collection, Box 17, Folder "Ugly American Correspondence, W. W. Norton."

42. EB to Eleanor Roosevelt, May 19, 1958, Burdick Collection, Box 17, Folder "Ugly American Correspondence, W. W. Norton."

43. EB to Alan Collins, November 1, 1957, Burdick Collection, Box 17, Folder "Ugly American Correspondence, W. W. Norton (Part 2)."

44. EB, "The Search for a Title," *San Francisco Chronicle,* September 28, 1958.

45. EB and William J. Lederer, *The Ugly American* (New York: Norton, 1958), 144–45.

46. Contract between William J. Lederer and EB and W. W. Norton, November 18, 1957, Burdick Collection, Box 17, Folder "Ugly American Correspondence, W. W. Norton (Part 2)." The book's original title was *Americans in the Far East.*

47. William Lederer to EB, Burdick Collection, Box 91, Folder "Lederer, Correspondence."

48. "In the Ugly American . . . we first wrote it as non-fiction . . . then decided, on advice of publisher to make novel . . . ALSO to avoid the American impulse to play angels-devils with history." EB, Lecture Notes, "The Political Novel," January 7, 1959, Burdick Collection, Box 65, Folder "The Political Novel Lecture."

49. The factual epilogue appears to have been drafted by Lederer, who had been opposed to its inclusion, arguing that if Harriet Beecher Stowe's *Uncle Tom's Cabin* had included a factual epilogue, the book would have had much less of an effect. William Lederer to Eric Swenson (Norton), June 7, 1958, Burdick Collection, Box 17, Folder "Ugly American Correspondence, W. W. Norton."

50. EB and Harvey Wheeler, *Fail-Safe* (New York: McGraw-Hill, 1962), 192, 277–80.

51. Ibid., 190. There was much discussion, and some controversy, over the accuracy of the scenarios in the novel. And yet the physicist Edward Teller read the book, asked to meet with Burdick and Wheeler, and when they all had dinner together at Burdick's house on October 15, 1962, Teller told them that the novel was basically fairly stated. See "Wheeler Recollections of the Dinner Discussion with Edward Teller," undated typescript, Burdick Collection, Box 40, no folder.

52. Ibid., 280.

53. *The Red Threat: President Orders Cuban Blockade,* newsreel, October 22, 1962, https://archive.org/details/1962-10-22_The _Red_Threat. Martin, *Adlai Stevenson and the World,* 719–21.

54. "U.S. Imposes Arms Blockade," *NYT,*

October 23, 1962.

55. Aldous, *Schlesinger,* 292–93. See also "The World on the Brink: John F. Kennedy and the Cuban Missile Crisis — Day 10," Kennedy Library, https://microsites .jfklibrary.org/cmc/oct25/.

56. Martin, *Adlai Stevenson and the World,* 728–37.

57. "The World on the Brink: John F. Kennedy and the Cuban Missile Crisis — Department of State Telegram," Kennedy Library, https://microsites.jfklibrary.org/ cmc/oct26/doc4.html.

58. Aldous, *Schlesinger,* 290.

59. Martin, *Adlai Stevenson and the World,* 728–37.

60. Paine Knickerbocker, "Gene Burdick Attacks a 'Lie' on 'Fail-Safe,' " *San Francisco Chronicle,* October 9, 1964.

61. Isserman and Kazin, *America Divided,* 78. David L. DiLeo, *George Ball, Vietnam, and the Rethinking of Containment* (Chapel Hill: University of North Carolina Press, 1991), 56.

62. Harold Faber to ELG, November 15, 1962, NYT Records, Daniel Papers, Box 33, Folder "Simulmatics Project."

63. JC to Solveigh Archer (*NYT*), Memo, Fall 1962, Coleman Papers, Box 48, Folder "Regression, 1962."

64. Harold Faber to Robert Garst, October

19, 1962, NYT Records, Daniel Papers, Box 33, Folder "Simulmatics Reports." Harold Faber to ELG, November 15, 1962, NYT Records, Daniel Papers, Box 33, Folder "Simulmatics Project." The "girls" were Marlene McClary, Grace Pecunia, Jackie Martino, and Helen Quill.

65. "Several of your people have made the point that Simulmatics was 80 per cent ready on Oct. 15 and that IBM was 20 per cent ready. I would say, rather, that Simulmatics was at best 50 per cent prepared." Who was even running the thing on Election Night? "When we started, Dr. Coleman was in charge; after the failure of the Oct. 15 test, Mr. Greenfield was in charge; in a few days, Mr. Bernstein was in charge." "IBM personnel backstopped Simulmatics personnel starting on Oct. 29," Harold Faber to ELG, November 15, 1962, NYT Records, Daniel Papers, Box 33, Folder "Simulmatics Project."

66. "Meeting Between New York Times and Simulmatics Corporation, Post Election Critique," November 9, 1962, p. 5, NYT Records, General Files, Box 123, Folder "Elections 1962."

67. "We did not catch the feel of the election quickly enough. Our early edition stories about the Republican gains were not correct. This was due to not getting fast enough returns in from key areas; not filter-

ing all significant returns to the main lead writer and failure to shift our approach with the shift in the emerging patterns of a very complex race." Harrison E. Salisbury, "Election Post Mortem," Memo, November 19, 1962, NYT Records, Daniel Papers, Box 33, Folder "Simulmatics Project."

68. Leo Egan and Clayton Knowles, "Joint Report to Mr. Adams on Simulmatics," November 8, 1962, NYT Records, General Files, Box 123, Folder "Elections 1962."

69. IP to Faber, November 8, 1962, NYT Records, Daniel Papers, Box 33, Folder "Simulmatics Reports."

70. "Man and the Machine," *Times Talk,* November 1962.

71. J. I. Henry to Harold Faber, Memo, November 19, 1962, NYT Records, Daniel Papers, Folder "Simulmatics Reports"

72. Harold Faber to Turner Catledge, Memo, November 21, 1962, NYT Records, General Files, Box 123, Folder "Simulmatics Reports."

73. JC to Harold Faber, November 7, 1962, NYT Records, Daniel Papers, Box 33, Folder "Simulmatics Reports."

74. Layhmond Robinson, "City Negroes Seen Behind Kennedy," *NYT,* November 3, 1960.

75. Daniel E. Slotnik, "Layhmond Robinson Jr., Who Paved Way for Black Journalists, Dies at 88," *NYT,* July 11, 2013.

76. Layhmond Robinson Jr., to Harold Faber, Memo, November 18, 1962, NYT Records, Daniel Papers, Box 33, Folder "Simulmatics Project." Pool, notably, does not mention his conversation with Robinson in his various notes: Pool Papers, Box 95, Folder "NY Times Analysis, 1962."

77. Tom Wicker, "Election Returns Bring into Focus the Personalities for '64," *NYT,* November 11, 1962.

78. ELG to Turner Catledge, Memo, December 12, 1962, NYT Records, Daniel Papers, Box 33, Folder "Simulmatics Project."

79. Joseph Alduino, Memorandum for File, June 3, 1963, NYT Records, Daniel Papers, Box 33, Folder "Simulmatics Project."

80. Harold Faber to Clifton Daniel, Memo, April 10, 1963, NYT Records, Daniel Papers, Box 33, Folder "Simulmatics Project." The original contract: ELG to Clifton Daniel, April 20, 1962, NYT Records, Daniel Papers, Box 33, Folder "Simulmatics Project."

81. ELG to Clifton Daniel, "Cost of NEW YORK TIMES Project," May 10, 1963, NYT Records, Daniel Papers, Box 33, Folder "Simulmatics Project."

82. A. Kuhn, "The New York Newspaper Strike," *International Communication Gazette,* May 1, 1964.

83. Joseph Alduino, Memorandum for File, June 3, 1963, NYT Records, Daniel Papers,

Box 33, Folder "Simulmatics Project."

84. NYT to ELG, June 6, 1963, NYT Records, Daniel Papers, Box 33, Folder "Simulmatics Project."

85. NYT to ELG, October 30, 1963, and Clifton Daniel to ELG, November 5, 1963, NYT Records, Daniel Papers, Box 33, Folder "Simulmatics Project."

86. IP, "Research Program on Problems of Communication and International Security," February 21, 1963, Pool Papers, Box 176, Folder "Eval of ComCom Project 1963."

87. IP, "Description of Research Program on Problems of Communication and International Security," Pool Papers, Box 20, Folder "Project ComCom."

88. J. P. Ruina (ARPA) to IP, July 23, 1962; IP to Ruina, October 5, 1962; and assorted charts and tables from the speech, Pool Papers, Box 32, Folder "Speeches — ARPA, 1962 December 12." IP had earlier reported to Simulmatics that in October 1961 he had presented a keynote address called "Cold War Modeling" at the Eighth Classified Military Operations Research Symposium, a meeting of about four hundred people able to conduct classified research. IP to Abelson, AB, JC, ELG, TBM, and others, Memo, October 23, 1961, JC Papers, Box 64, Folder "Simulmatics, 1961." He also reported learning at that conference that

"there is apparently a major simulation being done of post-nuclear recovery." He was interested in developing a bid for Simulmatics: "we need to get our foot in the door." It is not clear that this proceeded.

89. IP, "Research Program on Problems of Communication and International Security," February 21, 1963, Pool Papers, Box 176, Folder "Eval of ComCom Project 1963."

90. Tom Van Vleck, "Project MAC," Multicians website, https://multicians.org/project -mac.html.

91. SP, interview with the author, May 16, 2018.

92. Receipts for equipment and travel can be found in Pool Papers, Box 95, Folder "ComCom/Supplies, computer tape, and Xerox reimbursing."

93. Tom Van Vleck, interview with the author, June 1, 2018.

94. Ibid. Tom Van Vleck, "The History of Electronic Mail," originally posted February 1, 2001, http://www.multicians.org/thvv/mail-history.html. Errol Morris, "Did My Brother Invent E-mail with Tom Van Vleck?" Opinionator, *NYT,* June 17, 2011.

95. Aldous, *Schlesinger,* 272–74.

96. EB attended a black-tie gala premiere at the Trans-Lux East on April 10, 1963; his two tickets can be found in Burdick Collection, Box 18, Folder "Publicity — The Ugly

American Movie." The screenplay was a finalist for Best Written Drama of 1963, by the Writers Guild of America. See the certificate and correspondence in Burdick Collection, Box 18, Folder "General Correspondence — Ugly American Movie."

97. Ballantine Ale, "Who Is the Ale Man?" appeared in *Life* in 1961 and 1962. On the television campaign, see Ballantine Ale, Storyboard, 1962, Burdick Collection, Box 75, Folder "Ballantine Ale."

98. "Stanley Kubrick's Point of View," *Variety,* February 27, 1963. Anthony Gruner, *London Report,* February 25, 1963. A second draft of the *Fail-Safe* screenplay, by Walter Bernstein, dated March 15, 1963, and with edits possibly made by EB, can be found in Burdick Collection, Box 41, Folder "Fail-Safe Screenplay."

99. Harvey Wheeler, "The Background of Fail-Safe," undated typescript, Burdick Collection, Box 41, Folder "Fail Safe (Red State)" and Harvey Wheeler to EB, May 12, 1962, Burdick Collection, Box 41, Folder "Fail Safe (Red Alert)." Wheeler's account is supported by other evidence, including Hallock Hoffman to Harvey Wheeler, January 28, 1963, in which Hoffman provides a specific recollection of Wheeler first telling him the story idea on November 4, 1957; see Burdick Collection, Box 41, Folder "Fail Safe (Red Alert)."

100. Jim Truitt (*Washington Post*) to EB, October 26, 1961, Burdick Collection, Box 41, Folder "Fail Safe (Red Alert)."

101. The settlement statement was drafted in April 1963 and signed in July. See the settlement papers in Burdick Collection, Box 41, Folder "Fail Safe (Red Alert)." The settlement involved no payment of money from Burdick and Wheeler (or ECA or McGraw-Hill) to either George or Kubrick (or Columbia or to George's publisher). Instead, Columbia paid $300,000 to Burdick and Wheeler for the rights to *Fail-Safe.* The plagiarism charge was essentially dismissed. But Kubrick's nuisance suit delayed the film.

102. Peter Krämer, " 'To Prevent the Present Heat from Dissipating': Stanley Kubrick and the Marketing of *Dr. Strangelove* (1964)," *InMedia* 3 (2013), accessed online.

103. See, e.g., Richard L. Coe, " 'Fail Safe' Is a Spellbinder," *Washington Post,* October 16, 1964.

Chapter 9: The Four-Eighty

1. "Eugene Burdick, Author of 'The Ugly American,' " *Washington Post,* July 28, 1965. For details on EB's condition at the time, see Francis L. Chamberlain, M.D., to William C. Kuzell, M.D., October 10, 1959, Burdick Collection, Box 144, Folder

"Burdick, Medical."

2. EB to Donald Rivkin (Simulmatics), December 2, 1963, Burdick Collection, Box 87, Folder "Rivkin, Donald." On Rivkin, see "Donald Rivkin, International Lawyer, 77," *NYT,* August 4, 2001.

3. On "The Candidate" as the book's title at the time, see Edward Kuhn Jr. to EB, November 4, 1963, Burdick Collection, Box 61, Folder "McGraw-Hill."

4. "I almost dropped the whole project the day Kennedy was assassinated. I had first written the book with the idea that Kennedy would run through it the way Johnson does now. I concluded, however, that the basis of the book remained intact and there was no need to drop the book. In fact, the question of whether or not the incumbent President could be elected when he held the office for less than a year is an intriguing one." EB to Sonia Lilienthal (McGraw-Hill), May 18, 1964, Burdick Collection, Box 35, Folder "The 480 — Correspondence."

5. NBC Radio Network, November 22, 1963, available at https://www.youtube.com/watch?v=OC-Y8Pm8RDU.

6. EB to Donald Rivkin (Simulmatics), December 2, 1963.

7. Aldous, *Schlesinger,* 1–2.

8. Isserman and Kazin, *America Divided,* 103.

9. LBJ, "Address to Joint Session of Con-

gress," November 27, 1963.

10. Isserman and Kazin, *America Divided,* 110, 107.

11. Ken Burns and Lynn Novick, dirs., *Vietnam,* episode 5, " 'This Is What We Do' (July 1967–December 1967)" (PBS, 2017).

12. Aldous, *Schlesinger,* ch. 17.

13. LBJ, "Address to Joint Session of Congress," November 27, 1963.

14. John F. Maloney to William Lederer, September 7, 1961, William Lederer Papers, Special Collections and University Archives, University of Massachusetts, Amherst. Burdick and Lederer occasionally pitched pieces to *Reader's Digest* together; see the correspondence in Burdick Collection, Box 90, Folder *"Reader's Digest."*

15. William Lederer to IP, May 26, 1962 and IP to William Lederer, May 26, 1962, William Lederer Papers.

16. EB to Ralph Tyler (CASBS), April 15, 1964, Burdick Collection, Box 145, Folder XLIII.

17. The Hope Pictures arrangement was made in the summer of 1963. J. D. Burdick to EB et al., August 8, 1963, Burdick Collection, Box 50, Folder "Kamina, Inc., Correspondence #1." It appears that EB, Inc., agreed not to make it into a film, a TV or radio production, or sell rights to someone else until July 1, 1970. Martin Gang to Eugene Burdick, October 10, 1963, Burdick

Collection, Box 50, Folder "Kamima, Inc., Correspondence #1."

18. "I have checked the agreement of July 1, 1963 between Hope Pictures, Inc., and EB, Inc., which as you know was negotiated and executed without the knowledge or most of the brain trust. I mention this because the contract deals with 'The Candidate.' That contract gives Hope Pictures, Inc., the 'non-exclusive and perpetual right' to produce one feature based on the work which it has the right to exhibit in perpetuity in all media, including the theatrical film medium provided that it may not be exhibited theatrically until it has first been exhibited on television." Martin Gang to EB October 10, 1963, Burdick Collection, Box 50 "Folder Kamima, Inc., Correspondence #1."

19. EB to Martin Gang, October 21, 1963, Burdick Collection, Box 53, Folder "Gang, Martin."

20. Martin Gang to EB October 10, 1963, Burdick Collection, Box 50, Folder "Kamima, Inc., Correspondence #1."

21. EB to Martin Gang, October 21, 1963, Burdick Collection, Box 53, Folder "Gang, Martin." Martin Gang nevertheless pursued other possibilities. On April 14, 1964, he met with George Englund, who had directed *The Ugly American.* "George agreed with me that 'The 480' would be a fairly

expensive picture to make because of the important cast that would be available and a $3,000,000 figure would be a minimum. . . . He has a deal with MGM." Martin Gang to EB, Burdick Collection, April 14, 1964, and George Englund to Martin Gang, April 14, 1964, Box 35, Folder "The 480 — Correspondence."

22. D. B. Willard, "Book of the Day," *Boston Globe,* January 30, 1964.

23. Martin Gang to EB, October 10, 1963, Burdick Collection, Box 50, Folder "Kamima, Inc., Correspondence #1."

24. EB, *The 480* (New York: McGraw-Hill, 1964), 17–18.

25. "One big hot and very current item will be whether or not Goldwater can be defeated in the Republican Convention. I think if a couple of television commentators could be persuaded to say that THE 480 explores the strategy which might be used by the Eisenhower-Scranton-Nixon-Rockefeller people to beat Goldwater it would be an interesting theme during the first month of the book's publication." EB to Sonia Lilienthal (McGraw-Hill), May 18, 1964, Burdick Collection, Box 35, Folder "The 480 — Correspondence."

26. James H. Marshall (Simulmatics) to EB, December 5, 1963, Burdick Collection, Box 35, Folder "The 480 — Correspondence."

27. AB to EB, undated but c. 1963. AB sent

him materials, including "Project on the Compilation and Detailed Analysis of Political Survey Data: Progress Report," October 15, 1959, Burdick Collection, Box 34, no folder.

28. EB, *The 480,* 51–57.

29. *Saturday Review,* October 4, 1958. *Reporter,* October 16, 1958.

30. Edward Kuhn, Jr. (McGraw-Hill) to EB, March 31, 1964, Burdick Collection, Box 61, Folder "McGraw-Hill."

31. EB to Martin Gang, January 3, 1964, Burdick Collection, Box 50, Folder "Kamima Inc., Correspondence #1."

32. McGraw-Hill to EB, January 7, 1965, Burdick Collection, Box 51, Folder "General Correspondence, January–June 1965."

33. "Writers on Campus," *NYT Magazine,* February 16, 1964.

34. Edward Kuhn to EB, April 20, 1964, Burdick Collection, Box 52, Folder "General Correspondence 1964, January–May."

35. "Gentlemen: Will you kindly send me a copy of 'Simulation of the 1960 Presidential Election' which I understand is to be published in August." EB to MIT Press, April 29, 1964, Burdick Collection, Box 52, Folder "General Correspondence 1964, January–May." MIT sent Burdick a set of galleys on July 14. Howard Levin (MIT Press) to EB, July 14, 1964, Box 39, stapled to the cover of the galleys.

36. ". . . to be serialized in June in the *Saturday Evening Post.* First printing 50,000. $15,000 ad campaign; co-op ads; ad mats; posters." *Publishers' Weekly,* June 22–26, 1964, and in Burdick Collection, Box 35, Folder "The 480 — Correspondence."

37. "Simulmatics: How to Make an Instant President," *Sunday Times* (London), November 1, 1964.

38. Naomi Spatz, interview with the author, May 24, 2018; Ann Greenfield, interview with the author, June 9, 2018.

39. Sonia Levienthal (McGraw-Hill) to EB, June 10, 1964, Burdick Collection, Box 35, Folder "The 480 — Correspondence." "I'm enclosing the final release which I worked out with the Simulmatics people. They were here for lunch." The enclosed press release includes a disclaimer: "A spokesman for The Simulmatics Corporation points out that the firm disagrees with the conclusions Mr. Burdick expresses in 'The 480.' He emphasizes that Burdick himself says: 'The interpretation of The Simulmatics Corporation's work and its implications are entirely my own.' " (Press release, undated, page 2.)

40. Galouye, *Simulacron-3;* the quotation is from p. 9.

41. IP, "Simulmatics 1964 Preconvention Project," June 11, 1964, Pool Papers, Box

143, no folder.

42. EB, "The 480," Part 1 of 3, *Saturday Evening Post,* June 13, 1964.

43. Thomas D. Finney Jr. to Walter Jenkins, July 24, 1964; Richard Scammon to Walter Jenkins, Memo, July 28, 164; Walter Jenkins to Thomas Finney, July 31, 1964, Johnson Library, White House Central File, Box 352, Folder PR 7/1/64-8/9/64.

44. MIT Press, "Advance Information," undated but mid-1964, Burdick Collection, Box 34, with the galleys.

45. Edward Kuhn to EB, July 8, 1964, Burdick Collection, Box 35, Folder "The 480 — Correspondence."

46. EB, *The 480,* 78–79.

47. Ibid., 250.

48. "Eugene Burdick's THE 480: A Novel of Politics," Press Release, June 24, 1964, Burdick Collection, Box 34, Folder "The 480 — Correspondence."

49. Edward Kuhn Jr. to [Governor or Senator *********], June 1964, Burdick Collection, Box 34, Folder "The 480 — Correspondence."

50. EB to Sonia Lilienthal (McGraw-Hill), May 18, 1964, Burdick Collection, Box 35, Folder "The 480 — Correspondence."

51. EB to AS, November 13, 1958, Burdick Collection, Box 116, Folder "Nixon, Richard."

52. Orville Prescott, " 'The New

Underground' of Politics," *NYT,* June 24, 1964.

53. Victor P. Hass, "The New Kingmakers," *Chicago Tribune,* June 18, 1964.

54. The plot summary refers to "a fictional company, Simulations Enterprises, using (apparently without copyright release) the ideas of real-life Simulmatics Corporation." "Behavioral Science Fiction," *American Behavioral Scientist* 81 (1964): 40.

55. Ben H. Bagdikian, "Working in Secret," *NYT,* June 28, 1964. The article is a review of David Wise and Thomas B. Ross, *The Invisible Government.* Wise was the Washington bureau chief for the *New York Herald Tribune,* Ross a correspondent for the *Chicago Sun-Times.* Pool Papers, Box 95, untitled folder.

56. Sidney Hyman, "If Computers Called the Tune," *NYT,* June 28, 1964. Hyman further suggested that everything Simulmatics had done was documented in Larry O'Brien's campaign manual, a claim that was either ill-informed or disingenuous. O'Brien's primer, an update of the manual he wrote for the 1960 Kennedy campaign, includes, for instance, a page and a half on polling but does not cover the methods deployed by Simulmatics. [Lawrence O'Brien], *The Democratic Campaign Manual, 1964* (Washington, DC: Democratic National Commit-

tee, 1964), 31–32.

57. "Now, I enjoy being titillated by apocalyptic fantasies as much as the next man, and can swallow my full draught of science fiction without blinking. But it is disconcerting to learn right at the outset that Mr. Burdick himself seems not to know the difference between political science and political science fiction." Irving Kristol, "Computer Politics," *Observer*, October 4, 1964.

58. Rick Perlstein, "1964 Republican Convention: Revolution from the Right," *Smithsonian Magazine*, August 2008.

59. Burns and Novick, *Vietnam*, episode 5.

60. Isserman and Kazin, *America Divided*, 117–18. And see Fredrik Logevall, *Choosing War: The Lost Chance for Peace and the Escalation of War in Vietnam* (Berkeley: University of California Press, 1999).

61. Isserman and Kazin, *America Divided*, 134.

62. Earnest N. Bracey, *Fannie Lou Hamer: The Life of a Civil Rights Icon* (Jefferson, NC: McFarland, 2011).

63. Fannie Lou Hamer, Testimony Before the Credentials Committee, Democratic National Convention, August 22, 1964.

64. LBJ, Acceptance Speech, Democratic Convention, August 27, 1964.

65. Clark Kerr, *The Gold and the Blue: A*

Personal Memoir of the University of California, 1949–1967, vol. 2, *Political Turmoil* (Berkeley: University of California Press, 2003), 176. Mario Savio, *The Essential Mario Savio: Speeches and Writings That Changed America,* ed. Robert Cohen (Berkeley: University of California Press, 2014), 11.

66. "Percentage, by Department, of Faculty and TA's Observing Strike," 1964: twenty-nine out of thirty TAs in the Political Science Department did not cross picket lines. Free Speech Movement, Press Release, December 4, [1964]. "Of 637 demonstrators interviewed, 597 were students, which included 62 from Political science." Free Speech Movement: Pamphlets, Flyers and Bulletins, 1964–1965, Widener Library, Harvard University.

67. Margot Adler, "My Life in the FSM," Adler to her mother, December 2, 1964, at nine p.m., in Robert Cohen and Reginald E. Zelnik, eds., *The Free Speech Movement: Reflections on Berkeley in the 1960s,* ed. Robert Cohen and Reginald E. Zelnik (Berkeley: University of California Press, 2002), 119.

68. Ray Colvig, *Turning Points and Ironies: Issues and Events — Berkeley, 1959–67* (Berkeley, CA: Berkeley Public Policy Press, 2004), 93.

69. Mark Kitchell, dir., *Berkeley in the Sixties* (First Run Features, 1990).

70. Colvig, *Turning Points and Ironies,* 79. Fred Turner, *From Counterculture to Cyberculture: Stewart Brand, the Whole Earth Network, and the Rise of Digital Utopianism* (Chicago: University of Chicago Press, 2006), 2 and ch. 1.

71. Turner, *From Counterculture to Cyberculture,* 12.

72. Ralph Gleason, "The Tragedy at the Greek Theater," *San Francisco Chronicle,* December 8, 1964.

73. Kitchell, *Berkeley in the Sixties.*

74. EB to Robert Scalapino, December 16, 1964, cc to Clark Kerr with a cover note, dated December 19, 1964, Burdick Collection, Box 51, Folder "General Correspondence, January–June 1965."

75. In the margin on material Burdick collected in Berkeley that spring, he penciled in a single word: "Paranoid." Berkeley Citizens' Committee of Inquiry, April 22, 1964, Memo and Informational Flyer. And see David Andrews to Thoughtful Citizens Concerned About American Freedom, Memo, undated. Both in Burdick Papers, Box 116, Folder 46.

76. Robert Scalapino to EB, May 12, 1965, Burdick Collection, Box 51, Folder "General Correspondence, January–June 1965."

77. EB to Robert Scalapino, December 16, 1964, from the Savoy Hotel, London, and Scalapino to Dean William B. Fretter, Sproul Hall, with a cc to EB, January 21, 1965, Burdick Collection, Box 51, Folder "General Correspondence, January–June 1965."

78. Logevall, *Choosing War.*

79. Isserman and Kazin, *America Divided,* 137.

80. Martina E. Greene to EB, May 20, 1965, and EB to Greene, June 3, 1965, Burdick Collection, Box 51, Folder "General Correspondence, January–June 1965."

81. R. L. Sproull (ARPA), to Simulmatics, October 6, 1964, Pool Papers, Box 143, no folder.

82. EB to Bruce M. Polichar, June 1, 1965, Burdick Collection, Box 51, Folder "General Correspondence, January–June 1965."

Chapter 10: Armies of the Night

1. MS to Everyone, August 16, 1966, Shea Letters. MS, interview with the author, May 22, 2018.

2. MS to Everyone, August 16, 1966, Shea Letters.

3. Ibid.

4. Ward Just, *To What End: Report from Vietnam* (1968; repr., New York: PublicAffairs, 2000), 1. On Just, see Harrison Smith,

"Ward Just, *Washington Post* Reporter and Acclaimed Political Novelist, Dies at 84," *Washington Post,* December 19, 2019.

5. MS to Everyone, August 16, 1966.

6. Just, *To What End,* 8, 11.

7. Details on RAND come from Mai Elliott, *RAND in Southeast Asia: A History of the Vietnam War Era* (Santa Monica, CA: RAND Corporation, 2010).

8. Oliver Belcher, "Data Anxieties: Objectivity and Difference in Early Vietnam War Computing," in *Algorithmic Life: Calculative Devices in the Age of Big Data,* ed. Louise Amoore and Volha Piotuhk (London: Routledge, 2015), 130. Dan E. Feltman, *When Big Blue Went to War: The History of the IBM Corporation's Mission in Southeast Asia During the Vietnam War, 1965–1975* (Bloomington, IN: Abbott Press, 2012),14–16, 22.

9. Feltman, *When Big Blue Went to War,* 57.

10. IP to Jonathan Robert Pool, June 1966, Pool Family Papers.

11. MS to Everyone, August 16, 1966, Shea Letters.

12. IP to D. MacArthur, July 29, 1966, Pool Papers, Box 74, Folder "Correspondence 1961–66."

13. Gilbert W. Chapman (chairman of the board) and ELG, to Simulmatics Stockholders, September 5, 1967, Coleman

Papers, Box 65, Folder "Simulmatics 1966–1967."

14. Donald Fisher Harrison, "Computers, Electronic Data, and the Vietnam War," *Archivaria* 26 (1988): 18.

15. Ball, *The Past Has Another Pattern,* 174.

16. J. Peter Scoblic, "Robert McNamara's Logical Legacy," *Arms Control Today* 39 (2009): 58.

17. Mark Solovey, "Project Camelot and the 1960s Epistemological Revolution: Rethinking the Politics–Patronage–Social Science Nexus," *Social Studies of Science* 31 (2001): 178.

18. David Halberstam, *The Best and the Brightest* (1972; repr., New York: Ballantine, 1993), 405.

19. Elizabeth Dickinson, "A Bright Shining Slogan: How 'Hearts and Minds' Came to Be," *Foreign Policy,* August 22, 2009.

20. Simulmatics Corporation, *Dynamic Models for Simulating the Venezuelan Economy* (Cambridge, MA: Simulmatics Corporation, 1966). Simulmatics Corporation, *The Development of a Simulation of the Venezuelan Economy with the Simulmatics-CENDES Global Model* (New York: Simulmatics Corporation, 1965). And see the reports in the Simulmatics Corporation newsletters dated May 3, 1963, and July 26, 1963, Lasswell Papers, Box 39, Folder 544.

21. Simulmatics Corporation, "The Simulmatics-CENDES Economic Study of Venezuela: Background and First Report on Operations," October 1963, Burdick Collection, Box 35, no folder. Edward P. Holland (director, National Economic Systems Studies, Simulmatics Corporation), "Principles of Simulation," 1962, Burdick Collection, Box 35, no folder. On MIT CIS's collaboration with CENDES, see Center for International Studies Records, MIT Institute Archives and Special Collections, Box 7, Folders 27–30.

22. AB to June Bernstein, 1963, in the possession of Elizabeth Bernstein Rand.

23. Irving Louis Horowitz, ed., *The Rise and Fall of Project Camelot* (Cambridge, MA: MIT Press, 1967), 4–5, 49.

24. Michael Desch, *Cult of the Irrelevant: The Waning Influence of Social Science on National Security* (Princeton, NJ: Princeton University Press, 2019), 132.

25. Project Camelot, Working Paper, December 5, 1964, in Horowitz, *The Rise and Fall of Project Camelot,* 50–55.

26. Solovey, "Project Camelot and the 1960s Epistemological Revolution," 171–206; Philip Y. Kao, "Shelling from the Ivory Tower: Project Camelot and the Post–World War II Operationalization of Social Science," *Focaal: Journal of Global and*

Historical Anthropology 80 (2018): 105–19; Ron Robin, *The Making of the Cold War Enemy: Culture and Politics in the Military-Intellectual Complex* (Princeton, NJ: Princeton University Press, 2001), ch. 10; Ellen Herman, "Project Camelot and the Career of Cold War Psychology," in *Universities and Empire: Money and Politics in the Social Sciences During the Cold War,* ed. Christopher Simpson (New York: Norton, 1998), 97–134; and Rohde, *Armed with Expertise,* ch. 3. See also Michael T. Klare, *War Without End: American Planning for the Next Vietnams* (1970; repr., New York: Vintage, 1972), ch. 4, "Social Systems Engineering: Project Camelot and Its Successors"; and Gene M. Lyons, *The Uneasy Partnership: Social Science and the Federal Government in the Twentieth Century* (New York: Russell Sage Foundation, 1969), 167–69, 194–96.

27. Fulbright quoted in Marshall Sahlins, "The Established Order: Do Not Fold, Spindle, or Mutilate," in Horowitz, *The Rise and Fall of Project Camelot,* 77

28. Sahlins, "The Established Order," 75–76.

29. Subcommittee on International Organizations and Movements, House Committee on Foreign Affairs, *Behavioral Science and the National Security, Report No. 4,* December 6, 1965, 102–3.

30. IP, "The Necessity for Social Scientists

Doing Research for Governments," *Background* 10 (August 1966): 111–22.

31. Adam de Sola Pool, interview with the author, May 19, 2018. Susan Greenfield, interview with the author, July 27, 2018.

32. Elliott, *RAND in Southeast Asia,* ch. 2. And see Robin, *The Making of the Cold War Enemy,* ch. 9.

33. Elliott, *RAND in Southeast Asia,* ch. 3. DiLeo, *George Ball, Vietnam, and the Rethinking of Containment,* 75, 150–51.

34. Burns and Novick, *Vietnam,* episode 5.

35. R. L. Sproull (ARPA) to Simulmatics Corporation, October 6, 1964, with two appendices, Pool Papers, Box 143, no folder.

36. Elliott, *RAND in Southeast Asia,* ch. 2.

37. Burns and Novick, *Vietnam,* episode 5.

38. Just, *To What End,* 66–68.

39. Fritz Stern, *Five Germanys I Have Known* (New York: Macmillan, 2007), 247. And see Jean Pool to Adam de Sola Pool, January 28, 1967, Pool Family Papers.

40. Seymour J. Deitchman, *The Best-Laid Schemes: A Tale of Social Research and Bureaucracy* (1976; repr., Quantico, VA: Marine Corps University Press, 2104), 298–311.

41. AB to Seymour Deitchman, Memo, undated, Pool Papers, Box 67, Folder "Simulmatics Vietnam Correspondence." And see Simulmatics Corporation, "A

Proposal for Research on Urban Insurgency," January 1965 (Pool Papers, Box 145, Folder "Insurgency"). Deitchman wrote to IP that he'd forwarded AB's proposal to J.C.R. Licklider, head of ARPA's behavioral science program, noting, "He and I were much taken with the coincidence of the beginning of insurgency in Saigon." (President Ngo Dinh Diem's brutal suppression of Buddhists had led to protests that, at the end of 1963, led to his arrest and assassination.) Deitchman joked, "Are you rigging this?" Seymour Deitchman to IP, February 13, 1964, Pool Papers, Box 67, Folder "Simulmatics Vietnam Correspondence."

42. IP to Jonathan Robert Pool, July 26, 1966.

43. Simulmatics Corporation, "Simulmatics Efforts in Vietnam," February 1968, Pool Papers, Box 67, Folder "Simulmatics Vietnam Correspondence."

44. Simulmatics Field Team, First Progress Report, August 12, 1966, Pool Papers, Box 67, Folder "Simulmatics Vietnam Correspondence."

45. George Ball had gone to London to bring his body back. At Stevenson's memorial service, in Washington, Lyndon Johnson came up to a weeping Ball, put his hand on his shoulder, and said, "George, I never trust a man who can't cry for a friend." Ball, *The Past Has Another Pattern,* 152.

46. Frances FitzGerald, interview with the author, January 3, 2019.

47. Frances FitzGerald, *Fire in the Lake: The Vietnamese and the Americans in Vietnam* (Boston: Little, Brown, 1972), 362.

48. SP, interview with the author, April 27, 2018.

49. SP, interviews with the author, April 27, 2018, and May 16, 2018.

50. SP, "Dear World," undated but late fall 1966, Pool Papers, Box 145, Folder "Vietnam Correspondence."

51. Burdick and Lederer, *The Ugly American,* 181. Christie, *"The Quiet American" and "The Ugly American,"* 44–45. James Gibney, "The Ugly American," *NYT,* January 15, 2006. Louis Menand, "What Went Wrong in Vietnam," *New Yorker,* February 26, 2018.

52. Daniel Ellsberg, *Secrets: A Memoir of Vietnam and the Pentagon Papers* (New York: Viking, 2002), 4. On Ellsberg's time in Vietnam in 1966 and 1967, see chs. 7–11.

53. Tom Wells, *Wild Man: The Life and Times of Daniel Ellsberg* (New York: Palgrave, 2001), 248–49.

54. MS to Everyone, December 18, 1966, Shea Letters.

55. IP to Jean Pool, July 15, 1967, Pool Family Papers.

56. Cuc Thu Duong, interview with the author, June 11, 2018.

57. A. Ficks to IP, April 13 and 14, 1966, Pool Papers, Box 67, Folder "Simulmatics Viet Nam correspondence."

58. Lieutenant J. Stephen Morris, Memo for the Record, August 19, 1966, Project Agile Records, Box 49, Folder "Simulmatics — Problem Analysis, NN3-330-099-009." And see IP to Colonel John Patterson, May 18, 1967, same folder. My great thanks to Sharon Weinberger for help in locating the records of Project Agile.

59. Simulmatics uniform, in the possession of MS. MS describes the logo this way: "two tapes and a dollar sign of broken lances." MS to Everyone, December 23, 1966, Shea Letters.

60. Ward Just, interview with the author, January 2, 2019.

61. Lieutenant J. Stephen Morris, Memo for the Record, August 19, 1966, Project Agile Records, Box 49, Folder "Simulmatics — Problem Analysis, NN3-330-099-009."

62. L. A. Newberry (Research and Development Field Unit, Vietnam), Memorandum for the Record, December 17, 1967, Project Agile Records, Box 49, Folder "Simulmatics — Problem Analysis, NN3-330-099-009."

63. Cuc Thu Duong, interview with the author, June 11, 2018.

64. SP, interview, May 16, 2018. SP's dissertation attempted to explain the failure of

pacification. He argued that "if the war had been fought as though protecting the peasants were more important than killing Viet Cong the result would have been far less favorable to the Viet Cong than the peasant strategies have been." SP, "The Myth of the Village: Revolution and Reaction in Viet Nam" (PhD diss., MIT, 1969).

65. Rohde, "The Last Stand of the Psychocultural Cold Warriors," 232–50.

66. Naomi Spatz, interview with the author, May 24, 2018.

67. Quotations are from complete transcripts of the original 1966 interviews, in the possession of SP.

68. Slote's initial report is Walter H. Slote, "Observations on Psychodynamic Structures in Vietnamese Personality," undated draft, Pool Papers, Box 145, Folder "Vietnam Walter H. Slote." ARPA did not wish to release this report. Slote later published a version of it as Walter Slote, "Psychodynamic Structures in Vietnamese Personality," in *Transcultural Research in Mental Health,* ed. William P. Lebra (Honolulu: University Press of Hawaii, 1972), ch. 8. And see Ward Just, "Study Reveals Viet Dislike for U.S. but Eagerness to Be Protected by It," *Washington Post,* November 20, 1966.

69. Colonel William B. Arnold to Seymour Deitchman, August 22, 1967, Project Agile

Records, Box 49, Folder "Simulmatics — Problem Analysis, NN3-330-099-009."

70. Abraham Hirsch to Colonel John Patterson, June 30, 1967, Project Agile Records, Box 49, Folder "Simulmatics — Problem Analysis, NN3-330-099-009."

71. IP to Garry Quinn, July 30, 1967, Project Agile Records, Box 49, Folder "Simulmatics — Problem Analysis, NN3-330-099-009."

72. Susan Greenfield, interview with the author, July 27, 2018.

73. Garry Quinn to IP, August 14, 1967, Project Agile Records, Box 49, Folder "Simulmatics — Problem Analysis, NN3-330-099-009."

74. Garry Quinn to IP, July 24, 1967, Project Agile Records, Box 49, Folder "Simulmatics — Problem Analysis, NN3-330-099-009."

75. Ann Penner Winston, interview with the author, June 7, 2018.

76. Seymour Deitchman to Dr. Charles M. Herzfeld, ARPA/RFDU-Thailand, November 29, 1966, Project Agile Records, Box 49, Folder "Simulmatics — Problem Analysis, NN3-330-099-009."

77. Ward Just, interview with the author, January 2, 2019.

78. Just, *To What End,* 61.

79. Ibid., 71–72.

80. Isserman and Kazin, *America Divided,* 200.

81. Martha Gellhorn, "Civilian Casualties, 1966," *Ladies' Home Journal* (January 1967), reprinted in *Reporting Vietnam,* part 1, *American Journalism, 1959–1969* (New York: Library of America, 1998), 289–90.

82. MS to Everyone, August 29, 1966, Shea Letters.

83. Neil Sheehan, "Not a Dove, but No Longer a Hawk," *NYT Magazine,* October 9, 1966, reprinted in *Reporting Vietnam;* quotations from 299, 315.

84. MS to Family, September 24, 1966, Shea Letters.

85. MS to Everyone, August 16, 1966, Shea Letters.

86. Many of the interviews still survive. The interviews are filed in the Pool Papers, Boxes 131 and 132. E.g., "19 year old married seamstress and housewife named Mrs. Van, from Thuong Lac village in the Kien Phong Province, interviewed by Dan Grady with Cuc, Sanh, and Loc, September 7–13, 1966 at the National Chieu Hoi Center, Saigon. Had been a squad leader with the NLF for about two years." "Interview, September 13, 1966," in Pool Papers, Box 131.

87. Zalin Grant, "Vietnam by Computer: Counting Strength That's Not There," *New Republic,* June 13, 1968.

88. Simulmatics Field Team, First Progress Report, August 12, 1966, Pool Papers, Box 67, Folder "Simulmatics Vietnam Correspondence."

89. FitzGerald, *Fire in the Lake,* 339–40.

90. MS to Everyone, October 1, 1966, and October 15, 1966, Shea Letters.

91. MS to Everyone, October 28, 1966, Shea Letters.

92. MS to Everyone, October 15, 1966, Shea Letters.

93. Deitchman, *Best-Laid Schemes,* 313–14.

94. MS to Everyone, November 4, 1966, and October 15, 1966, Shea Letters.

95. Ibid.

96. ADG, diary entry, July 23, 1965, Grazian Archive.

97. MS to Everyone, November 15, 1966, Shea Letters.

98. Daniel Ellsberg, interview with the author, August 22, 2018.

99. MS to Everyone, November 16, 1966, Shea Letters.

100. Wendy McPhee, interview with the author, July 16, 2018.

101. Patricia Greenfield to Edwin Safford, undated but late November or December 1967, Greenfield Papers.

102. MS to Family, November 29, 1966, Shea Letters.

103. IP to John Vann, December 1, 1966, Neal Sheehan Papers, Library of Congress,

Box 27, Folder 2.

104. MS to Family, December 10, 1966, Shea Letters.

105. MS to Everyone, December 23, 1966, Shea Letters.

106. Alexis C. Madrigal, "The Computer That Predicted the U.S. Would Win the Vietnam War," *Atlantic,* October 5, 2017. And see Harry G. Summers, *American Strategy in Vietnam: Critical Analysis* (New York: Dover, 2007), 11.

107. LBJ, "Remarks in New York City upon Receiving the National Freedom Award," February 23, 1966.

Chapter 11: The Things They Carried

1. James Ridgeway, *The Closed Corporation: American Universities in Crisis* (New York: Random House, 1968), 64.

2. FitzGerald, *Fire in the Lake,* 342.

3. MLK, "Beyond Vietnam," speech, April 4, 1967.

4. MLK, Address on the Vietnam War, April 15, 1967. And see Simon Hall, *Peace and Freedom: The Civil Rights and Antiwar Movements in the 1960s* (Philadelphia: University of Pennsylvania Press, 2011), 105–6.

5. MLK, address outside the United Nations Building, April 15, 1967, black-and-white film, United Nations Audiovisual Library Archives, available at https://www.youtube

.com/watch?v=YpGFOiSTs3Q.

6. IP to JC, July 20, 1967, Pool Papers, Box 75, Folder "Correspondence May to August 1967."

7. IP to WMc, June 28, 1967, Pool Papers, Box 75, Folder "Correspondence May to August 1967."

8. Garry L. Quinn, Memo for Defense Contract Administration Services Region, New York, October 7, 1967, Project Agile Records, Box 49, Folder "Simulmatics — Problem Analysis, NN3-330-099-009." One security violation took place in May 1967: "The security markings appearing on SIM/CAM/5(A)/67, Brief Comparison of Simulmatics Chieu Hoi Report and RAND Study, are simply inconceivable to anyone remotely familiar with Defense security procedures." Garry Quinn to J. David Yates (Simulmatics), May 15, 1967, Project Agile Records, Box 49, Folder "Simulmatics — Problem Analysis, NN3-330-099-009." Another took place in June 1967; it involved sending a document that was not marked with any security labels and had been sent by Simulmatics accountant Shlomo Hagai, who had no clearance to even see such documents. Garry Quinn, Memo re Security Violation, June 6, 1967; William Hunt to Defense Contract Administrative Services Region, June 13, 1967; AB to ARPA, June 26, 1967, Project Agile Records, Box

49, Folder "Simulmatics — Problem Analysis, NN3-330-099-009." Greenfield eventually appointed Bernstein as Simulmatics' security officer, but this was too little, too late. ELG to F. B. Boice (ARPA), August 30, 1967, Project Agile Records, Box 49, Folder "Simulmatics — Problem Analysis, NN3-330-099-009."

9. F. B. Boice, Memorandum for the Record, December 12, 1966, Project Agile Records, Box 49, Folder "Simulmatics — Problem Analysis, NN3-330-099-009."

10. ADG, diary entries, January 21 and 23, 1967, Grazian Archive.

11. ADG, diary entry, undated but sometime in January or February 1967, Grazian Archive.

12. ADG, diary entry, January 28, 1967, Grazian Archive.

13. Victoria de Grazia, interview with the author, March 21, 2019.

14. Bob Dylan "The Times They Are a-Changin'," 1964.

15. Isserman and Kazin, *America Divided,* 195.

16. Jeremy de Sola Pool, interview with the author, May 23, 2018.

17. Ann Greenfield, interview with the author, June 9, 2018.

18. IP to Jonathan Robert Pool, various letters, July 1965, Pool Family Papers.

19. Tim O'Brien, *The Things They Carried*

(1990; repr., Boston: Mariner, 2009), ch. 1.

20. Marion Lee Kempner to his family, September 16, 1966, in *Dear America: Letters Home from Vietnam,* ed. Bernard Edelman (New York: Norton, 1985), 182.

21. Kim Le Cook, interview with the author, August 7, 2019.

22. Mary McCarthy, *Vietnam* (New York: Harcourt, Brace & World, 1967), 62.

23. IP to Garry Quinn, January 20, 1967, Project Agile Records, Box 49, Folder "Simulmatics — Problem Analysis, NN3-330-099-009."

24. Natalie W. Yates to ADG, February 8, 1967, and ADG, diary entries, August 1967, Grazian Archive.

25. ADG, diary entry, April 29, 1967, Grazian Archive.

26. Colonel John Patterson to Seymour Deitchman and Colonel M. E. Sorte, May 6, 1967, Project Agile Records, Box 49, Folder "Simulmatics — Problem Analysis, NN3-330-099-009."

27. Ward Just to Colonel John Patterson, Memorandum of Call, April 28, 1967, Project Agile Records, Box 49, Folder "Simulmatics — Problem Analysis, NN3-330-099-009." And see Colonel John Patterson to Seymour Deitchman and Colonel M. E. Sorte, May 6, 1967, Project Agile Records, Folder "Simulmatics — Problem Analysis, NN3-330-099-009."

28. ADG, diary entry, April 22, 1967, Grazian Archive.
29. Ward Just, "Swarms of Analysts Probing Vietnamese," *Washington Post,* April 30, 1967.
30. Colonel John Patterson to Seymour Deitchman and Colonel. M. E. Sorte, May 6, 1967, Project Agile Records, Box 49, Folder "Simulmatics — Problem Analysis, NN3-330-099-009."
31. "I hope you realize that since the first experience in Saigon, our people have all been thoroughly briefed on the problem of avoiding press relations. You know, as I do, that to some extent we are all in a fish bowl in Saigon. A careful reading of the latest Ward Just story makes it clear that his material was given him by someone who is critical of our kind of study." IP to Colonel John Patterson, May 18, 1967, Project Agile Records, Box 49, Folder "Simulmatics — Problem Analysis, NN3-330-099-009."
32. "The summer's events in Vietnam have generated a major conflict between the American Government and the press. It is a conflict of judgment over the course of the war." Richard Harwood, "The War Just Doesn't Add Up," *Washington Post,* September 3, 1967, reprinted in *Reporting Vietnam,* 484.
33. Ward Just to Frances FitzGerald, March 3, 1967, Frances FitzGerald Collection,

Boston University Libraries, Howard Gotlieb Archival Research Center, Box 25, Folder "Ward Just."

34. Zalin Grant, "Vietnam by Computer: Counting Strength That's Not There," *New Republic*, June 13, 1968.

35. Oliver Belcher, "Data Anxieties: Objectivity and Difference in Early Vietnam War Computing," in Amoore and Piotuhk, *Algorithmic Life*, 127–42.

36. Ward Just to Frances FitzGerald, April 7, 1967, FitzGerald Collection, Boston University Libraries, Howard Gotlieb Archival Research Center, Box 25, Folder "Ward Just."

37. ADG, diary entries, May 2 and May 4, 1967, Grazian Archive.

38. IP to SP, May 9, 1967, Pool Papers, Box 67, Folder "Simulmatics Correspondence."

39. Memorandum for Colonel Patterson, May 1967, Project Agile Records, Box 49, Folder "Simulmatics — Problem Analysis, NN3-330-099-009."

40. IP to Daniel Ellsberg, May 5, 1967, and IP to Richard Holbrook, May 6, 1967, Project Agile Records, Box 49, Folder "Simulmatics — Problem Analysis, NN3-330-099-009."

41. Robert S. McNamara, *In Retrospect: The Tragedy and Lessons of Vietnam* (New York: Vintage, 1996), 378.

42. IP to the Simulmatics Cambridge Office, July 14, 1967, Pool Papers, Box 75, Folder "Correspondence May to August 1967."

43. IP to Adam de Sola Pool, July 4, 1967, Pool Family Papers.

44. FitzGerald, *Fire in the Lake,* 339–40.

45. Colonel William B. Arnold to Seymour Deitchman, August 22, 1967, Project Agile Records, Box 49, Folder "Simulmatics — Problem Analysis, NN3-330-099-009." Pool strenuously disagreed with this characterization. See, e.g., IP to Philip Worchel, January 5, 1968, Pool Papers, Box 145, Folder "HES."

46. Colonel William Arnold to Mr. W. G. McMillan, November 21, 1967, Project Agile Records, Box 49, Folder "Simulmatics — Problem Analysis, NN3-330-099-009."

47. FitzGerald, *Fire in the Lake,* 362.

48. Frances FitzGerald, interview with the author, January 3, 2019.

49. IP to Jonathan Robert Pool, August 1967.

50. ADG, diary entry, July 21, 1967, Grazian Archive.

51. ADG to Seymour Deitchman, July 27, 1967, Grazian Archive.

52. ADG to Colonel W. B. Arnold, July 25, 1967, Grazian Archive.

53. IP to George Goss (CORDS), September 11, 1967, Grazian Archive.

54. ADG, Saigon Propaganda Release, July

28, 1967, Grazian Archive.

55. IP to Adam de Sola Pool, July 10, 1967, Pool Family Papers.

56. Elliott, RAND in Southeast Asia, ch. 4.

57. IP to Adam de Sola Pool, July 28, 1967, Pool Family Papers.

58. Ibid.

59. IP to Adam de Sola Pool, August 3, 1967, Pool Family Papers.

60. L. A. Newberry, RDFU-VN (Research and Development Field Unit, Vietnam), Memorandum for the Record, December 17, 1967, Project Agile Records, Box 49, Folder "Simulmatics — Problem Analysis, NN3-330-099-009."

61. Colonel William B. Arnold to Ambassador Locke, November 25, 1967, Project Agile Records, Box 49, Folder "Simulmatics — Problem Analysis, NN3-330-099-009."

62. L. A. Newberry (Research and Development Field Unit, Vietnam), Memorandum for the Record, December 17, 1967, Project Agile Records, Box 49, Folder "Simulmatics — Problem Analysis, NN3-330-099-009."

63. Simulmatics Corporation, "Communication and Attitudes in Viet Name, Report #2, The Extent of Radio Listening," undated, Pool Papers, Box 67, Folder "Simulmatics Vietnam Correspondence." Simulmatics Corporation, "Communication and

Attitudes in Viet Name, Report #5, Possessions and Desires of Vietnamese Villagers," undated, Pool Papers, Box 67, Folder "Simulmatics Vietnam Correspondence." Simulmatics Corporation, "Communications and Attitudes in Viet Nam, Report #8, Villagers' Contact with Television," undated, Pool Papers, Box 67, Folder "Simulmatics: Comm. & Attitudes in VN Report #8, Villagers Contact w/TV."

64. Jeffrey Whyte, "Psychological War in Vietnam: Governmentality at the United States Information Agency," *Geopolitics* 10 (2017): 19.

65. Lieutenant J. Stephen Morris to Garry Quinn, undated but September 1967, Project Agile Records, Box 49, Folder "Simulmatics — Problem Analysis, NN3-330-099-009."

66. Lieutenant J. Stephen Morris to Garry Quinn, July 31, 1967, Project Agile Records, Box 49, Folder "Simulmatics — Problem Analysis, NN3-330-099-009."

67. These include an extremely negative, unsigned, three-page evaluation in Pool Papers, Box 143, with a handwritten notation, by Pool, on the first page: "Exclude."

68. L. A. Newberry (Research and Development Field Unit, Vietnam), Memorandum for the Record, December 17, 1967, Project Agile Records, Box 49, Folder "Simulmat-

ics — Problem Analysis, NN3-330-099-009."

69. William Lederer, "Vietnam: Those Computer Reports," *New Republic,* December 23, 1967.

70. IP et al., *Hamlet Evaluation System Study* (Cambridge, MA: Simulmatics Corporation, 1968).

71. Philip Worchel to Colonel William G. Sullivan, January 2, 1968, Project Agile Records, Box 49, Folder "Simulmatics — Problem Analysis, NN3-330-099-009."

72. Lieutenant J. Stephen Morris to Colonel Edwin R. Brigham, August 21, 1967, Project Agile Records, Box 49, Folder "Simulmatics — Problem Analysis, NN3-330-099-009."

73. Colonel William B. Arnold to Seymour Deitchman, August 22, 1967, Project Agile Records, Box 49, Folder "Simulmatics — Problem Analysis, NN3-330-099-009."

74. Garry Quinn to IP, August 14, 1967, Project Agile Records, Box 49, Folder "Simulmatics — Problem Analysis, NN3-330-099-009."

75. IP to Robert Komer, January 16, 1968, Lasswell Papers, Box 40, Folder 547.

76. IP to Jean and Adam Pool, undated but summer 1967. Biographical information form, undated, Wayman Papers, Box 1, Folder 33. He wasn't supposed to go to Vietnam until the fall but went in the sum-

mer instead. "Father Hoc is in conversation with us about going to Viet Nam sooner than originally planned." IP to Garry Quinn, May 5, 1967, Project Agile Records, Box 49, Folder "Simulmatics — Problem Analysis, NN3-330-099-009."

77. IP to Jean and Adam Pool, undated but summer 1967, Pool Family Papers.
78. JH, photograph with biographical inscription, Wayman Papers, Box 2, Folder "Rev. Joseph M. Hoc, Sociologist Who Taught at Boston College," *Boston Globe,* February 2, 1988.
79. Canon Peppergrass, "Joe Hoc Leaves Us," *Catholic Worker* [?], April 2, 1954, unidentified newspaper clipping, Wayman Papers, Box 1, Folder 32.
80. Ibid.
81. JH to Dorothy Wayman, February 24, 1954, and May 22, 1954, Wayman Papers, Box 1, Folder 32.
82. JH, "Reds Destroy Fr. Hoc's Native Village," *Catholic Free Press,* March 11, 1955.
83. JH to Dorothy Wayman, September 17, 1965, Wayman Papers, Box 1, Folder 33.
84. JH to Dorothy Wayman, October 5, 1967, Wayman Papers, Box 1, Folder 33.
85. JH to Deitchman, February 1969, Pool Papers, Box 154, Folder "Hoc."
86. Jesse Orlansky (Institute for Defense Analysis), to Garry Quinn, October 22, 1968, Project Agile Records, Box 49, Folder

"Simulmatics — Problem Analysis, NN3-330-099-009." Joseph M. Hoc, "The Tet Offensive Viewed by the Vietnamese Villagers," September 1968.

87. JH, "Testing New Psychological Warfare Weapons in Viet Nam," May 1968, Pool Papers, Box 145, Folder "Hoc-New Psywar Weapons." See also JH, "Code Book for the New Psychological Warfare Weapons Study Viet-Nam Context," same folder. Hoc also subsequently prepared a report, "Viet Nam War Viewed by the Vietnamese Villagers," September 1968, Pool Papers, Box 145, Folder "Hoc — Viet Nam War Viewed by the Vietnamese Villagers, 9/68."

88. Garry Quinn to JH, October 2, 1968, Pool Papers, Box 145, Folder "Hoc."

89. Garry Quinn to JH, December 30, 1968, Pool Papers, Box 145, Folder "Hoc." And see JH to Seymour Deitchman, February 1 and February 5, 1969, and Seymour Deitchman to JH, February 24, 1969, Pool Papers, Box 145, Folder "Hoc."

90. IP to JH, June 23, 1969, Pool Papers, Box 64, Folder "Outgoing correspondence June to August 1969."

91. E. W. Kenworthy, "Thousands Reach Capital to Protest Vietnam War," *NYT,* October 20, 1967. William Chapman, "GIs Repel Pentagon Charge," *Washington Post,* October 22, 1967.

92. Jo Freeman, "Levitate the Pentagon

(1967)," https://www.jofreeman.com/photos/Pentagon67.html.

93. Peter Manseau, "Fifty Years Ago, a Rag-Tag Group of Acid-Dropping Activists Tried to 'Levitate' the Pentagon," *Smithsonian,* October 20, 2017; accessed online.

94. Norman Mailer, *Armies of the Night* (New York: New American Library, 1968), 15.

95. Robert G. Sherill, "Bastille Day on the Potomac," *Nation,* November 6, 1967. William Chapman, "GIs Repel Pentagon Charge," *Washington Post,* October 22, 1967. And see "The March on the Pentagon: An Oral History," *NYT,* October 20, 2017.

96. Wells, *Wild Man,* 283–84.

97. Dorothy Wayman to JH, November 1967, Wayman Papers, Box 1, Folder 33.

98. ELG, Memorandum for the Record, November 3, 1967, Lasswell Papers, Box 40, Folder 547.

99. Blackmer, *The MIT Center for International Studies,* 211. IP, "Notes from a Second Summer in Vietnam," September 1967, marked "For private circulation only; Not for quotation or attribution," Pool Papers, Box 145, Folder "I.P. File."

100. Ron McLaurin to Seymour Deitchman, December 13, 1967, Project Agile Records, Box 49, Folder "Simulmatics — Problem Analysis, NN3-330-099-009."

101. Colonel William Arnold to Mr. W. G. McMillan, November 21, 1967, Project Agile Records, Box 49, Folder "Simulmatics — Problem Analysis, NN3-330-099-009."

102. Ron McLauren (ARPA,) Memorandum for the Record, December 12, 1967, Project Agile Records, Box 49, Folder "Simulmatics — Problem Analysis, NN3-330-099-009."

103. Seymour Deitchman, Memorandum for the Record, November 29, 1967, Project Agile Records, Box 49, Folder "Simulmatics — Problem Analysis, NN3-330-099-009."

Chapter 12: The Fire Next Time

1. Ann Penner Winston, interview with the author, June 7, 2018.

2. Ann Penner Winston, e-mail to the author, June 10, 2018.

3. Ann Penner Winston, interview with the author, June 7, 2018.

4. Peter Shulman, interview with the author, June 11, 2018. Clarence Major, e-mail to the author, April 20, 2018.

5. Ann Greenfield, interview with the author, June 9, 2018.

6. Claude Brown and Arthur Dunmey, "Harlem's America," *New Leader,* September 1966.

7. R.D.G. Wadhwani, "Kodak, FIGHT, and the Definition of Civil Rights in Rochester, New York, 1966–1967," *Historian* 60 (1977): 59–75.

8. Isserman and Kazin, *America Divided,* 145. And see, broadly, Malcolm McLaughlin, *The Long Hot Summer of 1967: Urban Rebellion in America* (New York: Palgrave Macmillan, 2014).

9. Claude Brown and Arthur Dunmey, "Harlem's America," *New Leader,* September 1966. For a brilliant reconstruction and discussion of the relationship between the war on poverty and the war on crime, including the emphasis on prediction in both efforts, see Elizabeth Hinton, *From the War on Poverty to the War on Crime: The Making of Mass Incarceration in America* (Cambridge, MA: Harvard University Press, 2017).

10. Especially useful is ELG, "Statement to Stockholders, September 20, 1966," in which Greenfield notes the growth of the company between 1965 and 1966 and explains, "The seeds of our future growth were already germinating in 1964. They began to blossom in May and June of this year. For two years the Federal Government had been making historic decisions to sponsor two technological revolutions. The technological philosophy of government

domestic programs in many fields has been years behind that of the business community. Government attempts to influence political imbalances occurring around the world have suffered from a similar technological gap. About a year ago the Federal Government began coming to terms with its 'cultural lag' and decided to apply the most advanced computer and social science techniques in the areas of poverty and education as well as to problems of international political development. Simulmatics is recognized as a company ideally suited to introduce and apply modern technology to these areas. . . . A management decision was made to commit Simulmatics to a deep involvement in both the Great Society programs and the Government's political development activities abroad. . . . In the end, any distinction between innovation at home and abroad is bound to be artificial. . . . Simulmatics has been privileged to be in on the beginning of the technological revolution in the United States, and we confidently expect to participate in its extension around the world." Historians have called this turn, in which technical expertise that came out of defense-funded work was repurposed on behalf of Johnson's Great Society programs, a turn from "warfare to welfare." See Jennifer S. Light, *From Warfare to Welfare: Defense Intellectuals and*

Urban Problems in Cold War America (Baltimore: Johns Hopkins University Press, 2003).

11. McCarthy, *Vietnam,* 6.

12. Simulation games JC designed while at Simulmatics included Democracy, Ghetto, High School, and Life. Sarane Spence Boocock, "Johns Hopkins Games Program," *Simulation & Gaming* 25 (1994), 172–78. Constance J. Seidner, "Simulation and the Bottom Line," *Simulation & Gaming* 26 (1995): 503–10. Funding for this work was provided, in part, by the Carnegie Foundation.

13. "The Van Santvoord Years, 1925 to 1959," Hotchkiss Timeline, http://www.hotchkissmedia.org/communications/timeline/1925-1959.html. Harvard University Archives, e-mail to the author, August 13, 2019. Ann Penner Winston, e-mail to the author, June 13, 2018. Livingston College, Rutgers University, *Course Catalog, 1969–1970* (New Brunswick, NJ: Rutgers University, 1969). Marcellus Winston is listed as a member of the faculty of the English Department. Mr. Winston declined to be interviewed for this book.

14. Useful discussions include a 2017 forum of the *History of Education Quarterly,* especially Harvey Kantor and Robert Lowe, "Introduction: What Difference Did the Coleman Report Make?," *History of Educa-*

tion Quarterly 57 (2017): 570–78; Ethan L. Hutt, " 'Seeing Like a State' in the Postwar Era: The Coleman Report, Longitudinal Datasets, and the Measurement of Human Capital," *History of Education Quarterly* 57 (2017): 615–25; and Zoë Burkholder, "The Perils of Integration: Conflicting Northern Black Responses to the Coleman Report in the Black Power Era, 1966–1974," *History of Education Quarterly* 57 (2017): 579–90. See also Caroline M. Hoxby, "The Immensity of the Coleman Data Project," *Education Next* 16 (2016).

15. Nathaniel J. Pallone, "In Memoriam: Sol Chaneles," *Journal of Offender Counseling Services Rehabilitation* 15 (1990): v.

16. James H. Beshers, ed., *Computer Methods in the Analysis of Large-Scale Social Systems* (Cambridge, MA: Joint Center for Urban Studies, 1964).

17. IP and AB, "Study of a System of National Manpower Accounts and the Role of Longitudinal Data in Them," December 1, 1964, Coleman Papers, Box I: 77, Folder 8. ELG to DPM, August 13, 1964, Moynihan Papers, Box I: 77, Folder 8.

18. At the time, DPM lived at 57 Francis Avenue. See, e.g., DPM to Morton Jaffe (General Manager, Simulmatics), March 30, 1968, Moynihan Papers, Box I: 156, Folder 6.

19. Leonard Story Zartman (Eastman Kodak) to DPM, March 16, 1967; Leonard Story Zartman (Eastman Kodak) to ELG, June 9, 1967; and Harman Brereton (Eastman Kodak) to ELG, June 29, 1967, Moynihan Papers, Box I: 156, Folder 3.

20. Peter Shulman, interview with the author, June 11, 2018.

21. Simulmatics Corporation, Contract with DPM, July 28, 1967, Moynihan Papers, Box I: 156, Folder 6.

22. DPM to Norman Klein (Skidmore, Owings, & Merrill), November 17, 1967, Moynihan Papers, Box I: 145, Folder 1.

23. Simulmatics Field Team, First Progress Report, August 12, 1966, Pool Papers, Box 67, Folder "Simulmatics Vietnam Correspondence."

24. R.D.G. Wadhwani, "Kodak, FIGHT, and the Definition of Civil Rights in Rochester, New York, 1966–1967," *Historian* 60 (1977): 59–75.

25. Thomas Coleman, interview with the author, March 18, 2019.

26. Desmond Stone, "A Twenty-Five-Year Friendship Pays Off," *Times-Union* (Rochester), June 26, 1967.

27. Simulmatics Corporation, "The Sources of Unrest in Rochester," 1967, Moynihan Papers, Box I: 156, Folder 5. Peter Shulman, interview with the author, June 11, 2018.

28. Simulmatics Corporation, "The Sources of Unrest in Rochester," 1967, Moynihan Papers, Box I: 156, Folder 5. There are various drafts of this same report in this folder.

29. LBJ, "Speech to the Nation on Civil Disorders," July 27, 1967. And see Richard Valeriani, *The World & Washington,* NBC-TV, January 7, 1968, Kerner Commission Records, Johnson Library, Box 7, Folder 9.

30. *Report of the National Advisory Commission on Civil Disorders* (New York: Bantam, 1968), 362.

31. Memorandum, September 25, 1967, Robert Shellow Papers, Johnson Library, Box 4, Folder 10.

32. Simulmatics Corporation, Urban Studies Division, "Mass Media Study for Commission on Civil Disorders: Analysis of News Media Coverage of 1967 Urban Riots," proposal submitted to the President's Commission on Civil Disorders, original September 29, 1967, revised October 7, 1967, Pool Papers, Box 67, Folder "Simulmatics Media Study."

33. The sole published source on Simulmatics' work for the Kerner Commission is Thomas J. Hrach, *The Riot Report and the News: How the Kerner Commission Changed rage of Black America* (Amherst: University of Massachusetts Press, 2016), ch. 6, "Simulmatics Produces a Contradictory

Analysis."

34. Bruce Lambert, "In the Works: A Computer to Predict Riots, Revolutions," *Boston Globe,* September 6, 1965.

35. "Predict Riots in Nation's Capital," *Philadelphia Tribune,* March 26, 1966.

36. Maggie Bellows, "A City Is Watched by the Computer," *Austin American,* July 10, 1966.

37. *Face the Nation,* April 25, 1965, transcript, p. 19, James Farmer Correspondence Files, ProQuest History Vault.

38. FBI file on MLK, Cleveland, June 12, 1967, MLK, FBI File, Part 1, ProQuest History Vault.

39. Simulmatics Corporation, "News Media Coverage of the 1967 Urban Riots: A Study Prepared for the National Advisory Commission on Civil Disorders," February 1, 1968, p. 3, Kerner Commission, Johnson Library, Box 7, Folder 9. I have chiefly relied on the records in the Chayes Papers but note that most of Simulmatics' research for the Kerner Commission was lost when the company dissolved. In 1970, a Cornell sociology PhD student wrote to Pool inquiring after "all of the material on the research design as well as the computer print-outs," including "the coding scheme and frequency distributions from the content analysis." Carol Mueller to IP, July 17,

1970. Pool replied, "The Simulmatics Corporation is no longer in operation. The content analysis was in New York and the documentation that you seek is undoubtedly in their warehouse of files." IP to Mueller, July 22, 1970, Pool Papers, Box 67, Folder "Simulmatics Correspondence." So far as I have been able to determine, everything in this warehouse was destroyed.

40. Clarence Major, e-mail to the author, April 20, 2018.

41. IP, untitled document listing "questions . . . intended to probe attitudes toward the media," September 27, 1967, Robert Shellow Papers, Johnson Library, Box 4, Folder 10.

42. Simulmatics Corporation, Urban Studies Division, Exploratory Coding Categories, "Level III Analysis, Riot Reporting," Pool Papers, Box 67, Folder "Simulmatics Media Study."

43. Simulmatics Corporation, Urban Studies Division, News Media Study, President's Commission on Civil Disorders, Weekly Progress Report for week ending November 3, 1967, Pool Papers, Box 67, Folder "Simulmatics Media Study."

44. Kurt Lang to Sol Chaneles, November 29, 1967, Abram Chayes Papers, Harvard Law School, Box 285, Folder 4.

45. Kurt Lang to Abram Chayes, February 11, 1968, Abram Chayes Papers, Harvard

Law School, Box 285, Folder 4. And see Lang to Chaneles, February 8, 1968, same folder.

46. Hrach, *The Riot Report and the News,* 80.

47. Peter Shulman, interview with the author, June 11, 2018.

48. National Advisory Commission on Civil Disorders, Executive Confidential Meeting, November 6, 1967, p. 3722, Civil Rights During the Johnson Administration, 1963–1969, Part V: Records of the National Advisory Commission on Civil Disorders (Kerner Commission), ProQuest History Vault.

49. Victor G. Strecher, Report, November 3, 1967, Civil Rights During the Johnson Administration, 1963–1969, Part V: Records of the National Advisory Commission on Civil Disorders (Kerner Commission), ProQuest History Vault.

50. John P. MacKenzie, "Computer in Justice Department Ready for Riot-Watching," *Washington Post,* February 16, 1968, as filed in Pool Papers, Box 67, Folder "Simulmatics Media Study."

51. Blackmer, *The MIT Center for International Studies,* 211. IP, "Notes from a Second Summer in Vietnam," September 1967, marked "For private circulation only; Not for quotation or attribution," Pool Papers, Box 145, Folder "I.P. File." Pool sent a copy of these "Notes" to ARPA. "The

enclosed notes are something I am using to answer questions of friends about my impressions after the summer in Vietnam. I do not regard it as a product of the project, but I thought you might want to see it anyhow." Pool to Deitchman, September 21, 1967, Pool Papers, Box 75, Folder Correspondence August to November 1967. He circulated the "Notes" to everyone in the MIT-Harvard Vietnam Seminar (with two exceptions) and to certain colleagues in his department (again, with exceptions), as well as to ELG, ADG, and family members.

52. IP, "Notes from a Second Summer in Vietnam," September 1967, marked "For private circulation only; Not for quotation or attribution," Pool Papers, Box 145, Folder "I.P. File." Edwin E. Moïse, *The Myths of Tet: The Most Misunderstood Event of the Vietnam War* (Lawrence: University of Kansas Press, 2017), 94–101.

53. JH to Seymour Deitchman, February 5, 1969, Pool Papers, Box 145, Folder "Hoc."

54. JH, "Testing New Psychological Warfare Weapons in Viet Nam," May 1968, Pool Papers, Box 145, Folder "Hoc-New Psywar Weapons."

55. IP, travel expense memo, February 1, 1968, Pool Papers, Box 75, Folder "Correspondence January to May 1968." Joseph M. Hoc, "Viet Nam War Viewed by the Viet-

namese Villagers," September 1968, p. 18, Pool Papers, Box 145, Folder "Hoc — Viet Nam War Viewed by the Vietnamese Villagers, 9/68."

56. Walter Cronkite, "We Are Mired in Stalemate," Broadcast, CBS Special Report, *Report from Vietnam: Who, What, When, Where, Why?*, February 27, 1968.

57. IP to the Editor of the *NYT*, March 5, 1968, Pool Papers, Box 75, Folder "Correspondence January to May 1968."

58. *Report of the National Advisory Commission on Civil Disorders,* ch. 15. The Kerner Commission released its report at the end of February. Its chapter on the media largely set aside IP's content analysis, which had found the media coverage fair, accurate, and unbiased, in favor of the commission's own interpretation, which found much to worry about in practices like the use of "scare" headlines in newspapers, inflated figures for property damage, and various other misrepresentations and distortions. More broadly, the commission argued that "the news media have failed to analyze and report adequately on racial problems in the United States and, as a related matter, to meet the Negro's expectations in journalism." See Chayes to David G. Clark (University of Wisconsin), August 16, 1968, Chayes Papers, Box 285, Folder 4. Clark, a journalism professor, had written (Clark to

Chayes, August 13, 1968) to ask for a copy of the content analysis. Chayes responded he did not have it but that it was filed with the commission's archives. He added, "The study was conducted for the Commission by a social science research organization. As you will see by examining chapter XV of the Commission's report, the findings and conclusions of the Commission ranged well beyond the material in the content analysis. In fact, the content analysis played a rather limited role in the Commission's overall conclusions and recommendations on the media."

59. *Report of the National Advisory Commission on Civil Disorders,* v.
60. "Johnson Says He Won't Run," *NYT,* April 1, 1968.
61. Isserman and Kazin, *America Divided,* 237.
62. RFK, "Statement on Assassination of Martin Luther King, Jr.," April 4, 1968.
63. Norman Mailer, *Miami and the Siege of Chicago* (1968; repr., New York: Random House, 2016), p. 92).
64. Aldous, *Schlesinger,* 349.
65. Jean Pool, notes taken of a phone call with ELG, undated but c. 1969, Pool Papers, Box 143, no folder.
66. Patricia Greenfield to Edwin Safford, Christmas 1967, Greenfield Papers.

67. Pat Greenfield, "Making Do at P.S. 51," *Chelsea-Clinton News,* November 25, 1965.

68. "Mrs. Edward L. Greenfield, Civic Leader in Chelsea," *NYT,* June 23, 1968. Ann Greenfield, interview with the author, June 9, 2018.

69. Thomas Coleman, interview with the author, March 18, 2019.

70. Wendy McPhee, interview with the author, July 16, 2018.

71. Patricia Greenfield to Michael Greenfield, c. February 1968, Greenfield Papers.

72. Patricia Greenfield to Michael Greenfield, c. March 1968, Greenfield Papers.

73. Susan Greenfield, interview with the author, July 27, 2018.

74. "Mrs. Edward L. Greenfield, Civic Leader in Chelsea," *NYT,* June 23, 1968. Office of the Chief Medical Examiner, Accession Record, June 22, 1968, p. 336, Municipal Records of the City of New York. Ann Greenfield, interview with the author, June 9, 2018.

75. Autopsy of Patricia Greenfield, Case No. 68-5410, June 21, 1968, Municipal Records of the City of New York.

76. "In Pat's Memory," materials from the funeral of Patricia Greenfield, June 25, 1968, collection of Jennifer Greenfield.

77. JC to ELG, October 18, 1968, and WMc to ELG, October 23, 1968, Pool Papers,

Box 67, Folder "Simulmatics Correspondence."

78. JC to ELG, October 23, 1968, Pool Papers, Box 67, Folder "Simulmatics Correspondence."

79. DPM, on letterhead as director of the Center for Urban Studies at MIT, to ELG at Simulmatics in New York with a cc to IP, October 28, 1968, Box 67, Folder "Simulmatics Correspondence."

80. Susan Greenfield, interview with the author, July 27, 2018. Ann Greenfield, interview with the author, June 9, 2018.

81. Ann Penner Winston, interview with the author, June 7, 2018.

82. Ann Greenfield, interview with the author, June 9, 2018.

83. James Ridgeway to DPM, June 19, 1968, Moynihan Papers, Box 1: 156, Folder 3.

84. Ridgeway, *The Closed Corporation,* 64. Published in October as per Susan Jacoby, "Universities in Society," *Washington Post,* October 31, 1968.

85. J. Y. Smith, "H. R. Haldeman Dies," *Washington Post,* November 13, 1993.

86. Three pieces of evidence — one speculative, one circumstantial, and one forensic — suggest the possibility that the Simultron Project was, in fact, Ithiel de Sola Pool. First, the speculation: Pool's support for Humphrey in 1968 appears to have been quite limited, and by 1972, Pool would

684

publicly support Nixon's reelection campaign. Second, the circumstances: the Simultron memo (Simultron Project, Memo to H. R. Haldeman, May 6, 1968, Nixon Library, White House Special Files, Box 35, Folder 8) submitted to Haldeman on May 6 refers to an earlier "initial memorandum," which may have involved an earlier in-person meeting, and Pool was in Washington on May 4, 1968 (as per the date of travel documented in the Travel Account Memo, May 27, 1968, Pool Papers, Box 75, Folder, "January–May 1968"). Third, forensic: the Simultron memo was produced on the same model of typewriter Pool used (as opposed to the typewriters used by his secretaries, who noted their having typed documents by including their initials). I draw this conclusion from a report I commissioned from Mark Songer, a former FBI special agent and former FBI document analyst who is a court-qualified expert in forensic document examination. Songer reported that the May 6, 1968, Simultron memo was produced on a proportional-spaced typewriting machine, pica pitch, meaning ten characters per inch, using a Courier-style font design, a type style and design consistent with that of a Smith Corona Electra 120 model from 1965–68. Songer compared that document with a September 1968 memo by Pool to

Robert Nathan of the Humphrey campaign (Pool Papers, Box 204, Folder "Humphrey 1968"): that document, too, was produced on a proportional-spaced type-writing machine, pica pitch, meaning ten characters per inch, using a Courier-style font design, a type style and design consistent with that of a Smith Corona Electra 120 model from 1965–68. Furthermore, Songer observed limited misalignment and spacing characteristics in the Simultron memo to be consistent with the typed entries in the Humphrey memo. Mark Songer, e-mail to the author, November 20, 2019.

87. The Simultron Project, Memo to H. R. Haldeman, May 6, 1968; Patrick Buchanan, Memo to RN, May 8, 1968; and RN, handwritten note to Buchanan, n.d., Nixon Library, White House Special Files, Box 35, Folder 8.

88. Schulman, *Lyndon B. Johnson and American Liberalism,* 161. Keith T. Poole and Howard Rosenthal, "On Party Polarization in Congress," *Daedalus* 136 (2007): 104–7.

89. RN, "Order," Paid Political Program, broadcast March 7, 1968.

90. H. R. Haldeman, transcript of an oral history, conducted 1991 by Dale E. Trevelen, State Government Oral History Program, California State Archives, 317.

Chapter 13: An Octoputer

1. J.C.R. Licklider, Robert Taylor, and Evan Herbert, "The Computer as a Communication Device," *Science & Technology,* April 1968, 30–31.
2. IP, "Social Trends," *Science & Technology,* April 1968, 88, 101.
3. Richard M. Pfeffer, ed., *No More Vietnams? The War and the Future of American Foreign Policy* (New York: Harper & Row, 1968), 141.
4. Display ad for *Toward the Year 2018, NYT,* June 5, 1968.
5. Emanuel G. Mesthene, ed., *Toward the Year 2018* (Washington, DC: Foreign Policy Association, 1968). And see Jill Lepore, "Unforeseen," *New Yorker,* January 2, 2019.
6. Simulmatics Corporation, *Human Behavior and the Electronic Computer.*
7. Myron J. Lefcowitz and Robert M. O'Shea, "A Proposal to Establish a National Archives for Social Science Survey Data," *American Behavioral Scientist* 6 (1963): 27–31.
8. Richard Ruggles et al., *Report of the Committee on the Preservation and Use of Economic Data* (Social Science Research Council, 1965).
9. Discussions include Sarah E. Igo, *The Known Citizen: A History of Privacy in Modern*

America (Cambridge, MA: Harvard University Press, 2018), ch. 6, and Christopher Loughnane and William Aspray, "Rethinking the Call for a U.S. National Data Center in the 1960s: Privacy, Social Science Research, and Data Fragmentation Viewed from the Perspective of Contemporary Archival Theory," *Information and Culture: A Journal of History* 53 (2018): 206–10. See also Dan Bouk, "The National Data Center and the Rise of the Data Double," *Historical Studies in the Natural Sciences* 48 (2018): 627–36.

10. *The Computer and Invasion of Privacy: Hearings Before a Subcommittee of the Committee on Government Operations, House of Representatives, Eighty-Ninth Congress, July 26, 27, and 28, 1966* (Washington, DC: GPO, 1966).

11. Cited in Igo, *Known Citizen,* 226.

12. Vance Packard, "Don't Tell It to the Computer," *NYT Magazine,* January 8, 1967.

13. Edgar S. Dunn Jr., "The Idea of a National Data Center and the Issue of Personal Privacy," *American Statistician* 21 (1967): 21–27.

14. These and other letters from constituents can be found in the Cornelius Gallagher Collection, Carl Albert Congressional Research and Studies Center, University of

Oklahoma, Box 21, Folders 15 and 16; Box 28, Folder 195; and Box 29, Folder 24.

15. Rebecca S. Kraus, "Statistical Déjà Vu: The National Data Center Proposal of 1965 and Its Descendants" (presentation at the Joint Statistical Meetings, Miami Beach, FL, August 2, 2011).

16. Arthur R. Miller, "The National Data Center and Personal Privacy," *Atlantic*, November 1967.

17. Alan Westin, "The Snooping Machine," *Playboy*, May 1968.

18. "Privacy and Efficient Government: Proposals for a National Data Center," *Harvard Law Review* 82 (1968): 400–417, quotation on p. 410.

19. Emily Stuart, "Lawmakers Seem Confused About What Facebook Does — and How to Fix It," *Vox*, April 10, 2018.

20. All quotations from the hearings are from *The Computer and Invasion of Privacy: Hearings Before a Subcommittee of the Committee on Government Operations, House of Representatives, Eighty-Ninth Congress, July 26, 27, and 28, 1966* (Washington, DC: GPO, 1966).

21. Arthur R. Miller, *The Assault on Privacy: Computers, Data Banks, and Dossiers* (Ann Arbor: University of Michigan Press, 1971), 54–66.

22. Judy Kaufman and Bob Park, eds., *The*

Cambridge Project: Social Science for Social Control (Cambridge, MA: Imperial City, 1969). "Will Project Cambridge Go?" *Scientific Research,* September 15, 1969, cover.

23. J.C.R. Licklider, *Libraries of the Future* (Cambridge, MA: MIT Press, 1965).

24. M. Michael Waldrop, *The Dream Machine: J.C.R. Licklider and the Revolution That Made Computing Personal* (New York: Penguin, 2001). Katie Hafner and Matthew Lyon, *Where Wizards Stay Up Late: The Origins of the Internet* (New York: Simon & Schuster, 1998). Janet Abbate, *Inventing the Internet* (Cambridge, MA: MIT Press, 1999).

25. IP to Jerome Wiesner, November 18, 1968, Pool Papers, Box 64, Folder "Outgoing Correspondence September to November 1968."

26. IP to J.C.R. Licklider, Memo, November 27, 1968, Pool Papers, Box 64, Folder "Outgoing Correspondence September to November 1968." IP to J.C.R. Licklider, Memo, June 16, 1969, Pool Papers, Box 64, Folder "Outgoing Correspondence June to August 1969."

27. "A Proposal for Establishment and Operation of a Program in Computer Analysis and Modeling in the Behavioral Sciences," reports, memoranda, correspondence, and other materials from the Cambridge Project

Subcommittee, June–December 1969, Harvard University Archives.

28. [MIT President] Jerome B. Wiesner to Members of the [MIT] Faculty, May 5, 1969, MIT News Office, MIT Institute Archives and Special Collections, AC-0069.

29. Mailer, *Miami and the Siege of Chicago,* ch. 3.

30. RN, "Acceptance Speech," Republican National Convention, August 8, 1968.

31. Mailer, *Miami and the Siege of Chicago,* 137.

32. Commission on the Democratic Selection of Presidential Nominees, *The Democratic Choice* (Washington, DC: Democratic National Committee, 1968), 15.

33. IP to Robert Nathan, hand-dated "9/68," Pool Papers, Box 204, Folder "Humphrey 1968." See also [IP], "Achieving Pacification in Viet Nam," marked "Private Communication, Not for Publication," undated but, given its place in the folder, must be about March 19, 1968, Pool Papers, Box 75, Folder "Jan–May 1968."

34. Aldous, *Schlesinger,* 348.

35. "Shouldn't you search out opportunities to expose yourself to my rhetoric and wit?" Buckley wrote Schlesinger. "How else will you fulfill your lifelong dream of emulating them?" The exchange appears in William F. Buckley Jr., *Cancel Your Own Goddam Subscription: Notes and Asides from National*

Review (New York: Basic Books, 2009), 48–51.

36. Richard Todd, "The 'Ins' and 'Outs' at M.I.T.," *NYT Magazine,* May 18, 1969. Pool, unsurprisingly, refused to support the work stoppage. See the documents in Pool Papers, Box 73, Folder "Memos, 1968."

37. J. William Fulbright, "The War and Its Effects — II," *Congressional Record,* December 13, 1967, 36181–82.

38. "McGeorge Bundy Supports De-Escalation in Vietnam," *Harvard Crimson,* October 14, 1968.

39. Charlie Mann, "Bundy Discusses Possibilities for Implementing End to War," *Tech,* October 18, 1968. Noam Chomsky, e-mail to the author, April 22, 2018.

40. Peter Shulman, interview with the author, June 11, 2018.

41. Noam Chomsky, "The Menace of Liberal Scholarship," *New York Review of Books,* January 2, 1969. Michael Albert cites this same passage in *Remembering Tomorrow: From SDS to Life After Capitalism* (New York: Seven Stories Press, 2006), 99.

42. IP to Henry Kissinger, January 2, 1969, Pool Papers, Box 64, Folder "Outgoing Correspondence, December 1968 to February 1969."

43. Bonita Harris (IP's secretary) to Jean Keppler (Defense Science Board), January 14, 1969, Pool Papers, Box 64, Folder

"Outgoing Correspondence, December 1968 to February 1969."

44. IP to Howard Johnson (president, MIT), January 28, 1969, Pool Papers, Box 64, Folder "Outgoing Correspondence, December 1968 to February 1969."

45. Broadly, see Ellsberg, *Secrets.*

46. See the exchanges between William Parker and Noam Chomsky in Pool Papers, Box 177, Folder "Bill Parker."

47. IP, letter to the editor, and Chomsky's reply, *New York Review of Books,* February 13, 1969.

48. Richard Todd, "The 'Ins' and 'Outs' at M.I.T.," *NYT Magazine,* May 18, 1969.

49. "The First 70 Days: A Chronicle," *Technology Review,* December 1969.

50. David Deitch, "The Professor and the Military-Industrial Complex," *Boston Globe,* April 20, 1969.

51. Joseph Kashi, "Millikan Opens CIS Files; Denies Use of CIA Funds," *Tech,* May 2, 1969.

52. Adam de Sola Pool, interview with the author, May 19, 2018.

53. IP to the Political Science Faculty, Memo, May 5, 1969, Pool Papers, Box 69, Folder "Cambridge Proj — Publicity."

54. Memorandum, May 5, 1969, unsigned but in a handwritten note attributed to IP, Pool Papers, Box 69, Folder "Cambridge

Proj — Publicity."

55. "Research for the Counter-Revolution," *Hard Times,* May 12–19, 1969, in Pool Papers, Box 69, Folder "Cambridge Proj — Publicity."

56. Joseph Weizenbaum, "To the Editor," *Tech,* May 16, 1969. William R. Ferrell, Ronald C. Rosenberg, and Thomas B. Sheridan, "To the Editor," *Tech,* May 23, 1969.

57. E.g., John H. Fenton, "MIT Group Assails Computer Plan," *NYT,* May 7, 1969.

58. Richard Todd, "The 'Ins' and 'Outs' at M.I.T.," *NYT Magazine,* May 18, 1969. Pool wrote a letter to the editor complaining about its author calling him "a celebrated hawk" and asking, "Is there nothing else one can be but a hawk or a dove?" IP to the editor of the *NYT Magazine,* May 19, 1969, Box 64, Folder "Outgoing Correspondence March to May 1969."

59. Laurence Stern, "Ending of Secret Research Won't Cool the Campuses," *Tech,* May 14, 1969.

60. IP to John Jachym, May 16, 1969, Pool Papers, Box 67, Folder "Simulmatics Correspondence."

61. Leyland graduated Harvard Business School in 1969; he had a master's degree from MIT (1966) and a BA in political science from Harvard (1960). George Leyland, CV, 1969, Pool Papers, Box 67,

Folder "Simulmatics Correspondence."

62. IP, "A Work Plan for the Simulmatics Corporation," June 15, 1969, Pool Papers, Box 67, Folder "Simulmatics Correspondence."

63. IP in separate letters to F. Randall Smith, Tibor Fabian, and Sidney Rolfe, August 8, 1969, enclosing the report. Pool Papers, Box 64, Folder "Outgoing Correspondence June to August 1969."

64. IP, separate letters to John J. Jachym, Simon Ramo, and Charles M. Herzfeld, August 13, 1969, and to Simon Ramo, same date. Pool Papers, Box 64, Folder "Outgoing Correspondence June to August 1969."

65. IP to Harold Guetzkow, August 13, 1969, Pool Papers, Box 67, Folder "Simulmatics Correspondence." It seemed for a time as though Mathematica would buy Simulmatics. Wrote IP, "The financial group behind Mathematica is putting up $10,000 for expenses of the reorganization, $40,000 in cash, and 100,000 shares for settling of all Simulmatics debt. We are confident that we can clear the books that way." Pool to Randall Smith, September 8, 1969, Pool Papers, Box 64, Folder "Outgoing Correspondence September to December 1969." But, so far as I and anyone at Mathematica have been able to discover, that plan never materialized. David Roberts

(Mathematica), e-mail to the author, May 17, 2018.

66. "The First 70 Days: A Chronicle," *Technology Review,* December 1969.

67. Ibid.

68. SACC Teach-in, September 25, 1969, MIT Institute Archives and Special Collections, SACC Records, Box 1.

69. Greg Bernhardt, "150 Students Peacefully Disrupt CIS," *Tech,* October 14, 1969.

70. "CIS IS CIA," leaflet, 1969, MIT Institute Archives and Special Collections, SACC Papers, Box 2. Emphasis in the original.

71. "The First 70 Days: A Chronicle," *Technology Review,* December 1969.

72. "What About Academic Freedom?" undated brochure, possibly SACC, likely 1970, MIT Institute Archives and Special Collections, SACC Records, Box 2. And, on threats against Pool, see Victor K. McElheny, "MIT Professor Reveals Threat by Student Protestors," *Boston Globe,* November 1, 1969.

73. Paul Mailman, "SDS Confronts CIS," typescript, 1969, MIT Institute Archives and Special Collections, SACC Papers, Box 2.

74. Jeffrey J. Page, "Will Project Cambridge Blow MIT's Cool?," *Scientific Research,* September 15, 1969.

75. Yasha Levine, *Surveillance Valley: The Secret Military History of the Internet* (New York: PublicAffairs, 2018), 61.

76. Kaufman and Park, *The Cambridge Project.*

77. Chris Wallace, "Echoes of 1969," *Harvard Magazine,* March–April 2019.

78. Samuel A. Yohai to Harvey Brooks, November 1969, reports, memoranda, correspondence, and other materials from the Cambridge Project Subcommittee, June–December 1969, Harvard University Archives.

79. Jerome S. Bruner to Barrington Moore, November 17, 1969, reports, memoranda, correspondence, and other materials from the Cambridge Project Subcommittee, June–December 1969, Harvard University Archives.

80. Martin Perlmutter to Harvey Brooks, October 24, 1969, reports, memoranda, correspondence, and other materials from the Cambridge Project Subcommittee, June–December 1969, Harvard University Archives.

81. Steven J. Marcus to Harvey Brooks, October 23, 1969, reports, memoranda, correspondence, and other materials from the Cambridge Project Subcommittee, June–December 1969, Harvard University Archives.

82. Barrington Moore Jr. to the Cambridge Project (Committee), October 23, 1969, reports, memoranda, correspondence, and other materials from the Cambridge Project Subcommittee, June–December 1969, Harvard University Archives. Nor did Moore have kind words for the claims of Pool's initial proposal, which had cited content analysis during World War II as an example of the kind of payoff of this sort of work. Wrote Moore: "This observation does not match my experience of wartime Washington. Content analysis was largely abandoned before the war ended because it turned out to be useless for the understanding of large political problems."

83. David N. Hollander and Jeff Magalif, "175 March into Univ. Hall, Protest Project Cambridge," *Harvard Crimson,* September 27, 1969.

84. David I. Bruck, "Brass Tacks: The Cambridge Project," *Harvard Crimson,* September 26, 1969.

85. IP, letter to the editor of the *Harvard Crimson,* October 6, 1969, Pool Family Papers. He had the response published elsewhere. IP, "MIT Professor Scents McCarthyism in Attack," *Boston Herald Traveler,* October 26, 1969. "Practically all my time in recent weeks has been devoted to dealing with the student rebellion," Pool

wrote a few weeks later. Pool to Wilbur Schramm, November 18, 1969, Pool Papers, Box 64, Folder "Outgoing Correspondence September to December 1969."

86. On October 29, 1969, the advisory committee met at MIT and produced a statement explaining what the project was, and what it was not (this report indicates that "Harvard is deciding whether to take part"). "The purpose of the Cambridge Project is to develop better computer methods for behavioral science." "None of the work is classified." "There is no plan to set up a vast archive of data." [IP], "The Cambridge Project," *ACM SIGSOC Bulletin,* December 1, 1969, pp. 12–14.

87. Robert Elkin, "Rally, Sit-in Protest War Research," *Tech,* November 7, 1969. Later, when a poster featuring only Pool was pasted all over campus, it was ordered taken down. "Execomm Bans SDS Poster," *Tech,* April 27, 1971.

88. Blackmer, *The MIT Center for International Studies Founding Years,* 217. Lincoln P. Bloomfield, *In Search of American Foreign Policy: The Humane Use of Power* (New York: Oxford University Press, 1974), 6–9. And see the poster featuring only IP in the SACC Records, MIT Institute Archives and Special Collections, Box 2, Folder 76.

89. Parker Donham, "War Protest Week

Begins in Boston," *Boston Globe,* April 13, 1970.

Chapter 14: The Mood Corporation

1. Women's Coalition Strike Headquarters broadside, SY1970 no. 4, New-York Historical Society. Quoted in Anna Gedal, "The 1970 Women's March for Equality in NYC," March 10, 2015, http://behindthe scenes.nyhistory.org/march-for-equality-in -nyc/.
2. Naomi Spatz, e-mail to the author, March 13, 2019.
3. Paul Starr, "Seductions of Sim: Policy as a Simulation Game," *American Prospect,* https://prospect.org/article/ seductions-sim-policy-simulation-game. Jennifer S. Light, *From Warfare to Welfare: Defense Intellectuals and Urban Problems in Cold War America* (Baltimore: Johns Hopkins University Press, 2003), 90.
4. Peter Shulman, "Peter Shulman's War," http://peterswar.com/index.htm.
5. On TBM, see his obituary: Douglas Martin, "Thomas B. Morgan, Writer, Editor, and Lindsay Press Aide, Dies at 87," *NYT,* June 18, 2014. Also see an appreciation of his journalism: Alex Belth, "What a Novel Idea: The Essential Thomas B. Morgan," Esquire.com, November 1, 2016.
6. Kim Le Cook, interview with the author,

August 7, 2019.

7. IP to Cuc Thu Duong, January 26, 1968, Pool Papers, Box 67, Folder "Simulmatics Correspondence."

8. Charles F. Printz (Human Rights Advocates International) to MS, November 8, 1995, Shea Letters.

9. Ruth Washburn Cooperative Nursery School, "Miriam Emery Howbert, 1923–1989," https://rwcns.org/our-preschool/philosophy-history/.

10. John McPhee, e-mail to the author, July 25 and July 26, 2018. And for the investigation, see "L'Université du Nouveau Monde," November 20, 1971, available at https://www.rts.ch/archives/tv/information/affaires-publiques/8727432-l-universite-du-nouveau-monde.html.

11. Jennifer Greenfield, interview with the author, July 25, 2018.

12. "E. Greenfield, Founded Political Research Firm," *Chicago Tribune,* July 28, 1983. Michael Greenfield, "Obituary Notice: Edward Lawrence Greenfield," typescript, 1983, Greenfield Papers.

13. Susan Greenfield, interview with the author, July 27, 2018.

14. Levine, *Surveillance Valley,* 86. And see U.S. Judiciary Committee, Subcommittee on Constitutional Rights, *Army Surveillance of Civilians: A Documentary Analysis* (Washington, DC: GPO, 1972).

15. Daniel Ellsberg, interview with the author, August 22, 2018.

16. M. David Landau, "The Ellsberg File," *Harvard Crimson,* September 22, 1971.

17. The best chronicle of SP's experience is a report prepared by the American Political Science Association, "Initial Report on the Popkin Case and Related Materials by James D. Carroll to the American Political Science Association, Committee on Ethics and Academic Freedom, January 23, 1973, Pool Papers, Box 52, Folder "Popkin, Sam." But for more background and additional details, see the documents relating to my petition to unseal the grand jury records. Popkin's declaration is available at https://law.yale.edu/sites/default/files/area/center/mfia/document/ecf_no._10_-_popkin_declaration.pdf. These documents relate to my petition to unseal these grand jury records, submitted in 2018 by the Yale Law School's Media Freedom and Information Access Clinic. Those documents are all available online: https://law.yale.edu/mfia/projects/constitutional-access/re-petition-motion-order-directing-release-records-pentagon-papers-grand-juries. And for an account from the vantage of SP's lawyer: Mark S. Brodin, *William P. Homans Jr.: A Life in Court* (Lake Mary, FL: Vandeplas Publishing, 2016).

18. "Investigation Concerning Distribution

of the 'McNamara Study' to Newspapers Other Than 'The New York Times' During June and July 1971," FBI Report, August 18, 1971, *U.S. v. Ellsberg,* Attorney Files, National Archives at Boston. Saul Friedman, "How Ellsberg Switched from Hawk to Dove — and More," *Detroit Free Press,* August 1, 1971.

19. Robert E. Bowe, FBI Report on Daniel Ellsberg, November 12, 1971, *U.S. v. Ellsberg,* Attorney Files, National Archives at Boston. And see Julia Valenzuela, FBI interview, September 13, 1971, same file.

20. *Impeachment Inquiry Hearings Before the Committee on the Judiciary, House of Representatives, Ninety-Third Congress, 1025–26* (Washington, DC: GPO, 1974).

21. M. David Landau, "The Ellsberg File," *Harvard Crimson,* September 22, 1971.

22. See the American Political Science Association, Committee on Ethics and Academic Freedom, "Initial Report on the Popkin Case," January 23, 1973, Pool Papers, Box 52, Folder "Popkin, Sam." And see, e.g., "70 Educators Join Protest: Harvard, MIT Groups Back Popkin," *Boston Globe,* November 26, 1972.

23. Peter Shapiro, "Shifting Allegiances in Academia," *Harvard Crimson,* November 6, 1972.

24. Gordon Strachen to Len Garment,

Memo, June 20, 1972, Nixon Library, Contested Files, Box 13, Folder 16.

25. Recommended telephone call from Charles Colson to Ithiel de Sola Pool, Memo, October 5, 1972, Nixon Library, White House Special Files, Box 26, Folder 9.

26. Jeb Magruder to Clark MacGregor, October 13, 1972, Nixon Library, White House Contested Files, Box 40, Folder 3. On the letter, in retrospect, see Muriel Cohen, "Nixon's Academic Backers Still Defend Choice," *Boston Globe,* June 14, 1973.

27. Robert Reinhold, "Legal Obstacles Blocking Boston Grand Jury in Its Investigation of the Release of Pentagon Papers," *NYT,* November 1, 1971, p. 29.

28. *U.S. v. John Doe,* appeal of Samuel L. Popkin, 460 F.2d 328 (1st Cir. 1972). And see John B. Wood, "Professor Resisted Probe on Pentagon Papers; Supreme Court Denies Popkin Reprieve," *Boston Globe,* November 11, 1972.

29. Carl Bernstein and Bob Woodward, "FBI Finds Nixon Aides Sabotaged Democrats," *Washington Post,* October 10, 1972.

30. ABC News, "Elections '72," Broadcast, November 7, 1972, https://www.youtube.com/watch?v=HUQoQ_PiZ3U (which incorrectly gives the date as November 6).

31. Peter Jenkins, "Professor Gaoled in

Chains," *Guardian,* November 23, 1972.

32. Tom Wicker, "Liberty in Shackles," *NYT,* November 23, 1972.
33. Anthony Lewis, "Of Law and Men," *NYT,* November 25, 1972.
34. Bill Kovach, "Popkin Freed in a Surprise as U.S. Jury Is Dismissed," *NYT,* November 29, 1972. And see Declaration of Samuel L. Popkin, November 12, 2018, *In Re Petition of Jill Lepore,* available at https://law.yale.edu/system/files/area/center/mfia/document/ecf_no._10_-_popkin_declaration.pdf. See also American Political Science Association, Committee on Ethics and Academic Freedom, "Initial Report on the Popkin Case," January 23, 1973, Pool Papers, Box 52, Folder "Popkin, Sam." Popkin left his work on Vietnam behind after the publication of *The Rational Peasant: The Political Economy of Rural Society in Vietnam* (Berkeley: University of California Press, 1979). He returned to the study of voters, campaigns, and elections. See especially Samuel L. Popkin, *The Reasoning Voter: Communication and Persuasion in Presidential Campaigns* (Chicago: University of Chicago Press, 1991), and Samuel L. Popkin, *The Candidate: What It Takes to Win — and Hold — the White House* (New York: Oxford University Press, 2012).
35. John Wood, "Wife and Friends Say Pop-

kin Made a Martyr by U.S.," *Boston Globe,* November 23, 1972. The complete list of signatories appears in "70 Educators Join Protest: Harvard, MIT Groups Back Popkin," *Boston Globe,* November 26, 1972.

36. John B. Wood, "Move Comes as Surprise: Grand Jury Discharged, Popkin Freed," *Boston Globe,* November 29, 1972.

37. Robin Wright, "What Motivated Jailing of Popkin?" *Christian Science Monitor,* November 30, 1972.

38. Stewart Brand, "Spacewar: Fanatic Life and Symbolic Death Among the Computer Bums," *Rolling Stone,* December 1972.

39. "Scenarios for Using the ARPANET at the International Conference on Computer Communication," Washington, DC, October 24–26, 1972, Computer History Museum. And see, e.g., "Demonstration Heralds Next Wave: Connecting a Network of Networks," *Electronics,* November 6, 1972. For samples of these conversations, see https://web.stanford.edu/group/SHR/4-2/text/dialogues.html.

40. R. Buckminster Fuller, "Prime Design," *Bennington College Bulletin,* May 1960. From R. Buckminster Fuller, *Ideas and Integrities: A Spontaneous Autobiographical Disclosure* (Baden, Switzerland: Lars Müller, 2009), 329, quoted in Turner, *From Counterculture to Cyberculture,* 51–58.

41. Ibid., 104–8.

42. Brand, "Spacewar."

43. Turner, *From Counterculture to Cyberculture,* chs. 4 and 5.

44. O'Mara, *The Code,* 124.

45. NBC Nightly News, June 4, 1975, as reproduced in *Surveillance Technology: Joint Hearings Before the Subcommittee on Constitutional Rights of the Committee on the Judiciary and the Special Subcommittee on Science, Technology, and Commerce of the Committee on Commerce, United States Senate, Ninety-Fourth Congress, June 23, September 9 and 10, 1975* (Washington, DC: GPO, 1975), 3–8.

46. Ibid.

47. *Surveillance Technology,* 103–17.

48. John Klensin, interview with the author, May 21, 2018.

49. IP, "Social Trends," *Science & Technology,* April 1968, 101.

50. IP, *Technologies Without Boundaries,* ed. Eli M. Noam (Cambridge, MA: Harvard University Press, 1990), 60, 66. (Noam edited a manuscript that Pool had been working on at his death in early 1984.)

51. IP, *Technologies of Freedom* (Cambridge, MA: Harvard University Press, 1983), 226–51.

52. Brand as quoted in Lloyd S. Etheredge, "What's Next? The Intellectual Legacy of

Ithiel de Sola Pool," in IP, *Humane Politics and Methods of Inquiry,* ed. Lloyd S. Etheredge (New Brunswick, NJ: Transaction Publishers, 2000), 301–16.

53. Stewart Brand, *The Media Lab: Inventing the Future at M.I.T.* (New York: Viking, 1987), 18, 44, 214–19, 253, 267.
54. Ibid., 33.
55. Ibid., 183–84, 222–32.
56. Stewart Brand, "We Owe It All to the Hippies," *Time,* March 1, 1995.
57. Esther Dyson, George Gilder, George Keyworth, and Alvin Toffler, *Cyberspace and the American Dream: A Magna Carta for the Knowledge Age* (Progress and Freedom Foundation, 1994).
58. John Perry Barlow, "A Declaration of the Independence of Cyberspace," February 8, 1996.
59. IP, Questionnaire, undated, Pool Papers, Box 59, Folder "Contact Nets Diary."
60. John McPhee, "Link with Local History Lost," *Alamogordo* [NM] *Daily News,* April 10, 1998.
61. Wendy McPhee, interview with the author, July 16, 2018.

Epilogue: Meta Data

1. Jaron Lanier, "Jaron Lanier Fixes the Internet," *NYT,* September 23, 2019.
2. NM to AS, March 25, 1959, Stevenson

Papers, Box 38, Folder 7.

3. "The People Machine," *Newsweek,* April 2, 1962.

4. Lanier states this as an axiom: "Behaviorism is an inadequate way to think about society." Jaron Lanier, *Ten Arguments for Deleting Your Social Media Accounts Right Now* (New York: Henry Holt, 2018), 19.

5. Behavioral scientists, while admitting that there are no proven laws of human behavior, nevertheless often propose them. See, e.g., Aline Holzwarth, "The Three Laws of Human Behavior," behavioraleconomics .com, May 7, 2019, https://www.behavioral economics.com/the-three-laws-of-human -behavior/.

6. Ashlee Vance, "This Tech Bubble Is Different," *Bloomberg Businessweek,* April 14, 2011.

7. Although, notably, Google abandoned "Don't be evil" in 2015, and Alphabet Inc. adopted a code of conduct that urged employees to "do the right thing," http://blogs .wsj.com/digits/2015/10/02/as-google -becomes-alphabet-dont-be-evil-vanishes/.

8. Mathematicians-turned-businessmen Jeff Hammerbacher and D. J. Patil claim to have coined the term "data scientist" in 2008, and not long after, data science was also embraced both within and outside the academy as a new scientific method, a "fourth paradigm," following the earlier

paradigms of empirical, theoretical, and computational analysis. Thomas H. Davenport and D. J. Patil, "Data Scientist: The Sexiest Job of the 21st Century," *Harvard Business Review,* October 2012. Tony Hey, Stewart Tansley, and Kristin Michele Tolle, *The Fourth Paradigm: Data-Intensive Scientific Discovery* (Redmond, WA: Microsoft Research, 2009).

9. For an extremely trenchant discussion, see Francine Berman, Rob Rutenbar, et al., "Realizing the Potential of Data Science: Final Report from the National Science Foundation Computer and Information Science and Engineering Advisory Committee Data Science Working Group," December 2016, https://www.nsf.gov/cise/ac-data-science-report/CISEACData ScienceReport1.19.17.pdf.

10. For a distressing but illuminating summary, see https://www.cs.princeton.edu/~arvindn/talks/ MIT-STS-AI-snakeoil.pdf.

11. On scholarship that attempts to wrestle with data science's origins, see, e.g., Brian Beaton, Amelia Acker, Lauren Di Monte, Shivrang Setlur, Tonia Sutherland, and Sarah E. Tracy, "Debating Data Science: A Roundtable," *Radical History Review* 127 (2017): 133–48, and Rebecca Lemov, *Database of Dreams: The Long Quest to Catalog Humanity* (New Haven, CT: Yale University

Press, 2015).

12. Justin Peters, "The Moral Rot of the MIT Media Lab," *Slate,* September 8, 2019.

13. Michael L. Brodie, "Understanding Data Science: An Emerging Discipline for Data-Intensive Discovery," Proceedings of the XVII International Conference, *Data Analytics and Management in Data Intensive Domains* (DAMDID/RCDL 2015), Obninsk, Russia, October 13–16, 2015.

14. Predictive Analytics Market by Solutions (Financial Analytics, Risk Analytics, Marketing Analytics, Sales Analytics, Web & Social Media Analytics, Network Analytics), Services, Deployment, Organization Size and Vertical — Global Forecast to 2022, Markets and Markets, https://www.marketsandmarkets.com/Market-Reports/predictive-analytics-market-1181.html.

15. Paul P. Maglio, "Data Is Dead Without What If Methods," *Proceedings of the VLDB Endowment* 4, no. 12 (2011).

16. Stefano Rizzo, "What-If Analysis," in *Encyclopedia of Database Systems,* ed. Ling Liu and M. Tamer Özsu (New York: Springer, 2018), https://link.springer.com/referenceworkentry/10.1007%2F978-1-4614-8265-9_466.

17. Karim Amer, dir., *The Great Hack* (Netflix, 2019).

18. Berit Anderson, "The Rise of the Weap-

onized AI Propaganda Machine," *Medium*, February 12, 2017.

19. Charles Duhigg, "Did Uber Steal Google's Intellectual Property?," *New Yorker,* October 15, 2018.

ILLUSTRATION CREDITS

ILLUSTRATION CREDITS

11 Courtesy of Edwin Safford
25 Courtesy of Jennifer Greenfield
51 Courtesy of Fabst...
130 Courtesy of Sarah Neidhardt
163 Courtesy of John H. Kennedy Presidential Library and Museum
207 Courtesy of MIT Institute Archives and Special Collections
335 Courtesy of Maureen Shea
383 Courtesy of Kate Fahow Morgan
419 Courtesy of Ann Fenner Winston
455 Courtesy of MIT Institute Archives and Special Collections

ABOUT THE AUTHOR

Jill Lepore is the David Woods Kemper '41 Professor of American History at Harvard University and a staff writer at *The New Yorker.* A two-time Pulitzer Prize finalist, she is the author of a dozen books, including the international bestseller *These Truths: A History of the United States.* She lives in Cambridge, Massachusetts.

The employees of Thorndike Press hope you have enjoyed this Large Print book. All our Thorndike, Wheeler, and Kennebec Large Print titles are designed for easy reading, and all our books are made to last. Other Thorndike Press Large Print books are available at your library, through selected bookstores, or directly from us.

For information about titles, please call:
(800) 223-1244

or visit our website at:
gale.com/thorndike

To share your comments, please write:
Publisher
Thorndike Press
10 Water St., Suite 310
Waterville, ME 04901